BLACK, WHITE & GOLD

Goldmining in Papua New Guinea
1878-1930

BLACK, WHITE & GOLD

Goldmining in Papua New Guinea
1878-1930

HANK NELSON

PRESS

Published by ANU Press
The Australian National University
Acton ACT 2601, Australia
Email: anupress@anu.edu.au
This title is also available online at press.anu.edu.au

National Library of Australia Cataloguing-in-Publication entry

Creator: Nelson, Hank, 1937-2012, author.

Title: Black, white and gold : gold mining in Papua New Guinea, 1878-1930 / Hank Nelson.

ISBN: 9781921934339 (paperback) 9781921934346 (ebook)

Subjects: Gold mines and mining--Papua New Guinea--History.
                Gold miners--Papua New Guinea--History.

Dewey Number: 622.3420995

All rights reserved. No part of this publication may be reproduced, stored in a retrieval system or transmitted in any form or by any means, electronic, mechanical, photocopying or otherwise, without the prior permission of the publisher.

Cover design and layout by ANU Press.

First published 1976 by The Australian National University
Reprinted © 2016 ANU Press

Preface

Papua New Guinean communities living on islands in the Coral Sea, near creeks feeding the major rivers of the mainland, and in villages crowded along ridge-tops in the interior, gardened and hunted over land containing gold. Most of the men who came hungry for the gold were from Australia. They carried with them the skills to obtain it and the beliefs then common among Australian working men about foreigners and blacks. Most of the diggers believed that their guns and their brains made them superior to any 'coloured' men. Some also thought that they were physically superior, although that was harder to believe after 1902 when the first black American won a world boxing title. They did not doubt that they had a right to take the gold, and many thought that it was only a matter of time before Australians possessed the land.

Papua New Guineans had a variety of attitudes towards outsiders. Different communities put different values on skills in war, crafts, growing crops, trade, and the ability to control the non-material world. But the variety of their responses to miners was more than just a reflection of the fact that Papua New Guineans had worked out different methods of exploiting their environment and different sets of rules about how men should treat each other; communities were also separated by their experiences with foreigners. Some of the peoples living alongside alluvial land had for many years been meeting foreigners who came to fish for pearls, recruit men to work on boats and plantations, save souls and govern. Their beliefs and behaviour had been changed by those meetings. Others had only heard talk of foreigners when the miners and labourers came with their stores and began to dig and sluice.

So relationships between miners and villagers varied from one area to another. Some villagers retreated and the miners rarely saw them, some fought the miners and were shot in large numbers, some joined the police, some induced the foreigners to use their guns against enemy clans, some worked for the miners, some learnt to work gold for themselves (and still do), some extended their gardens and supplied the miners with tons of food, some village women decided to sleep with miners, and some were raped. The miners were gone from some fields within a few months; they stayed on others for forty years. On a few isolated fields the miners could not prospect or work unless they could buy food from the villages; on fields close to a port or navigable river the miners could be independent of local communities.

This book is about encounters between Papua New Guineans and miners from the first 'rush' in 1878 until 1930, four years after the alluvial miners finally found 'the big one'. It begins in 1888 with the north Queensland miners crowding to Sudest Island, New Guinea's first goldfield. It then turns back to look at the way of life and beliefs of the villagers and miners who met on Sudest. Chapters 2, 3 and 4 trace the relationships between foreigners and villagers as mining developed on Sudest, Misima and Woodlark Islands. Chapter 5 shifts to the mainland and the rush inland from Port Moresby in 1878 when no gold was found. Later chapters follow the defining of the alluvial areas along the northern rivers, near Milne Bay, in the Keveri Valley and on the upper Lakekamu River, and describe the meetings between peoples in those places. The last chapter is about the finding of rich alluvial and reef gold on Koranga and Edie Creeks and along the Bulolo Valley of the Morobe District. This book also attempts to tell some of the history of the miners and Papua New Guinea communities independent of their meetings. It is imperial history; not the history of colonial policy but of the behaviour of men on the frontier of empire.

Gold was the main export of British New Guinea and Papua until 1916 and it was again the most valuable export in 1938–39. In the Mandated Territory of New Guinea gold quickly replaced copra as the main export until by 1940 gold made up more than 80 per cent of the Territory's exports. Many writers have described the development of the Morobe Goldfield after 1926; here it is mentioned briefly. The history of some communities on other goldfields has been taken beyond 1930.

PREFACE

Gold-sluicing on Sharkeye Park's claim, Koranga Creek,
Morobe Goldfield, late 1920s
PHOTOGRAPH. H L. DOWNING

The information for this book has been obtained by reading, talking and travelling. Most of the reading was done in the National Archives of Papua New Guinea and in the library of the University of Papua New Guinea, Port Moresby; and in the Australian Archives, the National Library of Australia and the Menzies Library of the Australian National University, Canberra. Although most of the book is about miners who wrote infrequently, and villagers, labourers and policemen, nearly all of whom were illiterate, the written records are rich. The main defect in the written material is that so much of it is about people and not by them. This problem has been partly overcome by obtaining information from many sources about the communities described by outsiders. The writings of anthropologists, missionaries and linguists helped, and so did travelling and asking questions. This book has benefited from 'informed tourism' rather than 'field work'. In the Northern District and along the Waria River alone, communities from at least twenty language groups met the miners. Each community would provide valuable material for the student who earns the confidence of the people and has the skills to communicate with them. This book may assist those who attempt more detailed studies.

Fully-footnoted copies of the manuscript of this book showing sources of information and making additional comments will be placed in the library of the University of Papua New Guinea, the Menzies Library of the Australian National University, and the National Library of Australia, Canberra.

Canberra
1975   H. N. N.

Acknowledgements

When I went to the Territory of Papua and New Guinea in 1966 I knew little of the people who lived there. Out of choice and necessity I was soon pretending to teach some of them their history. I learnt slowly that many of the students came from communities which valued information about their past, retained much detail, judged the past actions of men and wondered about the road that they were on. After I had done some work we could share knowledge. I could tell a class when men from a particular area first joined the police force, what training they were given, what official powers they had, what dramatic incidents they were involved in, and what their white officers thought of them. The students could tell me why some men joined the police, what village people thought were the powers of the police, how the police behaved in the villages, what the police thought of their officers, and what happened to men who returned to the village after service with the police. I wish to thank the students who gave me the chance to learn about the history of the people of Papua New Guinea. In the notes attached to chapters I have mentioned only those students who gave particular help.

Kevin Green, Jim Gibbney, Moeka Helai and Bob Langdon helped me find written material. The staffs of the Library of the University of Papua New Guinea, the National Archives of Papua New Guinea, the Australian Archives, the Menzies Library of the Australian National University, the National Library of Australia, the Fryer Library of the University of Queensland, the Oxley Memorial Library of Queensland, the La Trobe Library of the State Library of Victoria, the Mitchell Library of the Library of New South Wales, the Australian Museum and the National Museum of Victoria gave courteous assistance.

Don Affleck, Elton Brash, Bert Brown, Leonie Christopherson, Fred and Nancy Damon, Don Denoon, Tom Dutton, Amirah Inglis, John Kolia, Diane Langmore, Jerry Leach, Nancy Lutton, Louise and Mekere Morauta, Alby Munt, Nigel Oram, Eddie Parr, Frank Pryke, Leo Pryke, Tony Pryke, Roger Southern, Jim Specht, Mary-Cath Togolo and John Waiko gave information on particular points. Barry Smith, Francis West,

Robin Gollan and Bill Gammage gave advice about the manuscript; and Stewart Firth and Ken Inglis made detailed comments on a completed draft.

Most of the book was written during 1973 and 1974 while I was a Research Fellow in the Department of History in the Research School of Social Sciences, Australian National University. I have done some revision since shifting to the Department of Pacific and Southeast Asian History. I would like to thank the staffs of both Departments for their hospitality. Jean Dillon, Jan Hicks, Janice Aldridge, Lois Simms and Christine Waring typed the manuscript. Maureen Krascum typed the footnoted copy of the manuscript placed in libraries in Canberra and Port Moresby. All were tolerant and skilful.

I also thank my wife, Janet, and my children, Tanya, Lauren and Michael, for putting up with me while I wrote or looked for lost pieces of paper.

For the presentation of this book I am indebted to the staff of the A.N.U. Press and the Graphic Design Unit, and to the Nauru Fund for a loan to help meet the cost of the drawings used in the text. Mr P. Belbin drew the illustrations appearing in the text.

Contents

Preface v
Acknowledgements ix
Chronology. xix

OUT OF COOKTOWN
1. A Meeting: north Queensland miners and Sudest Islanders 2

THE ISLANDS
2. Sudest: from protection to competition then isolation 18
3. Misima: warlike and civilised 28
4. Woodlark: a people free to walk about 49

OPENING THE MAINLAND
5. The Laloki: a beautiful country but a failure. 76
6. The South-east: a few fine colours and malaria 83

THE NORTHERN RIVERS
7. The Mambare: natives of the fighting variety. 90
8. New Ground: all golden country but very poor 112
9. The Yodda, Gira and Waria: unavoidable mishaps which
   constantly recur in warfare. 121

SIDESHOWS
10. Milne Bay: nothing very exceptional 176
11. Keveri: a magnificent valley and an intense interest in killing 183

THE LAKEKAMU
12. Two Ounces a Day and Dysentery: it grieves a man to lose one
    of them especially if he is a good boy 192
13. No Meeting: a salute of skewers 232

EDIE CREEK

14. On Gold: a quiet whisper ... up on the Bulolo old Shark-eye's getting gold. 254

Bibliography. 272
Index 285

Plates

Gold-sluicing on Sharkeye Park's claim, Koranga Creek,
 Morobe Goldfield, late 1920s..vii
Canoe, Louisiade Archipelago, after MacGillivray 1852 1
Ceremonial axe, Woodlark Island, after Seligman 1910.................17
Tree House, Koiari, after Stone 1880.................................75
Koiari ridge-top village and tree-houses, 1885 79
Hornbill headdress, Kumusi River, Australian Museum 1907 89
Head-dress with bird-of-paradise plumes from Kumusi River
 presented to the Australian Museum, 1907.......................103
Sketch of Neneba village and man from Neneba made by a member
 of William MacGregor's patrol, 1896............................117
Paddle handle, Massim, after Haddon 1894 175
The Abau detachment of the Armed Native Constabulary 187
Shield, Gulf of Papua, after Haddon 1894...........................191
A 'team' of labourers from Orokolo on the Lakekamu
 Goldfield, 1914..213
Two labourers bringing in a cassowary to feed the team, 1914.......213
Miners at Sunset Camp, Lakekamu Goldfield, 1914....................215
Labourers from Milne Bay with canoe that they have made
 to allow miners to prospect the tributaries of the upper
 Fly River, 1914. Frank Pryke is on the left....................222
Prospector trading with people on the upper Fly,
 Pryke expedition, 1914...222
Kukukuku warrior, Lakekamu Goldfield, 1914.........................241
Group of Kukukuku ('More of the nice boys'),
 Lakekamu Goldfield, 1914.......................................244
Patrol Officer Fred Chisholm trying to compile
 a Kukukuku vocabulary, 1914....................................244
Drum, Morobe District, National Museum of Victoria 1932 253
Miners' camp near the junction of Edie and Merri Creeks,
 'one of the picked spots'. Late 1920s..........................262
Frank Pryke's hut on Edie Creek 262

Salamaua's jovial billiard saloon keeper, Bill Cameron
 with a few of his patrons. June 1929. 263
Labourers sluicing with a monitor on Koranga Creek 264
Labourers carrying ore to the tram-line that serves the crusher,
 1938, Kupei. Bougainville, now the site of a giant copper mine. 264
Australian newspaper reports of the Morobe gold strikes. 267

Maps

Map 1 Papua New Guinea goldfields 1878-1930. The Waria
and the Laloki were not officially declared goldfields............3
Map 2 The South-East 4
Map 3 Sudest..8
Map 4 Misima Island 29
Map 5 Murua Goldfield, Woodlark Island 51
Map 6 D'Entrecasteaux Islands....................................66
Map 7 The Laloki...77
Map 8 The south-east mainland 84
Map 9 The northern rivers..91
Map 10 The Mambare 1895 98
Map 11 Milne Bay Goldfield......................................177
Map 12 Keveri Goldfield...184
Map 13 The Gulf of Papua 194
Map 14 Lakekamu Goldfield 200
Map 15 Morobe Goldfield...255

Maps drawn in Cartographic Office, Department of Human Geography, Australian National University

Tables

Table 1 Principal goldmining laws, Papua. 25
Table 2 Louisiade Goldfield European miners 26
Table 3 Louisiade Goldfield (Misima and Sudest) Production 26
Table 4 South-Eastern Division government officers 38
Table 5 Cuthbert's Misima Goldmine Ltd Monthly Labour Record
   February 1940. 44
Table 6 Murua Goldfield production and population 57
Table 7 Gira Goldfield production and population 124
Table 8 Yodda Goldfield production and population 125
Table 9 The Northern Division 126
Table 10 Papuan goldfields Total Gold Yield from first working
   to 30 June 1926. 172
Table 11 Milne Bay Goldfield production and population. 179
Table 12 Lakekamu Goldfield Death Rate. 205
Table 13 Lakekamu Goldfield production and population 210
Table 14 Principal goldmining laws, New Guinea. 256
Table 15 Morobe Goldfield Men employed in mining 257
Table 16 Morobe goldfield production (*ounces*). 258
Table 17 Sepik Goldfield. 259
Table 18 Indentured labourers employed in mining New Guinea 260
Table 19 Gold exports, Papua 266
Table 20 Gold exports, New Guinea 268

Chronology

1847   Members of the Society of Mary landed on Woodlark Island to establish the first mission station in New Guinea. It survived for eight years.

1871   Some Australian adventurers set out in the *Maria* to look for gold in New Guinea. The *Maria* was wrecked on the Queensland coast.

1873–4   Captain John Moresby mapped parts of the southern and north-eastern coasts.

1874   The Reverend William and Mrs Lawes of the London Missionary Society joined Polynesian mission teachers working in the Port Moresby area. The Lawes were the first Europeans to settle on the south coast.

1878   About 100 white miners prospected the Laloki and other rivers inland from Port Moresby, but found no gold.

1884   Britain established a Protectorate over south-east New Guinea and Germany claimed north-east New Guinea.

1888   In August David Whyte wrote to John Douglas, the Special Commissioner for New Guinea, to report that he and his party had found 142 ounces of gold on Sudest Island.

           In September British New Guinea became a Possession of the Crown and William MacGregor became the first Administrator (later Lieutenant-Governor).

1889   Gold was found on Misima Island. MacGregor proclaimed the Louisiade Goldfield which included both Misima and Sudest Islands.

1894   James Hurley, the leader of a prospecting party, was killed on the north-east coast.

1895   Early in the year Lobb and Ede found gold on Woodlark Island and the Murua Goldfield was proclaimed in November. In July George Clark, the leader of a prospecting expedition, was killed on the Mambare River.

1896	William Simpson's party found gold on MacLaughlins Creek, a tributary of the Mambare.
1897	Green, Fry, Haylor, police and prisoners were killed on the Mambare. Gold was found on the Gira.
1899	Gold was found on the Yodda. Gold was found at Milne Bay.
1901	Pryke and Klotz found gold inland from Cloudy Bay.
1906	British New Guinea became the Australian Territory of Papua. Crowe and Darling found gold on the Waria River.
1909	Crowe and Pryke found gold on the upper Lakekamu.
1914	Australian troops occupied German New Guinea.
1921	German New Guinea became the Australian Mandated Territory of New Guinea.
1923	Park and Nettleton registered claims on Koranga Creek in the Morobe District.
1926	Royal found gold on Edie Creek.

OUT OF COOKTOWN

Canoe, Louisiade Archipelago, after MacGillivray 1852

1

A Meeting

north Queensland miners and Sudest Islanders

By the end of 1888 nearly 400 Australian miners had pitched tents on the beach near Griffin Point, at the Four Mile and Nine Mile camps, and by claims scattered along the gullies of Sudest, the biggest island in the Louisiade Archipelago. The miners talked of Sullivan and his party who had taken 200 ounces in fourteen days from the west of the island, the seven men who arrived on the *Zephyr* and won 50 ounces in three days, and the diggers who turned up 300 ounces in one shallow gully. It was, they said, 'good looking gold' likely to 'go very nearly £4 per ounce'. But when men began returning to north Queensland early in 1889 none took fortunes with them. The Mercury carried twenty-three men who told the Cooktown customs they had 150 ounces; the *Lucy and Adelaide* brought twenty men and 240 ounces; the *Griffin*, twenty-three men and 241 ounces. A few men had made more than wages, but most who followed the rumours of rich gullies further on arrived to find that all the easy gold had been taken. They could re-work the creek beds or open up the terraces for a few pennyweight a day, or they could 'loaf on camps' hoping a rich strike would be made before their stores ran out. The talk on Cooktown wharf was that Sudest was for 'gully-rakers' and 'tucker men', those prepared to scratch a bare living; but of course you could never be sure. Already men were prospecting other islands.

Until November 1888 the miners found Sudest 'cool and pleasant to work'. The anchorage was sheltered by the small islands of Piron and Pana Tinani and the long ling of reef running off the Calvados Chain; and beyond narrow tangles of mangrove near the beach were patches of grassland, rounded hills, casuarinas and coconuts, timbered creeks and forested ridges. Fresh water was plentiful and travelling fair. The diggers shot pigeons and introduced a few head of cattle. One miner, more accustomed to distances in the Australian colonies, said: 'One hop, step and jump and you're over — a good running jump and you're at the other end.' Sudest was 40 miles long and 7 miles wide.

In November heavy rain began to fall, and for a week it did not slacken; it was 'not ordinary rain, but regular sheets of water'. Men could not light fires or dry clothes. 'You lie on your bunk and gasp', a miner wrote, 'for there is not a breath of air to stir the humid muggy atmosphere'. Miners suffered from malaria and dysentery. Later the north-west which brought the 'wet' came in sudden storms, ripping away tents and exposing the sick to lashing warm rain. Gaunt men came ashore at Cooktown, some still trembling from the effects of fever, and on each trip schooner captains reported that they had buried two or three men at sea. The boats from Cooktown and Cairns generally kept the five stores on Sudest well stocked, but not all the miners could pay for rations. With money short and 'liquoring expensive' even the storekeepers were 'not getting rich at a gallop'. And the 'few shady members of the spieler fraternity' who joined the rush found no men who were both rich and gullible.

Before the arrival of the miners about 1000 people lived on Sudest. Slightly built and brown-skinned, they had few clothes and many decorations. The women wore leafy skirts bunched on their hips and the men a pandanus leaf drawn tightly across the genitals. In his cushion of hair a man might carry a long-toothed wooden comb, a flower or scented leaves; he had shell ornaments to pass through the septum of his nose and hang from his ears; he wore necklaces and armlets of shell, bone and woven fibre; he tied streamers of pandanus to his ankles; and he carried his betel nut, lime gourd and carved spoon in a basket. The Sudest lived in small inland villages of only four or five houses; each house was a curve of palm thatch over a raised platform up to 30 feet

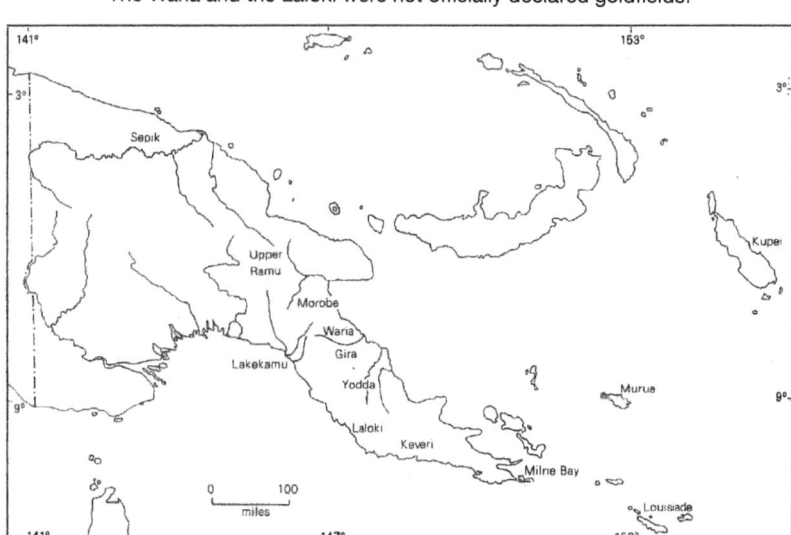

Map 1 Papua New Guinea goldfields 1878-1930.
The Waria and the Laloki were not officially declared goldfields.

long. The ground between the houses was swept clean, impressing many nineteenth-century European visitors with 'the cleanly habits of these savages'. The Sudest fished the lagoons and reefs with spears, lines and long woven seine nets. They cultivated yams, taro, sugar cane and bananas, made sago in the swamps, gathered fruit from their tree crops, and jealously husbanded their pigs.

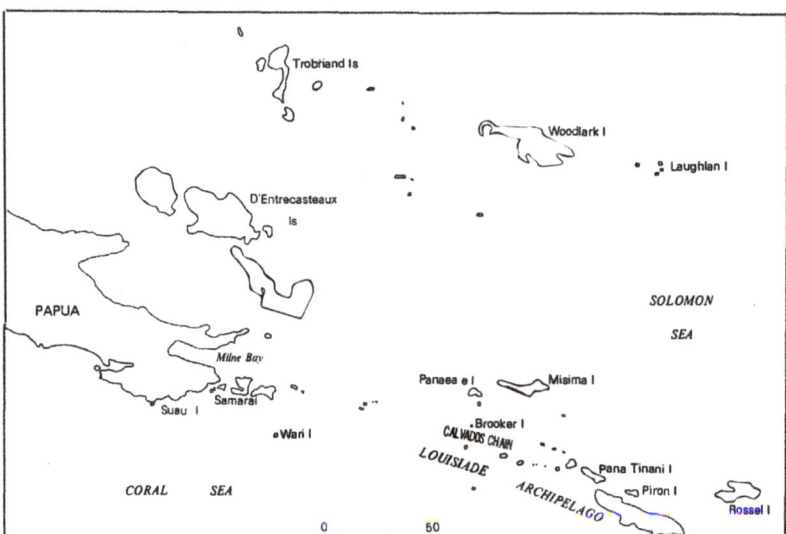

Map 2 The South-East

A sub-group of the Massim people, the Sudest shared many characteristics of a culture spread through the islands of Misima, Woodlark, the Trobriands, the D'Entrecasteaux and the Louisiade Archipelago, and to the mainland villages around Milne Bay. The English anthropologist Seligman, who visited the area in 1904, called the Massim 'merchant adventurers', and Malinowski, who began fieldwork among the northern Massim ten years later, described the people as 'daring sailors, industrious manufacturers, and keen traders'. 'Sailing about' was a common pastime and the construction of sea going canoes, *waga*, was probably the most highly developed craft of the area. The main centre of *waga*-making in the Louisiades was at Panaeate. There the prow and stern boards were carved, the hulls shaped, the side boards fitted, the outrigger fixed and the canoe painted and decorated with shells and given the name it would carry although its owners might change. The *waga*, propelled by an oval matsail, carried men from the Louisiades to Woodlark, the D'Entrecasteaux and Wari Islands. Able to use the many reefs and islets of the area as camping grounds, the traders normally did not spend the night at sea; and when foreign captains brought sailing boats into the area they too chose not to sail at night in waters so scattered with hazards. The men of some coastal communities spent two or three months a year away from their homes. Canoemen and villagers exchanged pots, stone axe-heads, shell armlets and necklaces, carved lime spoons, food and talk.

Before the arrival of the miners the Louisiade Islanders had encountered a variety of foreigners. Torres, Bougainville, D'Entrecasteaux, Coutance and D'Urville passed by without landing; and all except Torres gave islands and points names which continued to be used by later visitors. After the foundation of Sydney, boats going from Port Jackson to China sometimes threaded their way through the islands, and in the 1830s and 1840s whalers worked the Solomon Sea. By 1850 many islanders were accustomed to taking their canoes out to trade with the crews of passing ships; already some of the axes which men carried hooked over their shoulders were fitted with blades ground from hoop-iron, bolts, and fittings from wrecked vessels; and stories of how the Laughlan and Woodlark Islanders had killed more than twenty men from the whaler *Mary* in 1843 must have travelled the trade routes of the Massim.

When Captain Owen Stanley on a surveying voyage of the New Guinea coast in 1849 brought the *Rattlesnake* and the *Bramble* into Coral Haven, the Sudest had a chance to learn more of the foreigners, acquire their goods and test their strength. After their initial suspicions had passed and their 'violent gesticulations' had failed to persuade Her Majesty's boats to leave, villagers from Sudest and nearby islands traded frequently with the foreigners. They were 'greedy for iron', and when four or five canoes were alongside there was much 'squealing and shouting and laughing'. In one day the Brierly Islanders exchanged 368 pounds of yams for seventeen axes and a few knives; but for the month the *Rattlesnake* and the *Bramble* stayed in Sudest waters all men carried arms. The British sailors believed that the men who watched them were just waiting for a chance to attack, and the islanders who held the wooden swords and spears decorated with pandanus leaf pennants may have had the same fear. The Sudest heard the guns and saw the shot fall in the water when the sailors used their muskets against the canoemen, who took the iron used to anchor a tide-marker, but they did not know the sailors had fired to frighten and not to wound. After the *Rattlesnake* left Coral Haven three canoes from Pana Tinani came alongside one of the boats from the *Bramble*. In dull light just before sunrise they were seen by the watch and some men then came on board pretending to trade. Fighting broke out as the islanders on board grappled with seamen and those in the water attempted to drag the boat inshore and capsize it. Not deterred by the first musket shots, the islanders wounded two sailors before fleeing through the mangroves pursued by shot from a 12-pound howitzer. News of the encounter spread quickly among the coastal villagers, and men at Brierly Island tried to tell the crew on the *Rattlesnake* that someone at Pana Tinani had been killed. Not having heard of the fighting at Pana Tinani, the British sailors did not understand the strange pantomime, but they did agree to a request to demonstrate the power of their guns by shooting some birds. Later, when the *Bramble* reported the clash, her crewmen were unable to say how many of their attackers had been wounded or killed by gun fire, but as they had moved along the coast of Pana Tinani men on shore had followed them brandishing their spears and shouting challenges.

From the 1870s foreigners and conflict were more frequent in the Louisiades. The foreigners, coming in boats with names of peace and innocence — the *Annie Brooks*, *Pride of the Logan*, *Daisy*, *Emily*, *Alice Meade* and *Lizzie* — worked reefs and lagoons for pearl and bêche-de-mer, or they offered the villagers hoop-iron, axes, knives, calico, jews-harps, tobacco and guns in payment for copra, bêche-de-mer or labour. Many of them left no records of their voyages but the lists of the crewmen who died showed the variety of representatives from the family of man to pass through the Archipelago. Between 1878 and 1887 Chinese, Malays, Queensland Aborigines, South Sea Islanders, Australians, Americans, Englishmen, Frenchmen, an Indian, an African and a Greek were killed in the islands of south-east New Guinea. New Guinea saltwater men from other areas survived long canoe drifts, wrecks or massacres to live in the Louisiades. In 1887 John Douglas, the head of the administration of the Protectorate of British New Guinea, picked up three Torres Strait women whose husbands had been killed in 1878, and men from Torres Strait, New Britain and Manus lived on Brooker Island for various periods. The Solomon Island crew on the *Retrieve* killed their officers, burnt the boat and, taking Snider rifles with them, settled on Brooker Island where they became men of eminence. The Reverend Samuel MacFarlane, calling at Brooker in 1878 to see if there were any survivors from William Ingham's boat, the *Voura*, was greeted by 'the unmistakable Australian "cooey!" followed by the clear tones of a voice asking in English "Who are you?" 'He assumed both came from 'Billy', a Torres Strait Islander who had previously served on pearling boats; but he thought it prudent not to land and check.

No Christian missionaries had worked in the Louisiades before 1888, but the islanders' trading partners could give contradictory reports about the missionaries' behaviour and success. From 1847 until 1855 Catholic missionaries worked on Woodlark Island. At first valued as a source of trade goods, the missionaries were unable to convince the Woodlark Islanders that their explanation of the world was either intellectually more satisfying, likely to make men treat each other more generously and justly, or able to bring greater material rewards. For much of their time on Woodlark the missionaries were involved in an unproductive contest with the islanders, and when the missionaries left the islanders could believe that they had won. In 1877 the first Polynesian teachers of the London Missionary Society came to live on Wari Island and the Reverend James Chalmers established a mission station at Suau on the south coast near Milne Bay. Chalmers made his last visit to Suau in 1882, but several teachers, sustained by periodic visits from mission boats, became influential in the Milne Bay area where they helped interpret the outside world to the islanders. By 1878 at the latest Louisiade Islanders had encountered the missionaries at Wari.

When Commodore James Erskine proclaimed south-east New Guinea a British Protectorate in 1884 he failed to include the Louisiades within the British Empire. It was an oversight. In January 1885 Captain Cyprian Bridge arrived at Brierly Island to tell the people

> that the Queen had taken them under her protection, that they must give up fighting amongst themselves, cease to be cannibals — which they admitted they were occasionally — and on no account to injure white men but bring any grievances they might have before the first British officer who might come amongst them.

Bridge also conducted the 'customary ceremonies' at Pana Tinani, Rossel and Sudest, where the people were timid but 'stood the feu de joie better than could have been expected'. As evidence that they had joined the British Empire, Louisiade islanders had flags, copies of the proclamation, and medals, and one man, Rulitamu of Sudest, had taken Bridge's name. The ceremonies of 1885 were not followed by an increase in the number of government officers in the Archipelago; but other events had already occurred which greatly increased the turmoil in the area.

In January 1884 labour recruiters from Queensland entered the Louisiades. The crew of the *Lizzie* seized some Sudest men while they were asleep onshore and confined them in the hold until they were at sea, but most recruits were duped not kidnapped. Knowing a little of the ways of the bêche-de-mer fishermen, they went on board believing they would be 'sailing about' for a few months only. Kroos ('Sandfly') and Manboki ('Dixon'), who had served on bêche-de-mer boats, been to Cooktown and learnt Pidgin, may have knowingly agreed to work on the canefields, but most were like Tacomala of Piron Island. Although he himself had not worked for the bêche-de-mer fishermen he had seen their boats and he knew five Piron Islanders who had worked for them; without being able to 'hear' (understand) the recruiters he joined the

Map 3 Sudest

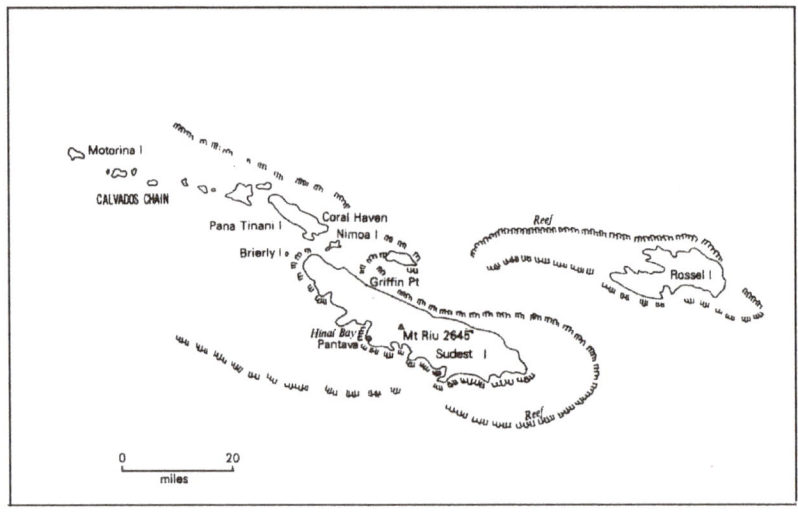

Ceara for Queensland. Many from Pana Tinani and Sudest went out to the recruiting boats with food expecting to be able to trade and were tempted to leave their canoes. Tagalita of Sudest and five others took fish out to the *Lizzie* and after receiving knives, tobacco and calico, all except one agreed to go and work for three months to earn more trade goods. Touinsi, speaking through an interpreter, told the Royal Commissioners inquiring into his removal from New Guinea waters that the recruiter had spoken to him and shown him an axe and other goods: 'I did not understand his talk ... he showed me tomahawk and knife; the tomahawk talked to me; I went in schooner.' In spite of the Queensland law preventing returning labourers from taking firearms with them, nearly all hoped that they would be paid a 'gun and box' (of trade goods). At sea many learnt for the first time that they would be away for three years; they had little idea of the work expected of them on the canefields until they were in Queensland. Some wept when they learnt of the 'gammon along me'. Later in 1884 the *Sybil* and the *Heath* were unable to get recruits in Sudest waters, probably because stories had spread of men being tricked and forced to go aboard earlier boats and because those men who were thought to be going away for only two or three months had not returned. The Commissioners having decided that the recruits were taken by deceit and violence, 405 men were returned to south-eastern New Guinea in June and July of 1885. At six landing places in Sudest seventy-one men and thirteen bundles of trade to compensate the relatives of dead labourers were put ashore; at all places except one the villagers would have nothing to do with those who returned their country men. Another twenty men were landed at Pana

Tinani, and a few others were taken to Panaeate, Misima and to islets near Sudest and in the Calvados Chain. In 1887 another group returned in the *Truganini*. Included among those going to Sudest was Siup of Rambuso village, who had gone on board the *Forrest King* in response to the recruiters' invitation: 'You like tomahawk? You come in the boat you get them.' He returned to Rambuso after three years in Queensland carrying an umbrella and wearing a feathered cap and a 'full dress suit of spotless white'. His relatives greeted him with demonstrations of affection. The crew of the *Truganini* passed among friendly people to picnic in a landscape of many greens. The change in attitude of the Sudest towards the carriers of returning men was a result of other events in the area, and not a response to Siup's finery.

A year after the first recruits returned Captain T. Mullins reported that a 'lad' left on Nimoa Island to look after a bêche-de-mer station had been murdered by people from an island in the Calvados Chain and his head sold on Misima. On Sudest, Mullins said, the villagers were 'disaffected, dangerous and threatening'. The degree of turbulence was soon known widely: a few days after Mullins wrote his report, Captain J.C. Craig and his crew of three Europeans and five Malays were murdered and his pearling boat, the *Emily*, was looted by Pana Tinani and Sudest Islanders. Some of those who died were shot by a man from Pana Tinani using Craig's Winchester rifle. It was the seventeenth incident in ten years in which foreigners had been killed in south-east New Guinea. Some of the attacks may have been made by men who had been ill-treated by labour recruiters, or by the relatives of men who had not returned, or by men confident that they now had the knowledge and power to defeat the traders. But many conflicts arose from particular disputes between islanders and traders. At Panaeate in 1885 the people said they had killed Frank Gerret because he had beaten a man to death, and Kasawai of Pana Tinani claimed that Craig was killed after promising rifles to two men and then supplying only one. The Acting Deputy Commissioner and Government Agent from Samarai, Henry Forbes, who visited Pana Tinani two months after the attack on the *Emily*, heard two other explanations. A Cooktown pearler, J.B. Robinson, working in the area thought that Craig had been attacked by men who believed that they had been underpaid; and Nimoa Islanders said that Godaw villagers of Pana Tinani stole a woman from Ewia village and gave her to Craig. Becoming afraid that the Ewia would attack them when the *Emily* left, the Godaw asked for the woman to be returned. Craig refused and they killed him and his crew to forestall Ewia reprisals. But Forbes could find only one woman who had been living with a member of the crew on the *Emily* and she, he thought, had been 'obtained with consent'. On his second visit Forbes decided that the story about the rifles was the most probable. The Craig case showed that while local people were likely to have specific grievances, foreigners were unlikely to know much about them.

The deaths of Craig and one of his crew, Walter Hollingsworth, left widows in Cooktown and cast 'great gloom' over the town. Within two years the miners were leaving Cooktown for Sudest taking with them their prejudices, their skills and a way of life. One of the early shanty owners on Sudest was 'late of the Royal, Cooktown' and Clunn's hotel with its iron lace balcony was taken in sections from Cooktown and put together in Samarai.

Unlike most other white Australians the Cooktown miners had already lived in a community where white men were in a minority. Established in 1873 to serve the Palmer goldfields, Cooktown was the port of entry for the 17,000 Chinese on the field by 1877. While the *Cooktown Courier* might deplore the 'hordes of Chinese which an idiotic Government permitted to swamp the best alluvial field ever discovered in Australasia' there were no vicious riots on the Palmer, and the *Courier* condemned the two drunks who helped themselves to the products of a Chinese fruit shop and pelted the women in charge. By the late 1870s the prosperity of Cooktown businessmen depended on the Chinese staying in the area: the business men opposed restricting Chinese immigration and spoke of the virtues of the Asiatic diggers.

The *Courier* recorded more violence between Aborigines and Europeans and showed more prejudice against those Aborigines who survived. On 20 May 1890 the *Courier* used the heading 'MURDERED BY THE NATIVES' for the third time that year. Under another common heading, 'THE BLACKS AGAIN', it told stories of miners and station hands being wounded, horses speared and the telegraph lines being cut so that tribesmen could obtain wire for spear points. Old diggers could recall the raids and counter raids of Battle Creek and Hells Gate where many white miners, Chinese and Aborigines had died, and others had sung the Old Palmer Song:

> I hear the blacks are troublesome,
> And spear both horse and man.

And while the *Courier* did not object to a little smoothing of the pillow for those Aborigines who escaped disease and dispersal, it did think such acts of charity should take place away from Cooktown. When the police shifted the 'blacks' out of the town area in 1890 the *Courier* commented: 'This action will be endorsed by ratepayers as the squalid niggers were a great nuisance.' It praised the decision to make part of the annual distribution of blankets at the mission, so preventing the Aborigines coming to town where they were 'always a nuisance'.

The north Queensland diggers also encountered Pacific Islanders. In Cairns, the second port of the New Guinea goldfields, men were debating whether the cane should be cut by white labourers or indentured Melanesians. To those who argued that the white man could not do hard manual labour in the heat of the canefields, the miners could reply that they had worked alluvial fields in the wet tropics for fifteen years. No miners wanted to change the Act of 1880 which excluded islanders from the Queensland goldfields. In Cooktown Melanesians working on luggers and schooners in the islands trade came ashore and unless they were like Whittens' Papuan crew who attended the Amalgamated Friendly Societies' Annual Sports in 'native dress', they attracted little attention. Cooktown's white community talked with heat about Melanesians when they killed the crews of boats which had often tied up at the Cooktown wharf; then the local press cried loudly for vengeance. The crew of the *Hopeful*, convicted of the murder and kidnapping of labour recruits, were 'martyrs' in Cooktown and the *Cooktown Courier* proclaimed with pride that no jury there would have found Captain Neils Sorenson guilty. Two historians have subsequently described Sorenson as a psychopath who bashed, murdered and kidnapped in the islands.

Believing Cooktown would be the main port for trade between the Australian colonies and New Guinea, Cooktown businessmen kept a watch on the way the government in Port Moresby regulated relations between black and white residents. When William MacGregor called at Cooktown on his way north to begin the administration of British New Guinea as a possession of the Crown, seventy 'leading and representative men' gave him a 'dejeuner' at the town hall. MacGregor had left a poverty-stricken farm in Scotland, graduated in medicine and served the Empire in Mauritius and Fiji before he was appointed to British New Guinea. Some citizens who dined with him in Cooktown were disappointed to learn that MacGregor was 'not disposed to recognize the superiority of the white over the "poor black"'. His early legislation concerned with land, labour and the supply of arms and drink to New Guineans provided further evidence for the north Queensland observers who believed that he was a 'niggerlover' introducing 'twaddling maudlin' protective policies to restrict the 'legitimate' interests of traders and miners in New Guinea.

In May 1888 David Whyte and nine prospectors left Cooktown on the *Juanita*. Whyte, captain of a pearler, had reported finding a gold-bearing reef on Pana Tinani and, encouraged by John Douglas, Cooktown businessmen helped meet the expenses of the expedition. By September 'knots of miners all over town' were talking of the 142 ounces of gold brought back by the men from the *Juanita*. Having found only hungry quartz on Pana Tinani, Whyte's party crossed to Sudest where they discovered alluvial gold in the Runcie River. Douglas had told Whyte that if he was unsuccessful on Pana Tinani he should try the big island, for the engineer of the *Truganini* had collected samples of quartz there in 1887. Already the *Griffin* had landed another party which had immediately begun to work 'good gold'. On 18 September the *Zephyr* cleared Cooktown with sixty miners and three days later another fifty left on the *Sea Breeze*. By the end of October 200 miners were on Sudest and more were planning to try 'the islands'.

There was no frenzied rush with clerks dropping their pens, sailors deserting their ships and grocers casting aside their aprons and order books. Most men who went to Sudest had worked on the Palmer, Hodgkinson, Etheridge, and Croydon fields of north Queensland: they were 'of the right stamp ... experienced, strong and willing workers'. They were proud of their skills as prospectors, miners, bushmen and pioneers; after them came the settlers and businessmen. They knew that men had died of fever on the abortive rush inland from Port Moresby in 1878; but the death rate had also been high in the early days of the Queensland fields; and if the climate was so harsh why did the missionaries, Chalmers and Lawes, look so 'sleek and fat' when they came to Cooktown? They had heard too of the savagery of the islanders, but their prejudices were modified by unexpected reports from early diggers who spoke of the Sudest as friendly and useful.

The Craig 'massacre' of 1886, the most violent in a series of clashes between traders and villagers in south-east New Guinea, was followed by widespread demands for harsh reprisals. The traders' cry that now no white man would be safe in the area was all the more shrill because the Pana Tinani had taken fourteen rifles, four revolvers and ammunition from the *Emily* before they soaked her stored sails in kerosene and burnt her. The *Sydney Morning Herald* called on the government to shoot some islanders:

> Killing a few pigs and burning a few huts, which is the usual punishment inflicted by the British authorities upon the aboriginal murderers of Englishmen, will not be regarded as sufficient punishment for the death of Captain Craig and his crew.

But as Queen Victoria's representative in Port Moresby, H.H. Romilly, pointed out, the government of the Protectorate of British New Guinea was uncertain of its powers to act and in the meantime 'the natives' could go on 'murdering away merrily'. A British man-of-war, H.M.S. *Diamond*, called at Pana Tinani but 'could do nothing'. Then Forbes fitted a gatling gun to the schooner *Coral Sea*, hired Nicholas Minister and his cutter, the *Lizzie*, at £12 per week, collected a force of forty-five men from Wari Island and the eastern Louisiades, and sailed for Pana Tinani. A bêche-de-mer trader already in trouble for seizing island men for work and women for pleasure, Minister led a force of irregulars ashore, each man wearing a red badge to distinguish him from other islanders. The unofficial report said that when Minister returned next morning Forbes leaned over the rail and asked him if he had made contact with the man suspected of leading the attack on Craig. Minister replied, 'Yes, there's the bastard', and handed up a basket containing a head. In his official report Forbes said Minister and his troop shot Dagomi, a 'noted cannibal and robber' and father of the man who shot Craig, wounded some other men, burnt three villages and recovered guns and ammunition. The Pana Tinani retreated into dense scrub from where they fired guns on the government party. On Forbes's instructions Minister also attempted a surprise raid on Popagania, a Sudest village and the home of Mutiana who was thought to have killed Craig. After visiting Robinson on his boat and a delay in which 'Everyone was drunk, from the mate in charge to the blacks & their gins', Minister left for Sudest. Finding the village deserted, he burnt the houses, cut down the coconut trees and 'ravaged ... gardens'. Forbes hoped that now the Pana Tinani 'marauders' had lost some of their guns other islanders would 'pay them back in their own coin, and help to reduce a tribe whose reputation is of the worst character'. He banned foreign traders from Sudest and Pana Tinani.

Responding to Forbes's encouragement or deciding for themselves to take advantage of the weakness of their enemies, Brooker Islanders, sometimes acting in alliance with other groups, launched a series of attacks on Sudest and Pana Tinani. The raiders carried guns and some of them had served in Minister's punitive force. By the time the miners came to the Louisiades there were only a few people, 'ill-fed, [and] miserable in appearance' on Pana Tinani. Their one village was built on swampy land in the middle of the island and spear points set in the undergrowth guarded the walking tracks leading to the houses. On Sudest the people had been forced inland, their gardens had been reduced in size, some tree crops had been destroyed and they owned no canoes for ocean voyaging. When Douglas visited Sudest just after the men from the *Juanita* began work at Runcie River, the Biowa villagers were lamenting the loss of six dead and three children carried off by raiders; and an early miner, C.L. Bourke, reported seeing 'a grand shindy' in which raiders almost wiped out one group. Douglas rescued the Biowa children and Whyte took them to their homes. But the arrival of more miners ended the attacks on Sudest.

The miners had taken their first gold from the land of a people already responding to a variety of outside forces. For fifty years the Sudest had used iron, they had tested their strength against European guns forty years before the arrival of the miners, about one hundred men from Sudest and Pana Tinani had sailed as labour recruits for Queensland, men from different parts of Sudest spoke some Pidgin, and it is impossible to tell how much the intensity and style of warfare in the area had been altered by the presence of foreigners and the use of guns. The Brooker Islanders may have been influenced by the Solomon Islanders who sailed with them; but the people of the Louisiades were headhunters before the arrival of any foreigners; and now that old alliances and trading partnerships had been broken, raiders carried firearms and violent foreigners had weakened some groups, it was likely that some islanders would exploit the situation to their own advantage.

In 1888 the Sudest were suffering at the hands of other islanders who had greater access to foreign goods and patronage, and therefore to wealth and power. Mutually suspicious and speaking different dialects, the Sudest were unlikely to unite against the raiders. They welcomed the miners because their presence gave protection and they probably hoped that later they could be exploited, enabling the Sudest to regain their position relative to other groups in the Archipelago. Had the miners arrived two years earlier they might have called the Sudest arrogant and savage; in 1888 they were largely unaware of the forces which now allowed them to report: 'the natives are all friendly'.

Much of the early information about mining is from the *Cooktown Courier* with a lesser amount from the *Cairns Post*. Brierly, Owen Stanley, MacGillivray 1852, Huxley 1935 and Wilcox wrote about the meeting between the Sudest and British seamen in 1849. The killing of the crew of the *Mary* is described in Ward, Vol. 4, pp. 8–9. Mission activity is recorded in L.M.S. Archives, MacFarlane 1888, Lovett 1903 and Laracy 1969.

Queensland Parliamentary Papers, 1887, Vol. 3, pp. 40–2 list conflicts between peoples of British New Guinea and foreigners. Bridge's account of his flag raising is in *Great Britain Parliamentary Papers*, 'Further Correspondence respecting New Guinea and other Islands in the Western Pacific Ocean', 1884–5, Vol. 54, pp. 100–5. The Report with minutes of evidence taken before the Royal Commission appointed to inquire into the circumstances under which labourers have been introduced into Queensland from New Guinea and other islands etc.' Q.P.P., 1885, Vol. 2, pp. 797–988, is a most valuable document as it records the evidence of the islanders. Additional information from 'Correspondence respecting the return of the New Guinea Islanders' Q.P.P., 1885, Vol. 2, pp. 1053–74; 'Return of Louisiade Islanders to their Native Islands', Q.P.P., 1887, Vol. 3, pp. 611–19; Romilly 1886, 1889 and 1893; Wawn 1973; Corns 1968; Bevan 1890. Two historians who have written about Sorenson are Corris 1973 and Scarr 1967. The violence between traders and islanders, is recorded in Royal Navy Australian Station, New Guinea 1884–8, microfilm, National Library of Australia; *Annual Reports* of British New Guinea; 'Massacres in British New Guinea (Correspondence respecting, and reports of Special Commissioner upon)', Q.P.P., 1887, Vol. 3, pp. 719–26, Records of the Protectorate of British New Guinea, C.A.O., G3–G29 (much of the correspondence is also printed in parliamentary papers); Mayo 1973; *Sydney Morning Herald*, 1 November 1886, and 10 February 1887; and *Pacific Islands Monthly*, December 1943, p. 43.

The experience of the diggers before leaving Queensland is taken from the Cooktown and Cairns newspapers. Jack 1921, Bolton 1963 and Holthouse 1967 provided general background. Binnie 1944 (the son of a mining engineer), Browne 1927 (a journalist), Corfield 1921 (a carrier) and Hill 1907 (a warden) have published their memories of Cooktown and the Palmer. The early development of mining on Sudest is recorded in the north Queensland newspapers; *Papuan Times*, 9 April 1913; Douglas 1888 and 1890; and see Chapter 2. Whyte's letter to Douglas of 24 August 1888 reporting the discovery of gold was printed in the *British New Guinea Government Gazette*, 1888, p. 49.

THE ISLANDS

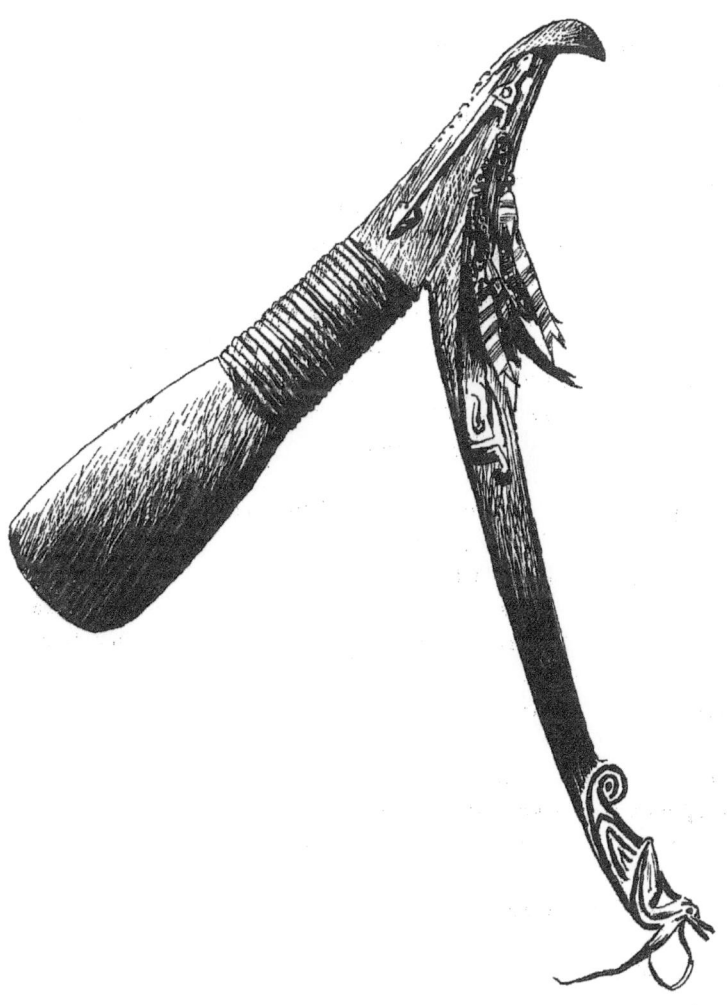

Ceremonial axe, Woodlark Island, after Seligman 1910

2

Sudest

from protection to competition then isolation

At first the Sudest and the miners did not compete for the resources of the island. The miners worked the gullies and coastal silts, 'blind-stabbing' in the shallow water at low tide to recover the alluvial washed out to sea; most of the Sudest stayed close to their villages on the ridges and worked their gardens on the slopes. A few men came down to sell fish and coconuts, but their gardens were too small to allow them to become the main suppliers of food to the miners.

The alliance between the Sudest and the miners was strengthened when the villagers decided to work for the foreigners. From the arrival of Whyte's party, men took sticks of tobacco for carrying and later they worked on the claims. The use of cheap labour to work alluvial claims was strange to diggers coming from Australian fields, and some found the practice repulsive. They thought it unjust for a miner to use a 'team' of labourers to work out a single man's claim quickly and move on to new ground; they wanted the alluvial fields to be the preserve of white men who began work as equals on their claims of uncertain value. MacGregor excluded Chinese from the goldfields of British New Guinea, but he and most miners eventually agreed that Papuans could choose to shovel alluvial, dig races, and look after sluice boxes on the goldfields of their own land. The Sudest continued the basic economic change which began when the first men had left their villages to work on visiting ships; they added wage earnings to subsistence farming. The Australian miners also underwent a subtle transformation; from independent workers to employers. Still thinking of themselves as battlers, depending on their skill and luck and keeping the corns on their hands, the miners had acquired new interests; wage rates, ration scales and the 'proper' relations between a white boss and a Papuan 'boy'.

The Sudest gave labour and food and tolerated the miners on their lands; the miners provided protection and a means of obtaining trade goods. These were basic and complementary bonds; but other factors caused minor clashes. The miners came without women and with false beliefs about the availability of 'native' women. Perhaps accepting a common belief among settlers of northern Australia, they thought that 'If you give [an Aboriginal] a stick of tobacco or say a handkerchief of a pretty description, or anything of that sort which appeals to his fancy, he will no doubt let you have the use of his lubra … '; or they knew of horsemen in outer areas who made a practice of 'running [an Aboriginal woman] down in the bush and collaring her'; or they were influenced by the fantasy that on the islands the women were gifted with erotic skills and always eager for sexual encounters. But the Sudest men would not pander and the women were restrained by customs and personal obligations not unknown among the wives of Cooktown. Disappointed miners threatened violence and the Sudest responded by hiding their women. A miner entering a village would find only men, and when MacGregor first visited Sudest in October 1888 he saw no young women in the first three or four villages he inspected. It was, he reported, 'only after they understood something of my official position that I was admitted to any intimacy and shown all the members of each family'. Three years later some villagers on the south coast still kept their women away from the puritanical MacGregor. The men of the area said that the miners were still asking for women but not getting them. Bingham Hely, Resident Magistrate for the Eastern Division, turned the miners' failure into a virtue. He wrote in his annual report of 1890: 'never since the commencement of the gold workings in this District have I heard of a single case of tampering with native women'. By 1893 some miners felt obliged to contradict the slightest suggestion that they were the sort of men who pursued island women. Lucien Fiolini, thought to have been an escapee from the French prison on New Caledonia, was killed on Rossel Island. On a tour of the south-east MacGregor met a deputation of six miners on Sudest who presented a list of the deficiencies of his government. They included a request that he withdraw a statement that Fiolini had been killed because he was after women; it reflected on all white men in the area. MacGregor chose not to withdraw.

While the miners won only a distant glance at Sudest women, the Sudest men coveted the possessions of the miners. In camps left unguarded during the day the Sudest had plenty of time to look around and take what they wanted; and the miners blamed the Sudest for everything they lost. Losses of stores and equipment were important to miners struggling to pay for their rations and they complained frequently about the thieving Sudest. Miners emptied their revolvers at 'these gentry'; a man believed guilty of theft on a previous occasion was found 'prowling round the camp again, and was promptly shot'; and J. Morrison, a storekeeper, said that one man caught stealing was tied to a tree and given a 'sound hiding'. John Cameron, a surveyor and prospector whom MacGregor appointed the first warden on Sudest, punished villagers by seizing hostages and cutting down coconut trees: he earned MacGregor's displeasure for acting illegally and the miners' praise for being 'firm handed'.

Two years after the opening of the field, when the number of white miners had fallen to about forty and both miners and villagers knew a lot about each other's behaviour, two Sudest men killed a miner. Gaiboa, a leading man of southern Sudest, had sent two men to sell some sweet potatoes and coconuts to William Bakem, an old miner working alone near Condé Point. When the men returned with half a stick of tobacco Gaiboa, incensed by the smallness of the payment, told the two men to kill Bakem. They returned, speared Bakem through the chest and hit him with an axe. One of the men, Tamana, visited Bakem again the next day, and finding him still alive, killed him. Immediately the miners learnt of Bakem's death they burnt some houses as a general punishment, but held the two murderers until a government officer arrived. In January 1891 at a sitting of the Central Court held on the government boat, the *Merrie England*, Judge Francis Winter sentenced Tamana to death and his accomplice to ten years' imprisonment. MacGregor reviewed the case in a dispatch to the Governor of Queensland on 28 January 1891:

> I have given full & careful consideration to all the circumstances of the case & have come to the conclusion that it is clearly my duty to direct that in this instance the law shall take its course. The natives of [Sudest] have been brought much into contact with white men & have been very frequently visited by Government Officers. It is impossible to believe that they do not know that it is a great & serious crime to kill a white man. There is nothing whatever in the case to justify or excuse the resort to violence ...

Careful to check that there was no 'injury to be avenged', MacGregor may not have known the basis of Gaiboa's deep sense of grievance. Bakem had violated the code regulating exchanges. He had not only underpaid Gaiboa; he had made a public declaration of his contempt for Gaiboa, and Gaiboa may have magnified Bakem's gesture by assuming that the old miner was a rich man. MacGregor ordered his officers to assemble people from different villages and Tamana was hanged before them. On his return to Sudest in June MacGregor was pleased to find that people at the other end of the island knew about the execution. The *Cooktown Courier*, which had predicted that MacGregor with his 'maudlin native policy and respect for his dear friends, the missionaries' would act against the interests of white Australians in New Guinea, had already begun to praise him for dealing out 'strict justice': soon after his arrival he had hanged four Papuans on the mainland for the murder of a white trader.

Some thieving, summary punishment, one spearing and one public hanging was for miners and islanders a gentle encounter. Five years after he found gold at Runcie River, Whyte told MacGregor that 'perhaps in no other country, placed in nearly parallel circumstances, has the commencement of settlement been attended by so few racial conflicts'. MacGregor was inclined to agree with him.

Lime spatula handle, Massim, after Haddon 1894

Early miners heard stories of a digger buying four ounces of gold from a villager for two sticks of tobacco, but the Sudest soon learnt the value of the metal in their ground. Men who could use the pan and sluice-box decided that it was better to work for themselves than be paid a few sticks of tobacco as labourers. By as early as 1891 Sudest were washing gold and selling it in the stores, and by 1895 the Sudest were obtaining most of the island's declining gold yield. After the diggers had worked the main gullies a lot of gold remained in widely scattered surface deposits. It was poor ground for the white miners, who were unwilling to work an area if they could not clear wages after paying for rations and labour. But for the Sudest, able to subsist on their gardens and with no obligations to be paid in cash, mining gave more independence and greater access to goods in the stores than the other ways open to those wishing to enter the cash economy: working as labourers, signing-on with the pearlers, or fishing for bêche-de-mer.

The Australian miners objected to meeting the Sudest as competitors. The natives, said an observer in 1894, 'rush the new patches, and with their keen eyesight they clean them out so thoroughly that a Chinaman could not live on the tailings. This enterprising feature of the natives is not relished by the white plodder after gold.' D.H. (Harry) Osborne, who arrived on Sudest in 1901, was told stories of Andy Jorgensen smashing the panning dishes of those Sudest who wanted to wash gold for their own benefit. In 1894 MacGregor heard rumours of diggers burning houses to intimidate Sudest who wanted to mine. John Graham, the Resident Magistrate who was sent to investigate, found insufficient evidence to take action. The miners complained that the Sudest paid no licence fees; but the Sudest were not to be excluded from the field by administrative ritual. They bought licences, although frequently a holder did not enforce his right to stop other villagers working on his claim.

As the total amount of gold which the Sudest placed in one and two ounce lots in the pans of the store scales became important, the storekeepers decided that they could not support their countrymen's demand for the exclusive right to exploit the goldfields of British New Guinea. They encouraged the Sudest to mine and spend: they stocked the tools, cloth, fishing lines, and ointment to cure skin diseases which the Sudest wanted, and for a while they acted in ignorance or defiance of the ordinance of 1888 prohibiting the sale of firearms to Papuans. Some departing diggers anxious to add to their stock of gold dust were also prepared to sell the guns that they had brought to protect themselves from the New Guinea blacks. One miner told MacGregor he had been offered 3 ounces of gold worth over £10 for his revolver. But from 1891 government officers supervised the area more closely and they confiscated the guns, rifles and revolvers owned by the Sudest. Most were handed over readily, perhaps because the raids which made them valuable to the villagers had ended. It was no longer possible for those communities who had guns to terrorise and rob those who did not. In the new order there were other ways of obtaining power.

Nearly two years before MacGregor defined the work of the village constables by regulation in 1892, he instructed William Campbell, Resident Magistrate for the Louisiades, to appoint 'rural police' who would be paid 'a few sticks of tobacco a week'. Jimmy Sudest of the Pantava area, who had worked in Queensland, and Iami of Griffin Point, the first police to hold office on Sudest, began the fashion of meeting visiting government officers to tell them of events on the island and to learn of the government's desires. When Iami reported to MacGregor in July 1892 he received a blue uniform, a belt, a knife and a pound of tobacco; and in return he assured MacGregor that he would have no trouble recovering a revolver from a nearby village. The people, he said, feared him because he was known as a friend of the government. Jimmy Sudest also gave evidence of his power and allegiance by handing in a revolver.

In 1890 MacLean and Samuelson began working a quartz reef on their claim, the Caledonian, about 2 miles across grasslands from the north-west coast. When MacGregor visited them in January 1891 they had put a shaft down 20 feet, taken out several tons of ore and spoke of floating a company to purchase machinery. While MacGregor watched, they washed some ground taken from the shaft and recovered a small nugget which they presented to him. Suspecting that they had put the nugget in the pan before testing the sample, MacGregor declined the gold but shared their tea, tinned meat and biscuits. MacLean and Samuelson shipped 3 tons of ore to Queensland for crushing, but the costs were high and the return was low. They struggled on for about two years, 'living in the greatest misery' and extracting a little gold by roasting the stone, pulverising it by hand, amalgamating the gold with mercury and then separating the mercury by straining it through a blanket. There were no villages close to the mine, but Sudest visited the miners, sold them food and stole some of their equipment. Miners still believed that somewhere in the islands there were reefs which would support a rich and permanent industry; but they now knew they were not on the Caledonian.

Inland from Hinai Bay on the south coast the Mount Adelaide reef seemed to offer a richer reward. After receiving a report from mining experts, the British New Guinea Goldfields Proprietary Co. Ltd appointed G.F.B. Hancock managing director and sent him north with a staff of twelve, a battery of stampers, sixteen working bullocks, two horses and twenty sheep. MacGregor, wanting to encourage forms of economic activity which were less transient than alluvial mining, supplied a gang of fifty-nine prisoners to build a road from the coast to the mine. The company agreed to pay the prisoners a penny a day, meet the costs of the wages of the overseer and the police guard, and supply rations. To tunnel, shovel the ore into the hoppers on the tramway and attend the stampers, the company recruited over 100 men from Rossel Island, Dobu and the Western Division. The Sudest knew the Rossel Islanders, having traded with them for *sapi-sapi* (shell beads), and they met other peoples who sailed to Dobu, but they knew nothing of the homeland and culture of the taller darker Kiwais from the Western Division, men they had seen infrequently as policemen and crewmen on boats. Before crushing had begun the overseer, nineteen prisoners and five of the seventeen Dobu recruits had died. The Papuans had suffered from beriberi caused by a deficiency in their diet. Most of the sixty-five Kiwais from the Western Division, led by Miserie, an ex-policeman, left the mine and set up camp near Pantava. Miserie told Alexander Campbell, the Resident Magistrate from Nivani, that they had been fed on rice and sago only, two of them had been beaten by a white overseer, and they feared they would die like so many of the other labourers. Hancock agreed to provide blankets and better rations but he defended his overseer, who, he said, had been threatened with an axe. The men returned to work: Papua New Guinea's first strike had ended. In June 1898 the Kiwais were paid off, and as their boat passed the government schooner they gave 'three cheers in good English style'. Campbell and his crew returned the salute.

The stampers at Mount Adelaide hammered for a few months only: the crushed ore freed little gold. The labourers were not replaced, the European staff abandoned their houses at the mine site and overlooking Hinai Bay, and by 1899 the Sudest were the only miners on the island.

As gold became hard to find the Sudest were caught in an economic trap. Some had become accustomed to going to the stores for the food, tools and clothing needed to satisfy their material needs and for the *tani* (strings of beads and shells) and other objects used in traditional exchanges. Men forced to shift a lot of alluvial for a few pennyweights of gold were persuaded to buy on credit, to buy 'belong book'; and for some storekeepers the 'book' was a means of forcing the Sudest to keep bringing their gold to the store scales. Villagers complained to Campbell: 'all time boy he afraid belong Mahony belong book'. John Mahony, a partner of Patrick Carvey, had been made a Justice of the Peace by MacGregor on the suggestion of the miners, and he now pretended to have the power to order the village constables to handcuff those who failed to pay their debts to his store. Carvey and Henry Burfitt, an employee, used more direct methods. Tomasi signed a statement to say that when he was unable to pay the 1 ounce, 10 pennyweight and 6 grains which he owed the store Carvey had taken a stick and thrashed him. Campbell believed him: he had heard similar stories before, Tomasi still had wounds on his scalp, and Campbell had noticed that while other traders went unarmed Mahony, Carvey and Burfitt carried 'either guns or revolvers in a very conspicuous manner'. Campbell, zealously keeping his books in order and seeing that all men obeyed the law, slowly collected the evidence to bring the traders to court. In January 1898 he asked Burfitt whether it was true that he had threatened to hang a man who had failed to pay a debt. Burfitt denied the charge and called on Wilsoni, one of his crewmen, for support. Wilsoni immediately told Campbell that he had seen Burfitt 'make fast fish line round neck of Sam'. Six months later on his next visit to Sudest Campbell saw Sam Manawah who said that Burfitt had come to his house and, finding him asleep, had seized him, banged his head on the floor and then pulled a fishing line tight around his neck. Manawah admitted owing 4 pennyweights to the store, but explained that he had been unable to pay as he had been collecting shell and afterwards had to work in his garden before he could again work for cash. Burfitt eased his anger by teaching the Sudest to repeat 'filthy and disgusting' statements about the government. Campbell imposed mild fines on Mahony, Carvey and Burfitt for assault and breaches of the labour and trading regulations.

The system 'belong book' continued but its abuse declined. The Sudest learnt that the law protected them from storekeepers who threatened to 'hammer' or 'make fast boy belong book', and by increasing their gardens they again became independent of the stores. Campbell compelled them to plant ten coconuts for every adult male and gaoled those who failed to make provision for their own independence. Some Sudest purchased their seed coconuts from Mahony and paid in gold. In 1900 Mahony and Burfitt asked if they could take some Sudest to a goldfield on the Papuan mainland where they would work 'on their own account'. Campbell thought Mahony and Carvey were only interested in enabling the Sudest to spend gold in their store and opposed the idea. He was supported by his superiors in Port Moresby; perhaps they acted only to protect the Sudest, but they may also have decided they had enough troubles on the mainland goldfields without taking the chance that white miners, their labourers or local villagers would clash with the Sudest.

Table 1
Principal goldmining laws*
Papua

The Gold Fields Ordinance, 1888
  adopted current Queensland legislation and regulations for the management of goldfields.

The New Gold Fields Ordinance of 1897
  gave the government the power to stop the granting of miners' licences to Africans and Asians.

The Mining Ordinance of 1899
  repealed previous legislation and adopted the Mining Act of 1898 of Queensland. The Queensland act was amended to fix the fee for a miner's right at ten shillings and to allow the granting of a larger reward claim for the finding of a new field.

The Mining Ordinance of 1907
  required any person about to mine on land 'owned and occupied by natives' to inform the warden, who assessed the probable damage, collected the money and held it for later payment to the owners of the land. The warden was to prevent any mining likely to cause 'substantial damage' until the owners of land and property gave their consent. The ordinance stopped the sale of miner's rights to Papuans; but all 'aboriginal natives' were now given the same powers as the holders of miner's rights except that they could not be employed to hold a lease or claim on behalf of another person.

The Goldfield Reward Ordinance of 1909
  provided for a reward of up to £1000 for anyone finding a new field able to support 200 miners of European descent for eighteen months. Members of prospecting parties subsidised by the government could not receive the reward.

Mining Ordinance, 1937
  repealed previous legislation. Although still based on the Queensland Act of 1898, Papua now had its own mining ordinance. The provisions of the 1907 legislation were included in the new ordinance.

* In both Papua and New Guinea many of the basic rules about the taking up and forfeiture of claims were set by regulations made under general provisions in the current mining ordinance.

After 1900 the Sudest could work with the few Malay, Greek, Filipino, Japanese and Australian traders who lived in the area; they could become labourers on Craig's or Mahony's plantations on Sudest or agree to serve in the Cosmopolitan Hotel in Samarai which Elizabeth Mahony had bought after the death of her husband, John; they could sign on to work elsewhere in Papua; or make copra or collect gum, shell or bêche-de-mer; but gold remained an important source of cash.

Table 2
Louisiade Goldfield
European miners

1888	7 October	Sudest	200
1889	8 March	Misima	89
1889	8 July	Sudest 300	Misima 400
1890	8 August	Sudest 70	50
1891	June	Sudest 38	Misima 38
1892	June	Sudest and Misima	65
1893	"	" " "	60
1894	"	" " "	38
1895	"	" " "	30
1896	"	" " "	20
1897	"	" " "	28

Table 3
Louisiade Goldfield
(Misima and Sudest)
Production

	ounces	£
1888/89	3850	14,387
1889/90	3470	12,440
1890/91	2486	8371
1891/92	1235	4332
1892/93	582	2236
1893/94	1128	3906
1894/95	728	2565
1895/96	600	2100
1896/97	560	1960
1897/98	600	2100

Sudest, briefly the main centre of foreign activity in British New Guinea, was visited infrequently by government officers and miners after 1902. Those foreigners who did go ashore at the old landings were surprised at the amount of Pidgin spoken (for the Sudest taught each other), the skill of the miners and the fact that some men had scales to keep a check on the pennyweights produced. On a horse supplied by Mrs Mahony and guided by a local 'boy', Assistant Resident Magistrate Henry Ryan in 1911 patrolled from Griffin Point past the old Four Mile camp. At Billy Bong creek he watched over twenty men exposing working faces on a hillside. They were 'working mates', each two men having their own claim. Further along the track at Jeneeta, a small village of seven houses, all the people, he thought, were 'gully workers'. Close to a settlement of four houses called Talk-Money another thirty men were working in Sago Gully. All the miners that he spoke to said that they were getting a little gold. One pair showed him 6 penny-weights which they had taken during the day; it was worth about £1. Until 1942 when the war forced the traders to leave, the Sudest continued to re-work the alluvial fields, sometimes striking patches which gave them incomes far higher than those obtained by any Papuans who signed on as indentured labourers.

MacGregor prepared legislation excluding Chinese from the goldfields in 1889 but it was not passed until 1898. Early, MacGregor said that the Chinese would have to be excluded because they would clash with Papuans: in 1898 he thought that the 200 Europeans on Woodlark Island would cause an 'immediate disturbance' if Chinese arrived.

The quotation describing relations between white men and Aboriginal women is from Mr Justice Dashwood, Government Resident of the Northern Territory, to the Select Committee on the Aborigines Bill, printed in Reynolds 1972.

MacGregor in his dispatches to the Governor of Queensland, his diary and *Annual Reports* provides much material on the early history of mining on Sudest. Other information is from the north Queensland newspapers; *Sydney Mail* 31 March 1894; D.H. Osborne, *Pacific Islands Monthly*, January 1944, pp. 34, 35; and *Queensland Parliamentary Papers*, A.G. Maitland, 'Geological observations in British New Guinea in 1891', 1893, Vol. 2, pp. 695–728. From the appointment of Cameron in October 1888 until Campbell left Nivani in 1902 there was normally a government officer in the Louisiades. Except for the sections appearing in the *Annual Reports*, much of the early correspondence from the resident officers has been lost. But from 1897 to 1901 there are full papers from Nivani and later papers from Samarai, Kulumadau and Bwagaoia include the reports of patrols to Sudest.

3

Misima

warlike and civilised

By October 1888 the miners had decided that Sudest was not rich enough to support all the men who were there or coming, and that if there was gold on one island then it was likely to be found on others, perhaps in greater amounts. As the ships leaving north Queensland were deep in the water, their holds packed with stores and the decks crowded with passengers, the diggers found it difficult to find boats to take them to islands beyond the reefs near Sudest. To assist the miners and to ensure that they did not 'disperse all over the Possession without authority or supervision' MacGregor agreed to accompany a party to Rossel Island. The miners selected twenty-one men to go aboard the government schooner and H.M.S. *Swinger* towed her through the passage to Rossel. For a week government officers and miners searched the creeks which cut the island's forested slopes. They found no gold, and although they visited several small villages and met two men who had been to Queensland, they saw few islanders. Back on Sudest the prospecting party was reduced to twelve and they again left with MacGregor for an extensive tour, testing ground on Pana Tinani, Misima, Normanby, Fergusson and Goodenough. No rich finds were made, but having washed 'rough colours' on Misima the diggers went back to work alluvial along creeks in the south-east of the island. One group returning to Sudest to collect stores reported finding 30 ounces, and other miners prepared to cross to the new field. Some paid £1 for their passage on canoes manned by Misima villagers and organised by Nicholas Minister. By March 1889 eighty men were on Misima and a storekeeper had set up business. A few were able to win an ounce a day, but no rich finds were made until Jimmy the Larrikin and Frenchy, rarely and more properly known as James McTier and Frank Rochefort, began working rich ground on 17 March 1889 on a branch of the Ana which flows into the sea on the north. They celebrated the day, a faith and a nationality by calling the branch St Patricks Creek. Miners looking at the coarse particles lying in front of the riffles in the sluice boxes decided that there was a reef shedding gold not far away. Ships leaving north Queensland began sailing direct for St Aignan (the miners and government officials at first called Misima by the name of a lieutenant who had sailed with D'Entrecasteaux on *La Recherche*, one hundred years before). Close to the beach at Siagara traders built four or five 'unsubstantial' stores of iron and thatch where they competed for the miners' patronage with rations and 'ordinary and better known medicines' at Cooktown prices. Before taking the 3-mile walk to the nearest mining area, many newly arrived men could talk of delays caused by headwinds and calms, and nights spent on hard boards, but thirty miners from Cairns could recount the danger of sailing on the schooner *Freddy* with Captain Don Smythe who (with willing help) broached the cargo, neglected his ship and his navigation, and had to beach the sinking *Freddy* about 300 miles off-course at Hula on the Papuan mainland. In spite of contrary winds and human frailty, within two months of St Patrick's day, 500 miners had landed on Misima. A year later fifty remained.

About one-third of the area of Sudest, Misima was higher, wetter and more densely timbered. James Cashman said it took half a day to clear a space to pitch a tent, and some miners had to move 10 feet or more of earth and boulders before reaching gold-bearing wash. By the middle of 1889 forty or fifty men were suffering from malaria, and many had decided that Misima was 'no place for miners'. It was certainly not an area for those wanting somewhere to scratch out a few pennyweights while they waited for news of the opening of a richer field.

Map 4 Misima Island

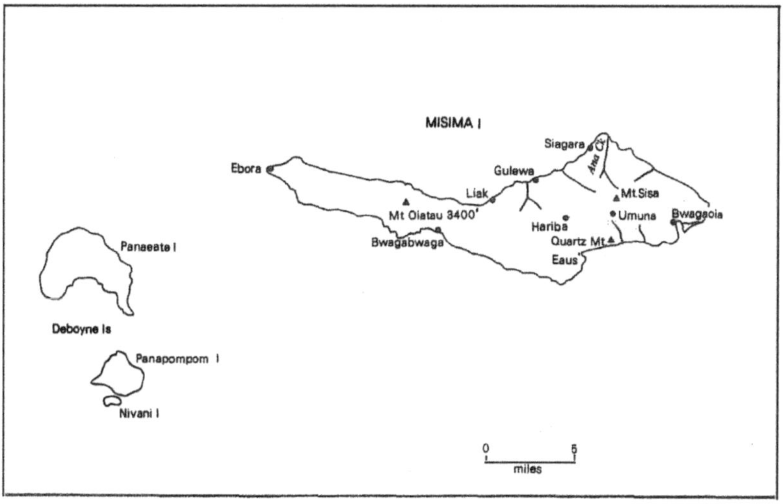

The miners thought the people of Misima 'warlike and civilized'. Between two and three thousand people lived on the island in 1888. They travelled frequently to Panaeate and other islands in the Calvados

Chain and regularly took their *waga* 100 miles north to Woodlark and west to Wari. Unlike most of the other Louisiade islands Misima had no large reef-protected lagoon. On the north and west coral limestone cliffs cut by fissures and ravines rose over 100 feet above the water; in the south-east there were shore reefs, sandy beaches and mangroves. The only sheltered anchorage was at Bwagaoia where ships could pass between the coast and the small lagoon enclosing Managun Islet. From their exposed coast it was difficult for the Misima to harvest the seas; but they were 'industrious cultivators of the soil'. Extensive, carefully worked gardens extended up the hillsides, lines of logs placed across the slopes marked individual gardens and stopped the topsoil from being washed away; fences kept the village pigs from the crops. Men from islands with little fertile land came to Misima to trade for betel nut taken from 'countless thousands' of palms, yams and other garden foods. All the villages except Hariba were near the coast, some containing up to fifty houses spread, in groups of four or five, for half a mile. Each house, like those on Sudest, had a thatch roof curving down from a central spine to put a cover over a raised platform. Looking like up turned whaleboats, they were well-designed to withstand the strong south-easterlies which blew in the middle of each year. Travellers on the tracks connecting the eastern villages had to use the bush ladders built against the steep cliff faces; in the west visitors went from one village to another by canoe. In high seas it was almost impossible to go from Ebora in the extreme west to any other part of the island. Some of the Misima were potters, but the best pots were brought by trading expeditions from Brooker and Panaeate Islands. To nineteenth-century Europeans the Misima appeared vivacious, industrious and healthy; the skulls that decorated their houses and the spears and shields that lay about were taken as evidence that they fought frequently and savagely. They suffered one obvious disability: many had *tinea imbricata*, a disease which made their skins dry and scaly. Europeans did not catch it, and could not cure it.

Partly protected from direct outside influence by the lack of safe anchorages, the Misima could keep informed of events in the area by their seamanship, their inclination to travel, and their close linguistic, trade and ceremonial links with peoples on other islands. Some labour recruiters came to Misima in 1884, and later John Douglas called one of the returned men, Molnos, 'an old friend'. In a lecture illustrated by slides, Douglas told the Brisbane branch of the Royal Geographical Society in 1888 how Molnos had returned from Queensland wearing 'regatta shirt, white trousers and straw hat'; but when he saw him on a later visit to the Louisiades he was naked, and Douglas remarked, 'What a beautiful young man I thought him'. For helping the government party return labourers to their home villages, Molnos was allowed to select from the products of European technology carried on the boat and he asked for soap; Douglas gave him a 'bar of yellow'.

In 1886 the people of Misima had a more dramatic confrontation with another group of foreigners, and this incident had a stronger influence on what foreigners thought about them. Lieutenant-Commander John Marx, looking for information about the killing of Frank Gerret in the Deboyne Islands, brought H.M.S. *Swinger* to anchor on the north coast of Misima. The people were cautious but eventually some men came on board and exchanged yams and fowls for tobacco. Hoping to increase the confidence of the Misima in the benevolence of Her Majesty's navy (and obtain a pig), Marx took a ship's boat to the beach while another boat, its crew armed with rifles, stood off covering the party onshore. The only man prepared to trade with Marx took tobacco in one hand and slashed at him with the other, cutting him on the head and arm with a bush knife. The sailors fired on the man, but they thought he escaped unhurt into dense bush close to the point of attack. Seeing other men concealed in the area, Marx believed his assailant had acted too soon, spoiling 'a plan for an attack on a large scale'. The only explanation Marx could give for the action was revenge: the Misima had explained through the interpreters carried on the *Swinger* that recruiters had been to the area, taken men away and not returned them.

Douglas, when asked by Rear-Admiral G. Tryon about punishing the islanders, advised that nothing be done in 'any retaliatory spirit' to the poor ignorant savage' who struck the blow; but for 'the future well-being of the Islanders themselves', for the safety of Her Majesty's subjects (and especially her servants), and for the protection of life and property, the islanders should be forced to surrender the guilty man on the promise that in due time he would be returned: 'only as a last resort should justice be vindicated by an act of war'. The 'outrage' against Captain Marx confirmed the 'very bad name' of the Misima. When Captain Francis Clayton visited 'Treachery Bay' in H.M.S. *Diamond* the people were defiant, and he had difficulty finding an anchorage. Eventually he 'fired a few shells to clear the bush, then with much difficulty owing to the surf, landed a boat's crew, burnt the small village, and destroyed two canoes'. The Misima may not have known that the shells and the burning were done for their improvement; but they must have learnt a little more about the behaviour and technology of some foreigners. Thirty-seven years later Resident Magistrate Louis Brown was surprised to find the people of Nigom village using an unexploded 6-inch naval shell as a canoe anchor; they explained that many years ago it had been fired at them from a man-of-war.

On the eve of the arrival of the miners in the Louisiades the Misima had some knowledge of white men and they were curious about them;

but while the Sudest and Pana Tinani were being harried and reduced, life in the Misima villages was little changed. The Misima could still believe that they could protect themselves. In the five months between the arrival of the prospectors on Sudest and the start of mining on Misima, they had time to learn something of the peculiar ways of the most numerous of foreigners, the diggers, and when they met, the Misima behaved much as the Sudest had done.

For payment of two sticks of tobacco a day they carried from the anchorages at Siagara and Bwagaoia to the mining areas, and a few worked on the claims. Siagara, a small village of only eight or nine houses, was overwhelmed by the rush to St Patricks Creek, and its inhabitants moved away. Hely said that the storekeepers paid the villagers in tobacco for their houses and as Siagara was only a temporary fishing village they had suffered no hardship. But MacGregor was less certain that they had been treated justly: he had seen Siagara before the arrival of the miners, at the height of the rush, and a year later when the Siagara had 'miserable houses' and lived as 'industrious beggars'. Other villagers, protected by distance and rough country from the direct impact of the rush, sold yams, coconuts, sago and breadfruit to the miners. While there were times when the Misima had only sago to sell and the miners complained that they could not do a full day's work on 'native food', many could not have stayed on the field without the produce of the Misima gardens. In the early weeks of the rush miners arrived more quickly than rations, and by 1891 only the *Wanganui* provided an unreliable three-monthly service from Cooktown to Sudest and Misima. At times the miners were forced to ask villagers for credit, and one man diverting a creek to bring it across an alluvial area explained to MacGregor that his labourers worked on 'tic'.

The villagers also visited the mining camps to steal. By the middle of 1890 'cases of pilfering' were 'everyday occurrences'. Perhaps it was inevitable that the Misima would take goods left unguarded on their lands, and that the frequency of their raids would increase when they learnt that it was difficult for the miners to protect their property or recover it from the villages; but one digger returning to Cooktown from Misima said that miners had started the cycle of thefts by taking village pigs and fowls. A man from Panapompom Island just to the west of Misima certainly acted in retaliation. William Campbell, the Resident Magistrate, confiscated a gun he had obtained from Nicholas Minister so the islander in turn stole tobacco from Campbell's store at Bwagaoia. MacGregor urged his officers to speed the building of a house at Bwagaoia 'calculated to inspire the native with respect', pursue all thieves relentlessly, and punish them severely. But when the miners at St Patricks Creek complained to MacGregor that Wagima of Kakoma, a village about 4 miles west of the mining area, had stolen tobacco he found it impossible to make an arrest. In Kakoma he asked for the 'king' and Harimoi was 'tendered'. Harimoi and the three elderly men who spoke with MacGregor admitted that Wagima had been stealing and they pointed to his house, but they could not or would not hand him to the government party. After MacGregor had detained one man by throwing him on his back and the police had seized another, he was able to extract a promise that Wagima would take a load of yams to St Patricks Creek to pay for the tobacco. He left doubting that Wagima would do so.

MacGregor found it equally difficult to impose the Queen's peace on Misima. In January 1892 he arrived at Bwagaoia where he met H. Neville Chester suffering from sandflies, malaria and a second theft of tobacco from the government store. Convinced that Chester had encouraged enormity by releasing the Panapompom thief and his accomplices after only three months in prison, MacGregor was determined to teach respect for the law and government officers. The Gulewa had recently raided Hariba killing two men, burning houses and cutting down coconut trees; they were, said MacGregor, the 'worst of a number of bad tribes' on Misima. Two miners, Sanderson and Simpson, increased the government's obligation to intervene by persuading the Hariba not to retaliate, but to wait until the government officers arrived.

But the government party could not land at Gulewa on the north coast and when MacGregor, Chester and six police tried to cross overland from the south coast, the Gulewa watched them struggling along the bed of a creek. The government force on its rush into a deserted village captured only one woman and two young boys. The woman was told that the government did not fight women and was released. MacGregor waited three days in Gulewa, feeding his men on the inhabitants' gardens, using their firewood, and attempting to bring the neighbouring villagers into an alliance against the Gulewa. Dogs and pigs returned to the village, and MacGregor wrote in his diary: 'the dingoes stick to the houses & not to their masters'. MacGregor had come to dislike the Gulewa. He wanted the subjects of the Possession to do as they were told or fight (and be defeated), but the Gulewa would neither submit nor fight. They refused to respond to his messages that the government wanted only the two leaders of the raid on Hariba and that they would be pursued until they surrendered. Having to go to Samarai, MacGregor left Chester with additional police and instructions to keep the Gulewa isolated, take as many male prisoners as possible, and exchange them for the two leaders. MacGregor took with him a policeman injured by one of the many spear traps that the Gulewa had left on the tracks around their village. Gulewa's neighbours drifted away from Chester's force, one prisoner cut his ropes with a shell and escaped, and Chester was left uncertain whether the Gulewa were living in rough country in the west or sheltering with other peoples. Six months later MacGregor reported that the 'Gulewa murderers have not yet been secured'. After a further lapse of two years MacGregor wrote that one, Kokove, had been sentenced to five years' imprisonment for killing a man from Hariba, but no mention was made of the other wanted man. Through the 1890s the Misima learnt what the government thought was good behaviour: public displays quickened their learning and their submission. Having been sentenced to gaol for theft, Babaga of Panaeate escaped and on his recapture was held on the government cutter in Bwagaoia Harbour. At night Babaga killed Constable Umi who was sleeping on deck and injured the Papuan coxswain with a tomahawk before jumping overboard. In spite of leg irons and Chester's attempts to shoot him, he reached shore. Babaga mixed freely with his own people and once he was seen fishing on the reef opposite the government residence, but it was two months before he was recaptured. He was hanged at Panaeate on 2 August 1893, 'dying without a struggle'; the first Papuan to be executed by the imperial government for an offence against another Papuan. On Sudest the islanders had seen that the penalty for killing a white man was hanging, now it was demonstrated that the police acting as servants of the government were to be given equal protection.

Among the forty or fifty men assembled to witness Babaga's death were two of his brothers. Babaga, MacGregor said, was a vicious man, feared by his neighbours for his violence. When questioned, his brothers said, 'i waisi ... ', it was good that he alone should die. When the Gulewa were asked to confirm that the *gavamani* was righteous they had argued that in the past the Hariba had killed them and it was unjust now to punish the Gulewa for retaliating. In 1892, three years after the miners had arrived, the Gulewa still believed that the strength of their group depended on their ability to raid and counter-raid. Within another two years the Louisiades were in 'a tranquil state'. The Misima then knew that the government was clumsy, but powerful and persistent. The village constables were hostages for their people: through them the government officers obtained information and gave directives.

By the end of the 1890s Alexander Campbell, the Resident Magistrate at Nivani, was unworried by clashes between warring parties or cultures on Misima. In court he dealt with cases of adultery, lying reports and petty theft. He celebrated distant occasions. Seventy days after the event he instructed the crew to dress the government boat and led the Armed Native Constabulary in three cheers for the Relief of Mafeking; a month after Queen Victoria's passing, he lowered the ensign to half-mast for seventy-two hours to mark her death. Campbell had time to order that all islanders visiting the government station have a bath once a day 'to instill some idea of personal cleanliness'. On patrol he checked to see if the villages were clean, the tracks kept in good order, enough coconuts planted, and whether there had been any sickness. Susuina in 1903 cut races and then robbed the miners' camps when they went to find out why the water had ceased to flow; but he was an exception. And while the way of life of the Misima had been changing the people had faced another group of foreigners who came with particular ideas of right and wrong for others to follow.

In 1891 Brother Samuel Fellows arrived at Panaeate to establish a branch of the Australasian Methodist Missionary Society. The people were 'very kind' and Fellows decided not to carry the Winchester rifle he had brought with him. 'Red-haired and emotional, sometimes on the mountain peaks and again in the valley', Fellows had gone to work in a Derbyshire steel mill at thirteen. Having furthered his education in Sunday Schools and shown his talent and dedication by lay preaching, Fellows became a candidate for the ministry after his migration to New Zealand. For three years he directed the work of the mission on the Deboyne, Calvados and Misima Islands. He learnt the Panaeate language, which allowed him to communicate with all people in the northern Louisiades, and he produced the first printed texts in Panaeate. Attracted by Misima's large population and productive gardens, he made his first journey along its southern coast two months after landing at Panaeate. Although sometimes sick with malaria and frustrated by winds which would not take the mission schooner, *Waverley*, in the right direction, he worked hard for a God who was close and whose message was clear and dominant. In his diary he wrote of his need for the 'atoning-cleansing blood' of his saviour: 'I do need Him in all His pity, tenderness & love, in all his willingness and power to save from sin.' The people of the Louisiades, he thought, lived a 'fairly enjoyable life', they worked 'systematically and laboriously' in their gardens, traded actively and kept agreements but they were superstitious, had many 'degrading and impure customs' and above all were unaware of their 'moral responsibility to God'. And he found they resisted his injunctions to change:

> They are as proud as the proudest Pharisee that ever lived, and as mean. This, with their inherent tendency to lying and deception, makes it an easy matter for them to deceive themselves with the idea that they are an exceptionally good sort of people, with whom it would be difficult to find serious fault.

Although he quickly gained some influence among them, he realised that 'The devil evidently [meant] to make a fight for his Kingdom among these people …'.

At the end of 1892 Fellows preached at twenty-three points on the Misima coast. Where landing was easy the crew of the mission boat carried Fellows's harmonium ashore and helped him to gather a crowd; where landing was difficult he alone took a dinghy through the surf while the crew held the *Waverley* in safe water. Believing his knowledge of the Panaeate language gave him 'an influence over them such as nothing else could have done', Fellows preached, prayed, repeated the liturgy and finished with two or three hymns. His flock was inclined to laugh and talk while he prayed but when they found how important the ceremony was to him they usually did as he told them and knelt and closed their eyes. In his sermons Fellows instructed the people to give up fighting, cannibalism and polygamy, and he spoke of Albert the Good as an example of a great man who had taken only one wife. He pleaded for a lessening of the women's share of the work, reverence during services, and an end to 'sabbath breaking' and 'the unrestrained sexual connection of the young people'. (For a time his devotion was tested by women who lay on their backs, put aside their grass skirts and called to him.) He told his congregations that God was their father and He loved them, but he also explained the 'way God will deal with [the] ungodly at Judgment & there was a sober earnestness on all faces, one woman shrieked when [he] described the thrusting down of Sinners into "prisons of fire"'. While some sermons aroused no interest or were interrupted by ribald comments, occasionally the audience was enthusiastic, and Fellows responded with greater passion and fluency.

> When I asked them if they knew the way to the Father's House in the heavens — they said in chorus — No. What a thrill of joy I had in telling them that I knew & would show them — teach them. I got thoroughly worked up & had a splendid time.

At the end of his first year on Panaeate Fellows had not gathered the 'spiritual fruits' he had 'longed & prayed for'; but he could soon record changes in behaviour: children went to school, congregations paid closer attention during services, the hymn, 'Pull for the Shore', promised to be 'a great favourite', and some young people were repeating the Lord's Prayer before going to bed at night. Soon after Fellows left the Louisiades in 1893, the propriety of the Panaeate sailing their canoes on Sunday when on long voyages had become an important question for both the islanders and the missionaries.

In 1892 two Samoan mission teachers and their wives landed at Bwagabwaga and a Tongan and his wife went to Alhoga further along the south coast. The villagers had previously told Fellows that they would accept the teachers and Fellows could sometimes bring them stores; but once the mission schooner had left, the teachers were largely dependent on their ability to exchange a few stores, tobacco, and knowledge for building sites, food, and a measure of village leadership. Kolinio, a Fijian who went to work at Liak on the north coast, strengthened his contract by marrying a local girl; but generally the mission insisted that the teacher went ready-married to his post. From the time the teachers took up residence on Misima the people obtained most of their knowledge of *tapwororo* (the church and its teaching) by watching and listening to a Tongan aristocrat, or a Fijian or Samoan commoner. They still heard men 'called' from the Midlands of England or suburban Australia but less often, more distantly and at times of greater ceremony. In the long term the South Sea Islanders may have been most influential in setting a style of behaviour for the Papuan teachers who worked under their direction and eventually replaced them. In villages where the South Sea Islanders preached and conducted schools, the Misima began the progression from catechumen to member-on-trial, baptism and full membership. Many were found guilty of fornication and forced to begin again; but by 1898 twenty-two people from Bwagabwaga had been received as full members, and the next year five men from Misima began training as mission workers. By 1920 twelve of the fourteen teachers on Misima were Papuans.

The Sudest, fewer in number, living in smaller villages, less able to support teachers and speaking a different language, saw little of the representatives of the Christian churches for fifty years after the arrival of the miners. On Sudest the government officers and police played an indirect part in stopping feuding on the island, and after the first few years of mining government officers landed infrequently at Griffin Point or Pantava. In 1932 Patrol Officer Ivan Champion reported that the inland hamlets of Sudest had not been visited since 1920; many people were listed as living in villages which had been shifted or no longer existed. Previous officers had made patrols by sending word of their coming so that all could meet them at the anchorages.

By contrast the Misima had faced many agents of change. The different foreigners had come at almost the same time and their influence, concentrated at particular points, had spread over the island. The Siagara had been thrust aside by the miners, the Gulewa had been harried by the government, and the Bwagabwaga had been hosts to Samoan and Tongan teachers. When Alexander Campbell came to Bwagabwaga in 1897, his first visit for fifteen months, he stayed at the teacher's house and he inspected the school where he listened to seventy-seven pupils sing a hymn, pray and repeat 'parrot fashion a few syllables on a blackboard'. Campbell might conclude that only eight of the pupils had 'any knowledge of reading and writing, and then it was of but a rudimentary nature; but he did not doubt that the Australasian Methodist Missionary Society, directed and interpreted by South Sea Islanders, was important in Bwagabwaga.

Fellows wanted the people to know that he was different from the government officers. He tried to persuade the escaped murderer Babaga that he should give himself up; but he also assured Babaga that he would not give the government officers any information. Some Gulewa prisoners captured by Chester's police 'laughingly told' the crew of the mission boat how they had been linked together by a rope around their necks and marched to the coast: the men wanted by the government for the attack on Hariba watched from close by. On the same day Fellows spoke with many of the Gulewa in a neighbouring village.

But while Fellows gave the Misima a chance to learn that the foreigners were divided, other members of the mission blurred the distinction between church and state. A teacher on Panaeate sent a group of men to Campbell to be punished for working on Sunday: they were relieved to find that while the government itself did not do business on Sunday it declined to punish those who did. Similarly Josephata, a Fijian at Bwagabwaga, insisted that the village constable from Panapompom take two couples charged with fornication, a man guilty of not attending divine service and witnesses to Nivani for trial. Campbell told them they had 'done very wrong' but they had broken no government laws and he could not punish them. Later Josephata asked for a village constable to be appointed to Bwagabwaga so that he did not have to spend so much of his time 'preaching government'.

Table 4
South-Eastern Division government officers

Officer	Date	Headquarters
J.B. Cameron	October 1888/89	Sudest
M.H. Moreton	1889	Siagara, Misima
	1890	Bwagaoia, Misima
W.T. Campbell	1890/91	"
N.H. Chester	1891	"
	1892/94	Nivani
J.W. Graham	1894	"
A.W. Butterworth	1894	"
W.T. Campbell	1894/95	"
R.J. Kennedy	1895/96	"
A.M. Campbell	1896/99	"
C.A.W. Monckton	1899 (January-April)	"
A.M. Campbell	1899/1902	"

Note: In 1902 the headquarters of the South-Eastern Division was moved from Nivani to Bonagai and then to Kulumadau on Woodlark Island. When mining again became important on Misima and declined on Woodlark the headquarters was shifted back to Bwagaoia in 1920 and remained there until 1942.

For the people of the Louisiades the distinction between church and government was puzzling, and their confusion must have been increased by the variety of men who spoke for the state. Robert Kennedy, Gold Warden and Resident Magistrate for the Louisiades in 1896, was charged with 'entertaining highly improper relations with native women', some of whom were procured for him by the crew of the government boat. At Nivani he lived with a Logea Island woman who was known to the people of the area as his wife. MacGregor allowed Kennedy to resign because he was young, previously of good character, and his mother was 'a very respectable resident of Brisbane'. Alexander Campbell, who had served in the customs departments of Fiji and Tonga, was temperate, fussy and just. He was concerned about the way recruiters completed indenture papers and the welfare of the men whose names and villages were listed on the forms. On special occasions he invited mission teachers and local villagers to the government station to see a magic lantern show and listen to his gramophone. Often acting to protect villagers and labourers, he was strongly prejudiced against 'the objectionable creation of the whiteman, the "whiteman's native"'. Faced with the '"over-civilized" type' capable of 'cunning, lying, and otherwise disreputable actions' Campbell felt he had to protect the 'honest whiteman'. To Chester at Bwagaoia the people were 'swine & niggers etc', and Fellows feared he would 'delight in shooting down the natives on the slightest chance'. Expressed in his dispatches to the Governor of Queensland, MacGregor's policies were rational and consistent: the governed in the Louisiades had to learn to live with the actions of men of different beliefs and appetites.

The missionaries and the government officers had come to change the islanders; but the miners were more numerous. By 1900 sixteen remained on Misima. They suffered from malaria, comforted by the belief that if they survived the first periods of fever later bouts would be less severe; they obtained food from the villages by purchase, exchange or begging; they built houses using local materials and skills; and they employed men and women from the villages to sluice for gold. They were less concerned about their sexual and racial purity than some of the early miners on Sudest. Robert Warren lived with a local woman and her kin had free run of his house. Alexander (Sandy) Grant and Charles Coppard, who came to the Louisiades in the first rush, formally married local women. Both accepted responsibility for maintaining and educating the children. Others took temporary 'wives' and tried to limit their association with the villagers. The most eccentric, the least capable of making a living elsewhere, the most integrated into village life, the most tolerant of the physical conditions, and two or three men who could reasonably hope to invest their earnings from gold in profitable trading and planting, stayed on. Among the best known were Jimmy the Reefer (James Carlow), who believed he had not slept for seven years and talked of the days when he had been a sculler of note on the Tyne; Carl Ernst, a German, confined to his bed with an ulcerated leg and fed by the charity of the villagers; and August Degen. 'Jumbo' Degen, 'a fine handsome man well over six feet', had been on Misima from the early days of the rush. He worked dressed in a bag, and even when new and unfrayed it reached only to his waist. Degen had fought for Prussia against Austria and France, and he asked Murray at their only meeting in 1908, 'How are the French?' Murray replied that as far as he knew they were all right. 'They are a bad lot the French', Degen said, 'they are like the natives; they should get a hammering every five years'. Murray 'understood his antipathy to the Great Nation but [thought] he was rather hard on the natives considering that he, like many others at that time on Misima was living almost entirely on native charity'.

The Misima fields were more difficult to work than those on Sudest and the Misima people had less need to mine than the Sudest, but from the late 1890s a few men worked old ground. At the same time those Misima who chose to sell coconuts received only one stick of tobacco for forty nuts. Although some Misima villagers had many palms, selling nuts was obviously a slow way to riches. Miners coming from villages a few hours' walk from the alluvial areas near Bwagaoia and along Mica, Cooktown and Ana Creeks could easily obtain food and they worked on their own land, or on land owned by people with whom they had close associations. But it was more difficult for men from distant villages to become miners. They had to search out the land-owner and then find some marriage, trade or totem relationship which they could use to begin discussion. After obtaining permission to work in an area, the miner would present rice or tobacco to the landowner. While most gold was taken by people from the villages on the edge of auriferous lands, eventually men from Ebora and the Deboyne Islands mined on Misima. Whereas on Sudest the gold was left to the villagers, on Misima both Papuans and Europeans mined until 1942. By about 1914 Papuan miners on Sudest and Misima were each obtaining about £500 of gold in a year.

In 1902 the 'once pretty station' at Nivani was abandoned: Mahony leased the government plantation and placed a 'Manilaman' there as overseer. Government officers on Woodlark or Samarai came in frequently to talk to the village constables, inspect a few villages and take away ten or so people to the lock-hospital on Eboma Island to join others suffering from venereal disease. The people of Panaeate told Campbell in 1905 that the disease had come recently from Wari and Tubetube. Campbell believed them for he had seen no sufferers during his years on Nivani, and he blamed Greek traders and pearl buyers for its introduction to the Louisiades. It was not brought by the early miners. Venereal disease probably did not reduce the population significantly on Sudest and Misima, but it may have contributed to severe depopulation on some of the smaller islands. In the eighteen years of little government supervision after 1902 Misima retained the interest of other foreigners: South Sea Island mission teachers and a white supervisor, about ten miners, a trader and one or two men with small leases hoping to establish plantations.

Clay Pot, made on Brooker and traded to Misima Island, 1974

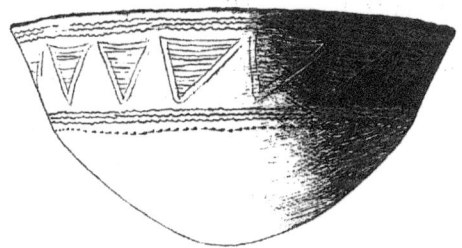

Sixteen years after their arrival on Misima, the diggers located the 'mother lodes' of much of the island's alluvial gold. Men pegged leases at Mount Sisa, Umuna and Quartz Mountain, and tried to raise capital, but the ores were too low grade, of uncertain extent or 'refractory', needing complex processes to free the gold from associated minerals. A little was recovered from test workings, and from the efforts of men who cut tracks and dragged heavy equipment to their leases. Then in 1914 Block 10 Misima Gold Mines (No Liability), controlled by Broken Hill Proprietary Limited, began extensive development works and eventually took over all leases on the Massive Lode at Umuna. Block 10 invested heavily, tunnelling for several thousand feet, installing crushing mills, cyaniding vats, sawmill, black-smiths' and carpenters' shops, electric lighting, staff accommodation and port facilities, and laying down 7 miles of tramway. The locomotive hauling the trucks on the 2-foot gauge tramline crossed twenty wooden bridges on the journey between Bwagaoia and Umuna, and 'some of the points of beauty to be seen *en route* ... would vie with similar spots in many locations'. At its height in 1921–2 Block 10 employed 63 Europeans and 512 Papuans to produce £56,508 of gold. Faced with further costs to maintain their gold yield, the directors closed the mine in September 1922. A local company employing about eight Europeans and 160 Papuans worked the leases until 1928 when Freddie Cuthbert, over seventy years old and already having made and lost a lot of money on Queensland fields, floated New Misima Gold Mines Limited to reintroduce larger-scale workings. The route of the old tramway was reformed to take motor trucks and new treatment plant was installed. During the 1930s steadily increasing amounts of gold were extracted from the Umuna mine. Before selling to Cuthbert's Misima Gold-mine Limited in 1935 New Misima had increased its dividend payments to 2s. a month on £1 shares. In the ten years before the mine closed at the start of the Pacific war, the Umuna mine had produced over half a million pounds worth of gold; it had been Papua's most valuable mine, and the most profitable enterprise conducted by Australians in the Territory. In some years it was probably the only Papuan company paying a dividend.

On Ana Creek workmen inched equipment up cliff faces to install a hydro-electricity plant, but soon after the machinery was set in motion the manager knew that returns were so low he would have difficulty recovering his salary. At Quartz Mountain and Mount Sisa companies worked intermittently to win little gold. In 1938–39 Gold Mines of Papua Limited earned £30,755 at Mount Sisa; it was the only year any group provided substantial evidence to support those who believed they could bring another Umuna into production.

Most traders, planters and alluvial miners adapted the way they worked to meet the needs and ways of the islanders. The Osborne brothers on Rossel, Mrs Mahony and Charles Arbouin on Sudest, employed local people, advanced money to those unable to find their £1 government tax, and paid men to make *sapi-sapi*, the shaped-shell beads used in inter-island trading. Alluvial miners were dependent on the goodwill of villagers for food and labour. When a miner handed trade goods to his workers to take to a neighbouring village to exchange for a pig, there was no equality between miner and labourer and villager; but there was interdependence. The companies came to reshape the land and its resources to make an efficient mine; adaptation was something for engineers and chemists faced with strange ores, and for inexperienced overseers directing indentured labourers. Most of the men recruited from Australia by the companies to occupy the company houses at Bwagaoia and the mine sites remained expatriates; the most frequent messages sent by the AWA transmitter at Bwagaoia were instructions to Australian bookmakers.

The companies planned to use local labourers to exploit the Misima lodes, but the demand soon exceeded the supply. Although the management of Block 10 hoped that the imposition of the head tax in 1919 would enable them to recruit 300 Misima for road work, most of the men on the island decided that there were easier ways of earning their tax money. By 1920 Block 10 employed over 600 Papuans and while the number of workers fell in the 1920s it rose again in the 1930s. In 1937 over 1000 Papuans worked as indentured labourers for Cuthbert's and the other companies. Most of the labourers were from the D'Entrecasteaux Islands, particularly from Goodenough. Known to recruiters and overseers as 'Gosiagos', they were more willing to work underground than other labourers. Much of the heavy surface work, cutting timber and roadmaking, was done by men from the Northern Division. They appeared on the monthly labour records as 'Tufis, Opis and Orokaivas'. Other labourers came from near Milne Bay and the south coast: the 'Baniaras, Buhutus, Suaus and Mailus'. Boatloads of recruits arrived regularly at Bwagaoia, signed-on at the Resident Magistrate's office, and walked up the road to the dormitories ('of 40 boys capacity') at Umuna; others, having completed their contracts and converted their wages into a box of trade goods from the Bwagaoia stores, went on board for the voyage home. Some men lost their trade goods on the voyage home: to other labourers who were better or luckier gamblers, or to women who offered delight at some of the island anchorages. A few men from the D'Entrecasteaux Islands and Milne Bay took wives with them; five women from Eaus village alone married labourers from the mines. A few other men from the south-east settled in the villages of Misima wives; one, Rupeni of Goodenough Island, was still living in Eaus in 1974. Generally Western, Gulf and Northern Division men did not marry Misima women. Labourers also took away memories of feasts on stolen pigs and garden foods; but during the twenty years that the indentured labourers made up about a quarter of the population on the island fights between 'sign-on boys' and villagers were rare.

In 1915, while 200 Misima men worked for the companies, other villagers earnt £674 washing gold at St Patricks Creek. In 1932 when the Umuna mine was paying 'handsome dividends' and few Misima worked regularly for the companies, the Resident Magistrate, Alexander Rentoul, stopped to talk to some men tending sluice boxes on a creek near the track to Mount Sisa. They told him that in two weeks two men could obtain gold worth £3, twice as much as they could earn by signing-on with the companies. Some villagers found it worthwhile to pay 3s. for a small bottle of mercury so they could amalgamate their own gold; their costs were lower if they stole the *silba* from the mines. With alternative ways of obtaining cash most Misima men chose not to be indentured to the companies. But they did do a lot of casual work, much of it by 'contract'. For a cash payment of specified goods villagers agreed to supply timber, thatch, sago or coconuts; or repair a road, or build a bridge. Having completed the 'contract' they were free to work in their gardens, travel or do government work.

When Block 10 began to exploit the Umuna lodes, the government returned to Misima. In 1920 white officers, police and prisoners began building the residency, the patrol officer's quarters, the office, barracks, post office, customs quarters, bond store, and two gaols. In five weeks of closer government supervision, four European employees of

Table 5
Cuthbert's Misima Goldmine Ltd
Monthly Labour Record[1]
February 1940

Origin	No.	Contract (months)
	454	12
Island boys[2] [D'Entrecasteaux Islanders]	13	18
Mailus	97	18
Baniaras	22	18
Orokaivas and Opis	75	18
Tufis	4	18
Buhutus	14	18
Misimas	5	12
Suaus	3	12
Casual employees	8	(not indentured)
	Total 695	

Distribution	
Mine	442[3]
Battery & treatment	62
Timber getting	41
Timber squaring	18
Trucks & roads	67
Engineers	7
Carpenters	9
Domestics	6
Cooks	40
Hospital	3
	Total 695

1. The monthly labour records are in C.A.O., C.R.S., G. 180, item 3, maps and charts.
2. The classification is from the labour records.
3. This group included all underground workers; 345 were from the 'Islands' and 74 from 'Mailu'.

Block 10 were fined for assaulting labourers, and one Papuan was gaoled for two months for assaulting a European. After hearing the complaints of three labourers who arrived at the station 'badly knocked about', W. R. (Dick) Humphries, the Resident Magistrate, wrote in the station journal, 'I have reason to believe that assaults have occurred here very frequently'. He told the mine manager that, 'in a friendly spirit', he should warn his staff not to take the law into their own hands. The manager replied that before the building of Bwagaoia station they had had little trouble with their labourers. Now many were refusing to work, and if they were sent before the Resident Magistrate for breaking their contract the punishment was so mild that it incited others to stop work. Humphries's sympathies remained with the labourers. He had seen a few men who had been battered; he had received a letter from a former accountant with the company offering to provide evidence of 'cruelty, flogging and overwork under a Contract System'; and he had gone underground to watch the Gosiagos at work. In his report to the Government Secretary Humphries wrote:

> To say that I was surprised at the nature of the work performed would not be fully expressing my feelings. In an atmosphere of heat and dust, lit only here and there with the light of a candle I caught glimpses of Island boys at work with the picks. Some of the cross-cuts were less than six feet high with a breadth much the same — a hole in fact, as the boys themselves describe it. Yet in these small dark heated spaces they do good work and the Underground 'boss' Mr Quintrill speaks very highly of them.

They worked a forty-eight hour week for 10s. a month and keep. They were probably the lowest paid free underground miners in the world.

Later few cases were heard in court and presumably few men were bashed at Umuna. The company apparently learnt that it would have to obey the regulations about keeping labour records, sanitation and diet, for later government inspectors recorded few complaints; and the labourers continued to sign-on for Block 10.

After 1920 government officers visited the villages more often, condemning dilapidated houses, lining the houses to make the villages more orderly, instructing the people to keep the tracks clear, telling the Hariba to move to the coast and then letting them shift back to the hills, collecting taxes, paying the baby bonus of 5s. to every mother with four children under the age of sixteen, checking the census and hearing adultery cases. Under the Native Plantations Ordinance villagers were instructed to clear land and plant coconuts in straight lines with careful spacing. Initially keen to have their own 'companies', later the villagers had to be compelled to maintain the plantations. The return from copra sales was low and the people, especially those on the north coast from Gulewa to Ewena, could not see the point of a few acres of carefully maintained trees when they had such an abundance of bearing palms that the uncollected nuts rotted on the ground. The land, gold, 'contracts', intermittent employment, and the loads of betel nut which they sold to traders and government officers to exchange for yams in the Trobriand and D'Entrecasteaux Islands allowed them to be both independent of the mining companies and critical of the government's modest plans for their self-improvement.

Government officers sometimes spoke of the Misima as a 'soft' people with a static or declining population. The officers did not read, or did not believe, the statistics collected by their colleagues. In 1913 and 1914 the Resident Magistrate had estimated that the population was about 2500. By the mid-1930s the officers knew that there were over 3000 in the Misima villages, and Native Medical Assistant Lahui Vai had calculated that 43 per cent of the population was under fifteen, clear evidence of an expanding population. Thirty years later the population of Misima was over 6000.

Paddle handle, Massim, after Haddon 1894

In January 1942, two days after the Japanese captured Rabaul, white government officers, the Reverend Harry Bartlett and most of the other Europeans on the island left for Samarai. Bulega, a young man from Siagara village, increased his influence. He preached of a new world order in which the black men would become white, they would have all the food and goods they needed, and those who were now white would work for them. Many of his followers gathered on Motorina Island in the Calvados Chain and decided they would kill the returning 'government'. In December Lieutenant R.G. Mader of the Australian New Guinea Administrative Unit (ANGAU), George Burfitt, Corporal Segaradi, Constable Gio and Kaioki of Suau went to Motorina to arrest Bulega, and, surrounded by people, they were killed without acting in their own defence. In other clashes men from the Calvados Islands killed an old Filipino resident and five islanders. ANGAU officers and police pursued those thought responsible for the murders through the islands of the chain, destroyed houses, damaged canoes, shot five people and arrested over 150. In 1893 the government had demonstrated its power and morality by hanging Babaga at Panaeate; fifty years later it reasserted both by a public hanging at Bwagaoia of eight men found guilty of murder. One man told the crowd assembled to watch him die that he was being killed because he had believed Bulega. Bulega, using a rope plaited from strips of his rami, hanged himself in the privacy of his cell. Another eighteen men were sent to gaol for three to ten years.

In their acceptance of Bulega's teaching and their violent rejection of the government's attempt to re-impose its rule, some people had shown they were dissatisfied with the old order, but many disruptive events had taken place between the departure of one lot of government officers and the return of another. Villagers, 'sign-on boys', two of the Bwagaoia police detachment and three of the Europeans who had stayed after the general evacuation had fought each other; abandoned stores and private houses were looted; the villagers had heard the 'terrific sounds of guns and bombs and ... aeroplanes' as the Battle of the Coral Sea was fought above and around Misima; the Japanese had established a float plane base close to the old government station at Nivani; someone had written 'Welcome Japan' on the trucks from the mine; Isikeli Hau'ofa, a Tongan mission teacher, and Kenneth Kaiw, a government clerk, had worked to establish peace on the island; villagers with no knowledge of the course of the war or the aims of the contestants had fled from their villages at the sound of low-flying aircraft; the Sudest had feared that inter-island raiding would begin again and wild rumours had swept the island, one of the most persistent being that the government would come and take all taxable men away to work. The killings in the Calvados Chain and the hangings at Bwagaoia were part of these traumatic events, but the men who responded to Bulega's dream that he could 'reverse the world' were also influenced by years when they had little gold and little power. Other Misima villagers, having known disorder and fear in 1942, cheered when they learnt that the 'government' was returning to Bwagaoia.

Much of the information on early alluvial mining is from the same sources used for the chapter on Sudest: MacGregor in his diary, dispatches and *Annual Reports*; other officers in *Annual Reports*; north Queensland newspapers; and J.P. Thomson and B.H. Thomson, 1889a, 1889b. There is very little early ethnographic material on the Louisiades except for Armstrong, and Tindale and Bartlett. The clash between Commander Marx and the people on the north coast is recorded in Royal Navy Australian Station, New Guinea 1884–8, Case 17, microfilm, N.L.A. The work of the early Methodist missionaries is taken from Fellows's diary and papers kept temporarily with his collection of artifacts in the National Gallery, Canberra; Bardsley; *Missionary Review*, and Minute Book of the Panaeate Station and Circuit, and Minutes and Journals of District Meetings, Boxes 13 and 20, United Church Papers, New Guinea Collection, University of Papua New Guinea. Mrs Amirah Inglis consulted the records of marriages at the Registrar General's Office, Port Moresby and provided details about marriages between white men and Papuan women. Other information about relationships between miners and villagers is from interviews on Misima in 1974. Murray recorded his meetings with eccentrics in his diary, 7 April 1908 and reminiscences pp. 18 and 19.

No comprehensive figures of gold produced by Papuans are available but sometimes the warden reported production in *Annual Reports*:

		£
Misima	1914/15	675
Sudest	1916/17	412
Sudest	1917/18	420
Sudest	1918/19	660

E.R. Stanley wrote his 'Report on the Geology of Misima, (St Aignan)' just as Block 10 was beginning work and he has a detailed account of the leases, and some photographs and maps. Other information about lode mining from Kulumadau and Bwagaoia Station Papers; and C.A.O., C.R.S., G180, item 3 (monthly labour records), and item 71 (correspondence between Humphries and Manager, Block 10 about the treatment of labourers 1920). Obituary notices of Charles Coppard and Freddie Cuthbert are in *Pacific Islands Monthly*, March 1945 and November 1948.

The events of 1942 are recorded in J.V. Barry, 'Commission of inquiry ... into the circumstances relating to the suspension of the Civil Administration of the Territory of Papua, February 1942', copies in Australian War Memorial, Australian Archives and New Guinea Collection, U.P.N.G.; transcript of evidence, C.A.O., C.R.S. A518, X 800/1/5; ANGAU War Diary, Australian War Memorial, 1/10/11 (includes two accounts of killing of Mader and arrest of Bulega's followers); Rex v. Bona etc., Australian War Memorial, trials of natives, 506/4; Rex v. Le Boutillier and Downey, Samarai 1948, folder 82, Supreme Court Building, Port Moresby.

Much information was also obtained on a visit to Misima in 1974. Three people who gave their time and knowledge generously were John Grant, Kenneth Kaiw and Alby Munt. In Canberra I was assisted by Epeli Hau'ofa, the son of Isikeli Hau'ofa.

4

Woodlark

a people free to walk about

Woodlark Island, over 40 miles in length and greater in area than Sudest, is lower and swampier than the other big islands of south-eastern Papua. Thick rain forest flourishes wherever the soil and drainage are adequate. The raised coral, mangroves, forest and small areas of garden lands of the west are divided from the east by the hills near Kulumadau in central Woodlark and the low Okiduse Range which rises at Mount Kabat in the north and culminates in a spear point of peninsula dominated by Suloga Peak. Inland from the mid-north coast and Guasopa Bay are extensive gardening lands.

In 1895 the beach opposite Mapas Island was covered in stone chips, a clearing about a mile inland was strewn with more fragments, and beyond that near an old village site on the flank of Suloga Peak were acres of chips. For many generations men had mined on Woodlark, taking stone from rock faces exposed in a gully on Suloga and working it until it became a tool, wealth and art. The hard volcanic rock was flaked by striking it with another stone, ground in sand and water, and then polished in water and the powder coming away from the stone itself. At the old village site on Suloga and at other places on Woodlark were large slabs of rock each with a circular depression made by men grinding and polishing. In the most valuable blades the polishing highlighted a network of lighter bands, the result of the irregular laying down of the original volcanic ash.

Suloga was the main source of stone blades for all the south-eastern islands. Blades were also traded to people on the mainland, going as far north as Collingwood Bay, and in the south they were picked up by Mailu Islanders and passed along the coast in a series of exchanges until they reached the Papuan Gulf 500 miles from the Suloga quarry. Different sized blades suited various tasks: felling trees and clearing scrub for gardens, adzing dug-outs, shaping the outside of logs for hulls, and cutting grooves and chipping holes to attach outriggers. Sometimes flaked blades were exported and polished by peoples on other islands. Even where polished blades were being used, as at the canoe-building sites on Panaeate, men were constantly grinding new cutting edges and re-binding loose blades to handles.

Finely banded stones, larger and thinner than any working blades, were objects of beauty and tokens of wealth. Men sought to acquire them in the competitive and ceremonial trading exchanges which linked the islands and the mainland. Men paid their debts in Suloga stone. Displayed before visitors from distant villages, they were tangible signs of the wealth and power of Trobriand chiefs. The islander who carried a finely ground axe hooked across his shoulder to a feast or a meeting with trading partners was displaying a valuable object; he was probably prepared neither for work nor war.

When iron was replacing stone as a cutting tool in the 1870s the inhabitants of the two villages who worked and distributed the blades suffered a 'big sickness'. The survivors abandoned their homes and their skills to settle in a small hamlet on Suloga beach. Without a knowledge of the techniques or magic which allowed other men to sail and plant with confidence, the Suloga were still a depressed community twenty years later when white miners came to Woodlark for another mineral of value, beauty and utility.

The Suloga blades, which had been selected and ground with the greatest care, retained their value. On Panaeate men who worked with iron tools, still loosely bound to wooden handles, accepted thirty or more stone blades for building a *waga*; after 1900 European traders paid from £5 to £10 for Suloga axes in one area to exchange them for shell and copra in another. The trader also benefited from the prestige which passed to the owner of a fine blade.

By 1850 Woodlark Islanders had traded, worked, fought and entered into religious dispute with Europeans to a degree unknown to other Papua New Guineans. In the 1830s they traded with whaling boats which came in search of food and water, and in 1840 some Europeans began calling the islanders' homeland after the whaler *Woodlark* which had anchored there in about 1832. In ignorance of the recommendation published in the *Nautical Magazine and Naval Chronicle* other foreigners and islanders continued to call it Murua.

On the advice of the captain of a whaler, Monseigneur Jean Collomb led a party of French missionaries of the Society of Mary to Woodlark in 1847 to begin the work of converting the people of New Guinea to Christianity. Aware that the islanders had killed all except one of the survivors from the wreck of the *Mary*, the missionaries held some Muruans hostage on the mission boat to secure the safety of those who went ashore. But the Muruans, anxious to trade for iron, competed to be hosts to the foreigners. The Marists chose to establish their mission on Guasopa Bay, a broad stretch of protected water backed by a gentle curve of sandy beach and flat forested land in south-eastern Woodlark; they called the 'beau et excellent' harbour, 'Nativité'. With the assistance of Pako, a Muruan who had been to Sydney and spoke some English, they purchased land and began to build. Collomb set himself the task of learning six new words of the Muruan language each day and he instructed his missionaries to patrol all parts of the island so that no infant need die without benefit of baptism. Three months after the missionaries arrived a baptised child died and Collomb forgot his fever to shed tears of joy for having been able to open heaven to another soul. But the Muruans were unmoved by the missionaries' assault on their beliefs. They would not accept that suffering and death had come to Woodlark because Adam had eaten of the forbidden fruit; and if the Marist God was so powerful then let him come down among them bringing his iron and displaying his axes. The Marists, who were often sick and unable to provide food for themselves, had no obviously superior material or spiritual power, while the Muruans could feel that in performing their rituals while gardening, trading and fishing they continually confirmed their relationship with the forces controlling the world. Having come into economic and ideological competition with the Marists, the Muruans were determined to cling to their own ways in spite of drought and disease. By 1849 the Marists had ceased active mission work to serve God by following a strict monastic life. The only convert was made by the Muruans: Brother Optat chose immediate pleasure on Woodlark rather than wait for eternal reward. He escaped the mission to meet Muruan women and commit 'improper familiarities ... in the sight of all'. His brothers in the church shipped him to Sydney in 1850.

Map 5 Murua Goldfield, Woodlark Island

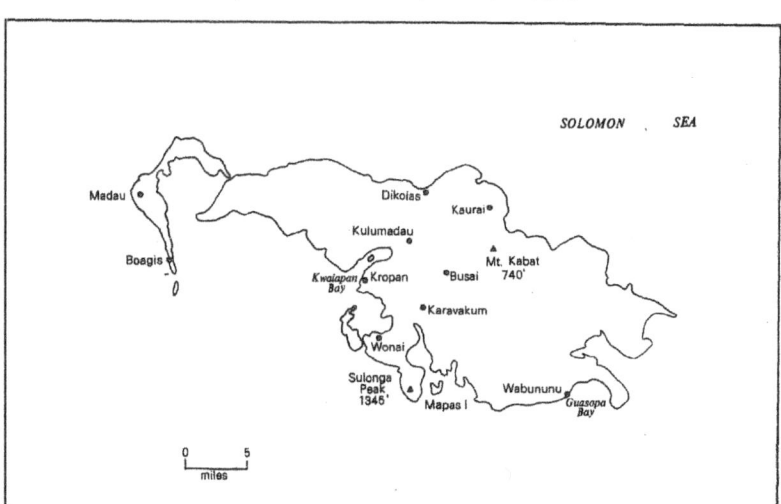

To shake the pride of the Muruans and demonstrate the superior ways of Europeans, the Marists took five Muruans and three Laughlan Islanders to Sydney in 1851. On their return the travellers led a movement to make another Sydney on Murua, and the missionaries believed that soon they would see a general acceptance of Christianity. But the Italian missionaries of the Missioni Estere di Milano who replaced the Marists in 1852 reaped no harvest. Drought and disease again afflicted the Muruans, and fighting broke out between villages; once more the missionaries became isolated from the islanders. In 1855 the Italians abandoned Woodlark. One priest, Giovanni Mazzucconi, who had been in Sydney recovering from illness, sailed on the *Gazelle* to rejoin his colleagues not knowing that they had already left for Australia. When the *Gazelle* ran on a reef outside Guasopa eight canoe loads of islanders came alongside, and feigning friendship to get on deck, killed all on board and looted the ship. The Muruan who told the crew of the schooner *Favourite* the story of the attack had been to Sydney and he chose to go away again on the *Favourite*. Individual Muruans were prepared for adventure and economic revolution; their resistance to Christian missionaries was not part of the conservatism of a community opposed to all change. And some Muruans may have thought they had paid a terrible price for their resistance; there had been about 2200 people on the island when the missionaries arrived but many islanders had died of disease by the time they left. Forty years after the Marists departed MacGregor asked Makavasi of Guasopa if he remembered the language of the missionaries: 'he promptly replied "travaillez comme ca"' (work like this).

The Muruans continued intermittent trading and fighting with visiting ships, and for the latter they were shelled by a French warship. Their relations with the outside world changed in 1880 when Wilhelm Tetzlaff ('German Charlie'), an agent of Eduard Hernsheim of Matupit, New Britain, established a trading station in the Laughlan Islands. The Laughlan Islanders were part of the Muruan community. They visited Woodlark frequently, many leaving the Laughlans with the southeast trades in November and returning home with the north-west in December, their canoes loaded with yams and other garden foods. Men from Woodlark visited Tetzlaff's station and an island woman who had lived with Tetzlaff later married a Guasopa man. Foreigners thought that Tetzlaff had acquired a position of influence among the Laughlan Islanders. When the Queensland labour traders came to the Laughlans in 1883 William McMurdo, the government agent appointed to see fair play on the *Stanley*, concluded that Tetzlaff had advised the recruits to desert. He punished all involved by burning village houses and Tetzlaff's station and destroying canoes. The Muruans attempted to retaliate in 1885 by seizing a boat from the *Victoria*, which was returning men from Queensland. A 'treacherous-looking lot' forced their way between the oars, grasped the gunwales and tried to drag the boat inshore, but the crew was able to take the boat clear without bloodshed; the two Muruans who had served a year with the Mourilyan Sugar Company waded ashore with their bundles. Few Muruans went to Queensland but many engaged in the copra, pearl and bêche-de-mer trades of south-eastern Papua.

Captain Bridge raised the Union Jack at Guasopa on 9 January 1885 before a gathering of 'the finest and most robust Papuans we had seen'. They were also hostile, indicating by mime that labour recruiters had tied men's hands behind their backs and taken them away. The Muruans, having been drawn towards the German commercial empire by Hernsheim, were now in the most north-western corner of British New Guinea.

Government officers had little reason to visit Woodlark until November 1889 when the traders Kickbush and Neilson were killed and their boat, the *Albatross*, was looted at Guasopa. According to evidence given at the trial of those accused of murdering the traders, Tetebra of Panemote Island, a crewman on the *Albatross*, had arranged the attack. Kwarma of Wakoia, who had been to Sydney, believed that his people living inland from Guasopa could attack the foreigners and be safe from any reprisal. Tetebra had then met the Wakoia and they had planned the raid. When news of the killings reached Samarai the Muruans were forced to take part in a sequence of events known to other people of British New Guinea. Assisted by Muruans who were hostile to the Wakoia, MacGregor led a small armed force inland. They quickly arrested most of those involved, but Mamadi, who was supposed to have taken a leading part, evaded capture. As a gesture of reconciliation Mamadi left a revolver and a Snider rifle with a 'bunch of sweet smelling herbs' tied to the trigger guard on a mat on the platform of his house. One of the captured men, Viviga of Wakoia, was found guilty of the murder of Albert Kickbush and sentenced to death. Against the advice of other members of the Legislative Council MacGregor directed that Viviga be hanged at Omdamuda, Guasopa Bay. MacGregor believed that plunder alone was the motive and although 'entertaining ... an extreme aversion to capital punishment' he decided that a public execution would be to the 'ultimate benefit of the natives'.

Having gone to Woodlark because of the killing of Kickbush and Neilson, government officers became involved in other events. They learnt that the Wakoia had attacked the villages close to Guasopa Bay killing thirteen people on one raid and surprising and killing fifteen women in their gardens on another. One group from Guasopa had abandoned their homes and taken refuge among the more numerous Wamana people of the north coast. The Wamana, having had more frequent contact with foreigners, accepted the coming of the government without any display of opposition; but the Wakoia, less involved with the traders and fearful of retribution for their part in the attack on the *Albatross*, left their villages and carried their axes, ebony spears, and painted shields when they learnt that government parties were in the area. A clash was avoided when Viviga was arrested in 1891 but in 1892 MacGregor sent word to Boiomea, a Wakoia leader, that he had arrived 'to fight if they wished to have fighting, or to make peace with them all except the murderer Mamadi, if they wished to have peace …'. Although suspicious that MacGregor wished to imprison him, Boiomea chose peace. Wamana, Wakoia and Guasopa met and MacGregor arranged for the Guasopa refugees to return to their own lands. MacGregor gave shirts and tobacco to the 'chiefs' hoping that in taking the 'clothes' a man also accepted an obligation to speak for the government. Six months later Bingham Hely reported that 'the natives everywhere came off to [the government boat] with the greatest confidence' and he believed that a 'lasting peace' had been established between the Wamana and the Wakoia.

Before the arrival of the miners the Muruans were caught in further strands of the loose government net by the appointment of a village constable at Guasopa, and the old people of the Laughlans in 1893 had 'burst into a wailing lament' as the first young man left to join the Armed Native Constabulary. His mother put a mat and two coconuts in the dinghy which took him to the *Merrie England*. The next year more young men joined the constabulary.

Apart from the occasional visit of a naval officer or a punitive explosive shell, the 'government' had followed the miners to Sudest, accompanied them to Misima, and preceded them to Woodlark.

Lime spatula, Woodlark Island, 1974

In 1895 the traders Richard Ede and Charlie Lobb, who had taken over Tetzlaff's trading station on the Laughlans, found gold on Woodlark. Lobb, a Cooktown miner, went to New Guinea in about 1890, and Ede kept him supplied on prospecting trips around the islands. Four months after making the strike, Ede and Lobb formally applied for a claim from which their labourers recovered 20 ounces of gold per week. Lobb drowned in 1897 when he was knocked overboard by the boom on his cutter; Ede remained on Woodlark as a planter and trader until 1942 when he went to Australia shortly before he died. Another find was made by the Papuan collectors employed by the English naturalist A.S. Meek. He turned digger to work a claim and win £250 worth of gold in six months. When MacGregor arrived in November 1895 he found about a dozen men mining on a spur of Suloga Peak, another dozen working on a creek at the head of Suloga Harbour, and a few prospecting other areas. Hoping to discourage competitors, some diggers told MacGregor that 'the whole field was utterly worthless', but MacGregor decided that they had been 'doing fairly well, some of them very well'; and he proclaimed the Murua as British New Guinea's second goldfield. Hearing of more finds in the creeks of the wet forested country between the Okiduse Range and Kulumadau, men prospecting the mainland or turning over old ground in the Louisiades left to try the Murua. By the end of 1895 about fifty diggers were on the new field. The number of miners on Woodlark continued to grow, and then in February, March and April of 1897 there was a rush, with steamers arriving every fortnight from Queensland. Shipping agents in Sydney said that the demand for passages was as great as during the early days of Coolgardie. Australian interest in gold in New Guinea grew on rumours, scraps of contradictory news and occasional reports which sober men could believe. In March 1897 they learnt that the *Ivanhoe* had arrived in Cooktown from Woodlark with 800 ounces consigned to Burns Philp and 32 passengers who were prepared to declare that they had another 300 ounces. There was certainly gold on Woodlark. In fact during the first three years, at least 32,000 ounces went from Murua through Queensland ports; not much compared with the richest Australian fields but more than three times the amount taken from Sudest and Misima from 1888 to 1890. Early diggers at places they named Elliotts Creek, Mackenzies, Colemans, Reillys and Big Ben Creek, Yanks, Skippers, New Chum and Floggers Gully made tucker and some had a full 'shammy', a soft leather bag of gold. Most of the alluvial workings were near Busai, with smaller centres at Suloga, Wonai, Okiduse, Karavakum and Kulumadau.

One man who sank test holes in many gullies on Woodlark was C.A.W. Monckton. Whit Monckton, the son of a New Zealand doctor, was twenty-three. He had gone to New Guinea in 1895 for adventure and, armed with a letter of introduction from the Governor of New Zealand to Sir William MacGregor, had hoped to obtain a position in the government service. Sir William had no vacancies and Monckton went to Woodlark. After a brief stay on the island Monckton recruited two labourers in Samarai and returned to mine. Nearly twenty years later when he was farming in New Zealand he listed his prospecting equipment:

> For food we depended on a small mat of rice of about fifty pounds weight carried by one boy, and as many sweet potatoes, yams or taro we could pick up from wandering natives. The other boy carried a pick and shovel, tin dish, crowbar, axe and knife, and three plain deal boards with a few nails, comprising our simple mining equipment, together with a sheet of calico, used as a 'fly' or tent, to keep the rain from us at night. My pack consisted of a spare shirt, trousers and boots, rifle, revolver, ammunition, two billy cans for making tea and boiling rice, compass and matches, and last but not least a small roll case of the excellent tabloid drugs of Messrs. Burroughs and Welcome.

On the rare occasions that panning revealed gold, they assembled the sluice box, set it with a fall of one inch in twelve, put stones or nailed vines across the bottom to trap the gold, and diverted creek water through it. Monckton found enough gold-bearing ground to shovel into the sluice box to pay his expenses; but many who came in the rush of 1897 did not.

In January MacGregor wrote to the Governor of Queensland asking him to correct the 'many sensational, probably misleading items of news' in the Australian press; but the number of diggers on Woodlark continued to increase until April when there were about 450 on the island. Alexander Campbell, attempting to administer the new field by making schooner trips north from his headquarters on Nivani, thought that Murua could support about 170 men. Many of the early diggers, coming from other parts of New Guinea, were accustomed to the climate and knew what stores, trade goods and equipment were needed to work an island field; but during 1897 the number of men on the island without money or stores or hope increased. The diggers and the constant rain turned Busai into a swamp of churned mud. The storekeepers found it almost impossible for their labourers or mules to carry goods from the anchorage in Suloga Harbour to Busai. A correspondent of the *Sydney Daily Telegraph* giving The Latest Information in March 1897 said: 'The sun burns down on the soaked ground, the rotting vegetation, the putrid beach, and the unwholesome mangrove swamp, and the air seems to be alive with fever bacilli.' He thought that alcohol had established its reputation as an enemy of fever, and with a bit of quinine the sick could pull through. Not all men, he conceded, had a constitution strong enough to survive the cure. At least fourteen miners died in March; one, Charles Rayman, with no money and his two mates dead from fever, put his gun in his mouth and blew his head off. Forty men left for Cooktown on the *Clara Ethel* in March: 'most ... crawled to the vessel stricken with fever and dysentery', two died before the ship cleared the harbour and five more were buried at sea. Campbell tried to auction the possessions of those who died on Woodlark, but found little to sell: 'when a man dies on the field be his claim rich or poor, there is nothing in the camp a few minutes after death'. William Page and Patrick Finnigan, charged with stealing 11 ounces of gold, a revolver and a watch from James Henderson, said in their defence that they had thought Henderson was dead. They were celebrating their luck in a licensed store when Henderson walked in and accused them of theft. Still drunk when he faced Campbell, Finnigan suffered an additional penalty for contempt of court. MacGregor called the early arrivals 'industrious workmen of good character', but many of those arriving in 1897 were men without mining skills and Campbell reported that they included about 'a dozen very bad characters', some of whom had served long prison sentences in the Australian colonies.

Table 6
Murua Goldfield
production and population

Year	White Miners (30 June)	Papuan Labourers (30 June)	Production (ounces)
1895/96	190	760	12,000
1896/97	400	1600	20,000
1897/98	160	640	10,000
1898/99	62	248	5000
1899/1900	76	304	6000
1900/01	150	600	7500
1901/02	100	400	7000
1902/03	120	480	8500
1903/04	125	500	9000
1904/05	100	400	9689
1905/06	80	294	10,527
1906/07	69	227	5296
1907/08	69	227	5296
1908/09	70	252	6339
1909/10	48	252	9780
1910/11	89	343	8632
1911/12	89	403	9447
1912/13	93	420	12,147
1913/14	39	347	9182
1914/15	58	450	7170
1915/16	48	318	6790
1916/17	36	300	4527
1917/18	11	120	3052
1918/19	16	35	405
1919/20	3	30	400

Although many men left after April 1897 the Murua did not decline as rapidly as the Louisiade fields. At the end of June 250 miners remained, 160 were still there in June 1898, 62 in 1899, 76 in June 1900 and 180 in November 1900. Men stayed because the alluvial field was richer, and others arrived because reefs had been found. Campbell granted the first reward claim for a reefing prospect at Colemans Creek near Busai in April 1897, but it was not until the testing of the Ivanhoe reefs at Kulumadau in 1899 that Australian capitalists were prepared to invest in the Woodlark mines.

Three companies were formed to exploit the Kulumadau reefs: the Woodlark Island Proprietary Goldmining Company of Sydney, the Woodlark Ivanhoe Goldmining Company Kulumadau of Adelaide, and the Kulumadau Woodlark Island Goldmining Company of Charters Towers. With the assistance of a surveyor, a court and assessors the companies bought out the claim holders and began recruiting men, importing machinery, developing plant sites and sinking shafts. The distance between the mines and a port was reduced to a mile and a half by a track cut from Kulumadau to Bonagai on Kwaiapan Bay. When the Proprietary Company decided that too much of its capital was being taken by Australian labourers and reduced wages from £5 to £4 a week, the men stopped work. The company brought in more workers from Queensland and they joined the strikers. The mine manager appealed to Campbell for police to protect the company's property, but Campbell decided that the company's property was in no danger and to bring in the Armed Native Constabulary would 'have been the signal for an otherwise orderly and well conducted body of strikers to have become a disorderly mob, and possibly both police and mining property might have fared badly'. The company agreed to pay £5 and the ten stamp battery was ready for the official opening by the Lieutenant-Governor, George Le Hunte, in April 1901.

By the end of 1901 the three companies were working the Kulumadau lodes, and at Suloga another company had installed an engine and pump to sluice for gold. But the warden thought that the method of extraction used by the Suloga company was unimportant because, 'if the gold is not there it cannot be dug up'. The Suloga company, whose stocks had soared 'like a rocket', did not earn enough to pay for its lease. At Kulumadau the Proprietary Company had forty Papuans working on the tram line between the mines and the port, another sixty cutting timber for shafts, building and fuel, and the three companies employed sixty-five Europeans and 130 Papuans at the mines; but the crushing mills were often silent. There was gold at Kulumadau, but it would not support three companies using extravagant methods. The Ivanhoe abandoned its leases in 1903 and the Proprietary Company closed its mine in 1905; in the previous year it had paid a dividend of 3d. on each £1 share. The Kulumadau

Woodlark Island Goldmining Company took over the Proprietary Company's leases in 1907 and maintained production until 1918. In most years the Kulumadau produced gold worth between £10,000 and £20,000. In 1915 it employed 28 Europeans and 253 Papuans, and when it closed in May 1918 it had to dismiss 11 Europeans and 120 Papuans.

At other old alluvial centres hopeful leaseholders made many trial crushings and attempted to form companies. In 1905 Karavakum (or Bonivat) looked like the 'coming locality'. Test crushings from the Woodlark King lease having given high yields, the owners installed a mill, concentrator and cyaniding plant. Over the next ten years the Woodlark King produced over £50,000 of gold. Although it was one of the richest single leases on Murua, the Woodlark King did not tempt investors to support large-scale development. At the Little McKenzie, another Karavakum lease which was briefly 'one of the most promising properties on the island', the owner installed a battery from the old Ivanhoe mine and for a few years it again crushed stone for little return.

Where the alluvial miners took thousands of ounces of gold from the old silts and many creeks of the Busai area, the reefers found numerous leaders and deposits of 'good stone'. The first crusher to work at Busai on the Federation lease in 1906 was powered by a waterwheel, but it was soon replaced by Fred Weekly and partners' Murua United battery. Weekly crushed stone for leaseholders at £1 a ton for small lots, most of the stone being bagged and lumped to the mill on Papuan shoulders until tramways were built to the most important mines in about 1911. George Jones's Federation lease, the most productive at Busai, yielded over 1800 ounces before 1914.

The leaseholders were hampered by water and their inability to raise the capital to open new lodes and introduce more efficient methods of recovering gold. Kulumadau had an annual rainfall of over 160 inches but surface water disappeared quickly in the porous coral and limestone soils. The level in the companies' dams fell quickly during brief dry spells and the mills were forced to stop. At other times the underground miners spent more time bailing than digging. The Kulumadau shafts went below 500 feet but at Busai and Karavakum where the mining areas were lower and the leaseholders could not afford to install efficient pumps, deep mining was impossible and even shallow shafts were flooded.

After May 1918 there were frequent attempts to open old mines. Most gold was won in 1938 and 1939 by a company cyaniding the sands from the Woodlark King. The alluvial miners had never completely abandoned Murua. In 1901 when the first crushers began operating the alluvial miners recovered 7500 ounces, and in 1913 eighteen white miners and their 'teams' of labourers were washing gold on the island. They reported finding 695 ounces but the warden conceded that the alluvial miners were still as reticent as they had been to MacGregor in November 1895. In the 1920s and 1930s men kept returning to Woodlark, where they set gangs of Papuans to rework old ground while they prospected for reefs and poked around old shafts. By 1940 the miners had taken over 200,000 ounces from lode and alluvium. Until the revival of the Louisiades in the 1930s when the Umuna mine was paying a monthly dividend to its shareholders, Murua was Papua's richest goldfield.

Paddle handle, Massim, after Haddon 1894

In 1901 Kulumadau was a crude settlement of tents, huts of bush material and a few new buildings of sawn timber and galvanised iron. It was inhabited by white company miners and construction workers, independent reefers and alluvial miners, and gangs of Papuan labourers and villagers who had come to look, sell or work. Although the death rate was much lower than it had been in the early days of alluvial mining, Kulumadau was still considered 'naturally an unhealthy locality' where 'fever' was persistent and outbreaks of dysentery common. Three stores licensed to sell liquor and another, 5 miles away at Busai, served the 150 white residents on the island. The Reverend James Walsh of the Methodist mission complained of men, on the Sabbath, gambling with dice, playing billiards, drinking and singing obscene parodies on 'The Holy City', 'I will arise' etc. It was worse one Friday night when Gus Nelsson celebrated an expansion of his business by giving free drinks to all: 'Saturday the men drank harder. Sunday was worse. Yesterday was worse still, and rioting was strong. Today the language is worthy of a sewage farm, and still the drinking goes on.' Two years later the majority of members in the Commonwealth Parliament briefly agreed that they should transform the way of life of most white men in British New Guinea.

In August 1903 Samuel Mauger, a Melbourne hat manufacturer eager for governments to pass legislation to protect men from evil, moved that a section be added to the Papua Bill prohibiting the sale of alcoholic liquor to all residents except when authorised by a doctor. Ardent prohibitionists and those who believed that 'The taste for drink is an overpowering passion with the black' and likely to be the cause of their destruction combined to pass Mauger's amendment. The Acting Administrator, Christopher Robinson, wrote to inform the parliament that the miners were 'accustomed to the use (and abuse also, for that matter) of alcoholic liquors' and he doubted the capacity of any government to change their ways. The miners who responded to Alfred Deakin's request for the views of every class told him that they would 'certainly not do without their grog'. W.E. Buchanan, who spoke as 'an old digger, with some 30 years' experience', seven of them spent prospecting and trading in New Guinea, said that the miners would 'certainly not for one moment brook any interference by the present native police; any such action would, without doubt, mean a very unhealthy experience for the native force'. And Buchanan 'with all courtesy' went on to tell the 'Federal politicians' to 'do something ... towards improving the disgraceful conditions of the unfortunate natives of your own land whom you have allowed to become utterly depraved by drink and opium'. In Western Australia and Queensland, said Buchanan, the laws against supplying liquor to Aborigines were a 'screaming farce, and the open and permitted degradation of the native women a crying shame'. Buchanan spoke from a position of some righteousness. Very few men had been charged with supplying Papua New Guineans with liquor, none with selling it to them, and convicted Europeans had been fined heavily. The Papua Act was passed without the prohibition clause: Deakin was able to satisfy Mauger and his supporters with provision for a poll giving the white residents the chance to reduce the number of licences to sell liquor. The white residents voted not to close any licensed store or hotel; and while some miners were prepared to give a medicinal nip to an exhausted labourer, they rarely allowed a Papuan to drain the bottle for pleasure, companionship or escape. The miners maintained their own reputation for hard drinking. In 1906 the three Royal Commissioners appointed by the Australian government to make recommendations about the future of Papua took evidence in Sheddon and Nelsson's Kulumadau store 'still redolent of fried tinned sausages'. The first comments addressed to the Commissioners were the incoherent ravings of a 'wildly drunken son of toil'. Some miners responded to the chairman's call for order and the drunk, the chairman noted, 'was dragged by his head into oblivion'.

Government officials, public spirited men and white women increased sobriety and civic order in Kulumadau. In 1901 one or two white women lived in the Kulumadau area, seven were on the island in 1905 and by 1912 the 'single ladies of Kulumadau' could conduct a 'ball' at Nelsson and Sheddon's store, with proceeds in aid of the Kulumadau General Hospital. A rival storekeeper (who had just arrived from the Lakekamu Goldfield where all women were scarce) reported that sixteen white women came to the ball and 'a lot more stayed away'. He accused all the white women of selfishness, but circumstances may have warped his judgment. Angelina 'the town whore' who 'could take the *Kia-Ora* [a trading boat] or even a larger size' had venereal disease; at least the storekeeper said that several of her customers had copped a load. For those with other interests or prepared to postpone sensual pleasures, there were meetings of the Hospital Committee, the Progress Association and the Amalgamated Workers' Union. The Kropan villagers were paid to 'erect shelters and seats for the convenience of travellers' on the Bonagai-Kulumadau road. Convinced that there would be a large white population on the island in the near future, the government extended its services. Roads were improved, a school for more than twenty white children was opened in 1913 and a wireless transmitter able to contact Australia began operating in 1915. But the white population had reached a peak of 179 in 1913.

When war broke out in Europe white men showed great enthusiasm for the European Armed Constabulary, which served the Empire by mounting guards at 'Saltwater' and the wireless station. They were instructed not to resist an 'attack in force'. Confident that in Woodlark's thick bush a group of thirty armed whites and ten Papuans would be able to resist 'any ordinary landing party', they asked for more precise instructions. The Government Secretary told them that the shelling of the township could be considered an attack in force, and in that event they should cease defending the wireless station. A month after news of the war reached Woodlark strange lights were seen entering Kwaiapan Harbour and the alarm was raised; but the lights were only the burning torches used by a Muruan fisherman. The German raider, *Wolf*, sunk the Burns Philp boat, *Matunga*, off the Laughlans in 1917; but by that time the European Armed Constabulary had handed in their rifles. Some white men sailed from Woodlark saying they were 'leaving for the front', prices rose, ships called less frequently and the mines did not flourish. The closing of the Kulumadau mine in 1918 and the shifting of the government station to Misima in 1920 ended white small town life on Woodlark. 'Doctor' John Taaffe, who had cared for victims of accident and disease since his arrival on Woodlark in 1896, lived for another year to look after government business and be generous to all men.

Clay Pot, Panaeate, after Tindale and Bartlett 1937

In 1897 James Gallagher, a miner, returned to his camp and was unable to find a tent he had left to dry. He asked the labourers of another miner, Thomas Scott, what had happened to his 'calico' and one said that he thought that Pumpkin (or Nagevagum) of Suloga had taken it. Firing a shot from his Winchester on the track just outside the village Gallagher entered Suloga and asked for Pumpkin. Eventually he appeared, and told Gallagher that 'House belong you stop house belong Scott'. He then accompanied Gallagher back to Scott's camp, and produced the tent. After a brief argument Pumpkin said, 'Suppose you want to fight By Christ I fight plenty belong white man I no afraid belong him'. Gallagher gave Pumpkin a poke with his rifle and told him to shut up. The next day Pumpkin and his father, Lakapu, fought Gallagher at his claim, slashing him on the back and wrist with knives before Scott came to Gallagher's aid. In retaliation Gallagher's brother Mat and two other miners burnt Lakapu's house. At a preliminary hearing Lakapu and Pumpkin claimed that they had gone to sharpen their knives on Gallagher's grindstone but he had misinterpreted their actions and attacked them. Sentenced to imprisonment on Nivani, Lakapu advanced to the position of warder, but returned to the labour gang when it was discovered that he had been spending his nights with a woman on Panapompom and returning to Nivani early in the morning. Lakapu died at Nivani in 1900. His son, although a 'young ruffian' who 'had to be taught his position' and strapped for attempting 'to have connection' with another man's wife, survived his term under government supervision.

Considered too young when arrested to face a charge before the Central Court, Pumpkin had served his time as a mandated child under the control of the resident magistrate. He returned to Suloga having suffered less than many of those who had remained in the village. There had been some prostitution. In 1899 Campbell cautioned two men at Suloga: Robert Lewis, a miner, who slept with a girl under twelve in his bunk, and Parakota of Suloga who attempted to obtain a shilling by falsely promising to lend his wife to a miner. 'Some alleged white man' introduced venereal disease in 1898 and it spread rapidly. After 'much sickness and many deaths' caused by unrecorded diseases the people of Suloga abandoned their village in 1901 and camped on the beach. Campbell wrote in the Nivani station journal:

> One thing appears to be pretty certain with regard to these people, and that is that the number of deaths during the past six years from diseases introduced by the whites, is very much larger than it could have been from all causes in the old savage days during a like period of time.

Eventually some people moved back and others left Suloga Peninsula to go north to Wasilasi and east to Unamatana. In 1915 Suloga was again abandoned and the village burnt after three people had died and others had suffered from dysentery.

Tudava, the first Suloga village constable, was recognised as a leader by his own people, and before his appointment in 1897 he had worked for the alluvial miners and learnt some English. Sometime after Tudava's death in 1915 Pumpkin took over the leadership of those people remaining in the Suloga area. He was most effective in directing their efforts to earn cash. By 1915 he had worked gold claims and his people had cleared 60 acres on Mapas Island for a copra plantation. When copra prices collapsed in the 1920s and 1930s the Suloga collected trochus shell and continued to work old alluvial areas. In 1941 they were still making sago to feed the men washing for gold, then almost the only product the islanders could sell. Pumpkin had died in about 1935. Some Muruans thought he had been poisoned, for he came from a community in which powerful and wealthy men were envied and in danger.

Lakapu and Pumpkin acted differently from all other Muruans. They were the only ones to attack and severely wound an early alluvial miner. Pumpkin showed greater determination than most Muruans to earn more cash than he needed to pay his £1 annual tax, buy tobacco and make infrequent purchases in the stores. But Pumpkin's attitudes and skills were not unusual. Frequently miners and government officials described the Muruans as independent, and sometimes they called them truculent and even 'overbearing'. Europeans in the 1920s, expecting to meet the simple savage, spoke disparagingly of Misima men who knew some English and 'seem[ed] very sophisticated, smoking Derby tobacco in clay pipes'. They were more disturbed by the Muruans' ability to select and reject from the new ways paraded before them. The Muruans faced numerous bearers of a culture who were confident, aggressive and given power by their technology, political organisation, and place in the economy, yet they were often able to choose the sort of relationship they wanted with the foreigners.

The Muruans knew some of the early miners, having met them as traders before 1895, and they were able to speak a little of their language. By the end of 1896 the people near Suloga Harbour were conducting a 'considerable industry' of their own and selling their gold in the calico and bush-built stores. The Muruans did not contest the right of foreigners to seize the richest patches, but some of the miners objected to their countrymen employing Papuans at 10s. a month to work ground outside their claims, and buying miners' rights for Papuans so that they could hold claims until their employers were ready to work them; and a few miners said Papuans ought not to work on any claims unless they received a 'due share of the results'.

If Muruans shared Campbell's distaste for the woman from a northern village who was 'prostituting herself all over the place', or for other women who formed casual relationships with Europeans, they did not allow it to influence their behaviour towards the diggers. They worked on the alluvial fields, carried for the storekeepers and replaced the miners' tents with the sago palm huts which the miners thought were healthier and more comfortable.

The Muruans were also prepared to work for the government, sometimes they came to the government station offering to work, but usually a government officer went to a village to recruit men for a particular task. Gangs of Muruans cut the tracks from 'Saltwater' to Kulumadau and on to Busai and Karavakum, bridging creeks, clearing timber, digging drains and carrying coral to surface tracks; they did constant work repairing sections washed away in heavy rains and cutting back long grass and undergrowth; they cleared the Nasai Point quarantine area; brought in bundles of sago leaf for thatching; they loaded and unloaded the punts which met the steamers in Kwaiapan Bay; worked on the construction of the wireless station; and when the timber for the Kulumadau school arrived they carried it from the landing and cut and transported timber for the stumps. At times the government employed up to eighty 'free' labourers. Most came from the northern villages of Dikoias and Kaurai, some from Guasopa, Wakoia and Kavatana in the east and Kropan in the south, and a few from Madau in the west. Normally they agreed to work for a short period for payment in tobacco. On 14 November 1911 fifty men from Kaurai, nearly all the able-bodied men from the village, came to Kulumadau and began cleaning up the cemetery; on 18 November they were paid 8 pounds of tobacco.

But some Muruans who worked for the government at Kulumadau were not free. Infrequently groups were gaoled for failing to carry out a government officer's instructions. In 1915 twenty-one Dikoias men were sentenced to four days' hard labour for failing to keep village tracks clear, and in 1917 thirty-three men from Kaurai were given fourteen days for not cleaning their village. Their experiences of prison work did not reduce their enthusiasm for 'free' labour. In March 1912 nineteen Kaurai men served seven days for refusing to carry for a government patrol and at the end of their sentence agreed to stay in Kulumadau to work on the road as paid labourers. And as free labourers they still ran the chance of being thrashed by road overseer Hamilton. To feed the prisoners and labourers maintained on the station, government officers bought food in the villages, supplementing the supply by taking frequent trips to draw on the yam gardens of the Trobriands.

Although some Suloga people worked as casual labourers for the reefers, Muruans would not sign-on as indentured labourers for the Kulumadau companies. In 1903–4, when the mines and stores employed 500 indentured men, only one Muruan signed-on; in the next year no Muruan signed a contract. Planters taking up leaseholds increased the demand for labour, but the Muruans still refused to sign-on. On Misima in the 1930s most indentured labourers came from outside the island, but there were always some local men prepared to accept the same conditions and they often left the north coast village of Bagalina to work at Umuna. On Woodlark when the Kulumadau companies' demand for labour was at its greatest Muruans from all villages, some of whom worked for the alluvial miners and the government, decided not to work as indentured labourers.

Nearly all the men who indicated to the government officer at Bonagai or Kulumadau that they understood the contract of service to work as a mining labourer on Woodlark came from the D'Entrecasteaux Islands. In December 1899 a hurricane swept through the archipelago, destroying tree crops and damaging gardens. It was followed by severe drought. Some Goodenough Island people, 'mere skeletons' desperately trying to keep alive, exchanged children who were then killed and eaten, or plundered gardens where a few root crops still survived. Many volunteered to follow earlier recruits to Woodlark where the companies were preparing for their first crushings; and government officers passed men who would normally have been rejected as unfit. The rains fell, the gardens recovered, but the rate of recruitment did not decline.

Map 6 D'Entrecasteaux Islands

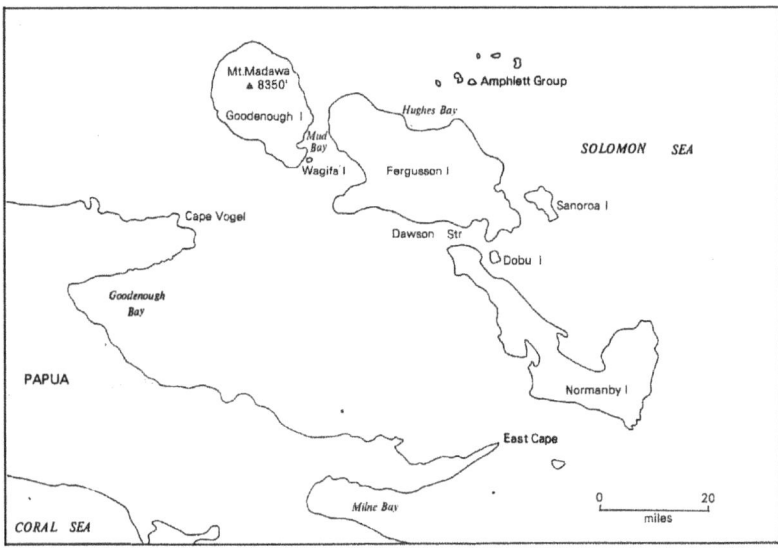

Coming from a competitive and abstemious community where all men worked hard in their gardens and strong men demonstrated that they had slight appetite for food or sex, the Goodenough Islanders knew the virtues of hard work; more than other Papuans they were willing to do the monotonous labouring tasks which the foreigners wanted other men to do. Young men went away before marriage and came back with greater prestige and the wealth to seal a marriage contract, or they left when their wife was pregnant for the first time. Some signed-on several times, coming home for only two or three months between contracts to help make new gardens and meet other obligations. By 1940 all men in many villages had been away to work and no woman had been more than a few miles from her home. In the Trobriand Islands where the people made different judgments about human behaviour, some women were prepared to sail as companions to white recruiters or live in miners' camps, but the men generally refused to work on alluvial fields or down the shafts at Kulumadau and Umuna.

The ordinary labourer on Woodlark was paid 10s. a month, and a few with special skills or those who became 'boss-boys' could earn up to 30s. Sometimes the companies paid a £2 bonus to men completing a twelve-month contract. White recruiters in 1900 were paid £5 a head for men landed in Woodlark and 10s. to return the 'time-finished' labourers to their homes. The recruiter took his boat into an anchorage and his Papuan crewmen went ashore to persuade men to come away with them. Sometimes a returned labourer or a leading man in a coastal village acted as his agent. Many early recruits from inland Goodenough and Fergusson came from villages unknown to government officers where the people still engaged in warfare and cannibalism. Government officers taking the first patrols inland met men who had worked in many parts of the Territory and spoke a little English and Motu. William Bowden, investigating the killing of some Fergusson Islanders on Goodenough in 1910, was showered with stones and spears. He asked an interpreter to call out that he had come to talk and not to fight. He did not have to wait for a translation of the reply: 'You policemen, you Government, we fight you', and more stones came down. Government officers were surprised that they could not recognise returned labourers until they spoke; nothing in their appearance or manner distinguished them from those who had stayed at home.

When men from the interior passed through coastal villages to go away as labourers they changed relationships between inland and coastal peoples. Because of their location coastal people had been the first to meet Europeans, and some had been able to exploit their position, distributing goods obtained from the foreigners and claiming power from the patronage of government officers, missionaries or traders. Labourers returning to inland villages in south-eastern Goodenough were likely to lose all or some of the contents of their boxes of trade; and some coastal peoples tried to preserve their position by preventing inland peoples making direct contact with foreigners. In 1903 Campbell called at Upaai on Hughes Bay, Fergusson Island. Men who had been friendly to him in the past and acted as guides when he had led a patrol which clashed with the inland Maiadoma now told him that the Maiadoma were talking loudly of destroying any government party that entered their territory. They had threatened to eat the police and use Campbell's 'skull to decorate the best house in their chief village'. The Upaai advised the government to 'kill em all'. Campbell agreed that if the Maiadoma had said such things they should be punished severely, but before he left the beach a leading man of Maiadoma came out to the government boat and, handing over a basket of yams for the 'master', accused the Upaai of causing all the trouble:

> From days long gone by they and we were always enemies. The coast people and the hill people. We used to kill each other one tribe taking payment for every man killed by the other tribe. The government came and the coast people first knew it, and they now are always threatening us that they will bring the government to eat us up. It is true my men took their spears to you when you came here last, that was because the Upaai people said you were coming to kill us all.

> You took two men away that time, the brother of one of those men is now standing there. We have come to say we do not want any more fighting, we have come to ask you that we the Maiadoma people may be protected from the threats of the Upaai people. Before you were led to our villages by them because we had killed a man belonging to their tribe, but we did this in payment for one of our men killed by them.

The Maiadoma also complained about Nicolas of Upaai, who had spent several years working for traders and now acted as an agent for recruiters. He threatened violence against the inland peoples and deceived his employers by keeping the tobacco which they had given him to pay intending recruits. Nicolas explained its loss by saying that hill people had stolen it from him. Campbell believed the Maiadoma: they told a plausible story and Campbell's old prejudices led him to favour the 'straightforward' inland men against Nicolas, 'one of the most objectionable of natives "the whiteman's nigger"'. Other officers followed Campbell's actions on his first patrol to Maiadoma: advised, guided and assisted by coastal people they punished unruly hill people.

Most D'Entrecasteaux Islanders tolerated the conditions on Woodlark, but some did not. By January 1897 the storekeepers were complaining about seventeen Dobu and Normanby carriers on the Busai track who stole a boat and deserted. Other islanders followed. After investigating a case in 1900 of thirteen Normanby men who stole a whaleboat, stocked it with tobacco and other goods, and sailed for home, Campbell decided that they 'have so frequently stolen boats, and escaped punishment, that they looked upon it as the right thing to do whenever they get tired of work, or become homesick'. A group of Dobuan labourers made sure they would not be pursued by stealing all rowlocks and oars, including those in the store. Sometimes deserters gave themselves up or government officers persuaded them to surrender by threatening to destroy a canoe unless they came forward; but until each man's name was entered in the village book, and there was a network of village constables through the islands, government officers found it difficult to capture deserters. Men facing a term in a prison gang and then returning to their old employer escaped by signing-on under another name and going to work in a different area.

Men accustomed to ocean voyaging who found little pleasure in mining or carrying may have taken any unguarded boat as an invitation to desert; others had specific grievances. Some Goodenough Islanders said that they had been mistreated by their overseer and had decided to take a boat when three labourers were killed by a fall of earth. And desertion was common in 1902–3 when thirty-seven indentured labourers died of beriberi, dysentery and German measles. The death rate reached its height in that year; over the next three years only eighteen labourers died out of a work force of over 400. In 1913–14 fifteen men died and four deserted from the 608 labourers on the island. The death rate was low compared with some mainland goldfields, but some white miners complained about the accident rate. Of the fifteen who died in 1913–14 six died as a result of injury, the highest in any one year. With many inexperienced men employed underground by struggling companies normally free of expert mining inspectors, a high accident rate was likely. The Kulumadau companies may have been lucky to avoid a major disaster.

Before the arrival of the miners the Laughlan Islanders told MacGregor that they did not want a mission teacher to live among them. MacGregor assumed that they had been influenced by the trader Tetzlaff; but in rejecting the missionaries the Muruans were confirming a decision of their fathers. They did not want the missionaries to return. In 1899 the Guasopa would not sell land to foreigners unless the buyers agreed not to let missionaries use it. A year later E.J. Glew of the Methodist Mission visited Woodlark with the hope of establishing a station there, but 'the natives [did] not seem to view the matter with favour'. They explained their opposition to Campbell:

> Those of the Dukoeasa [Dikoias], Kaurai, and Kropan villages called upon me and requested me to prevent the missionary settling at their respective villages. They stated that if he did establish a station in their villages that they would leave them and build new towns elsewhere. They appeared to have particularly strong views upon this subject. When asked for their reasons for objecting to have a missionary stationed amongst them they stated that now they were free so long as they did not commit any offence against the law, but once the missionary came they would no longer be able to 'walk about' i.e. be free.

At first the Methodists avoided Muruan hostility by offering guidance to the Europeans and indentured labourers in Kulumadau, but later they built a station at Guasopa. For over fifteen years European missionaries worked on Woodlark, but the Muruans chose not to believe them or the South Sea Island and Papuan teachers who came to Woodlark before 1940. No other small group of Papuans received so much attention from Christian missionaries and took so little notice of them.

The Misima had agreed to be hosts to mission teachers, who influenced a change in their beliefs and behaviour; the Sudest had shown no enthusiasm for the missionaries and the missionaries had spent little time there; the missionaries had worked hard on Woodlark but by 1940 their only consolation was that written by Xavier Montrouzier at Guasopa in 1851; they had suffered, sown in tears, and the islanders had resisted the grace of God.

The uniformity of the Muruan decision not to be Christians or indentured labourers was in contrast to the uneven impact of mining on the island. The Suloga, already suffering an economic revolution following the end of quarrying, lived on the beach where the diggers landed and on land where they mined and built their stores; those Muruans living in the main centres of population in the north and east were within a day's walk of the mining areas but distance gave them a buffer; in the extreme west the people of Boagis were little disturbed. Related by language to the Panaeate, they were seamen and traders, often 'sailing about' for two or three months of the year. They did not mine for themselves or work for foreign miners. In 1931 when they displayed their wealth to a visiting patrol officer it consisted of 270 necklaces and arm-shells soon to be taken to trading partners in other islands, the necklaces continuing their circle through the islands by going south and the arm-shells west. Forced to find cash to pay taxes, the Boagis looked to the beach or the sea; they worked shell or copra.

World War II touched Misima when production at Umuna was at its height; it broke across Woodlark when the white population had declined to six miners, three planters and traders, and one government officer. Two miners who were living with local women decided not to obey the general order to leave the area in January 1942. They were joined by Lieutenant P.J. Mollison, recently a patrol officer and now an officer in the Royal Australian Navy and a coastwatcher. Early on 23 June 1943 villagers told him that soldiers were landing on Guasopa beach. Hearing the troops speak English Mollison went forward and learnt that they were the advance party of Alamo Force, instructed to occupy Woodlark, code-named Leatherback. Within a few weeks nearly 8000 troops and 17,000 tons of equipment and stores had been landed at Guasopa. Most of the troops were from the United States 112th Cavalry Regiment, an artillery battalion and support units which had trained and assembled in Townsville. Lieutenant-General Walter Krueger, commander of the United States Sixth Army, recalled that a few men from the Australian New Guinea Administrative Unit had 'served as a cushion between troops and natives'. Twenty-four days after the arrival of the first troops the Guasopa airfield, built close to the site of the old Marist mission station, was ready for use; and a week later a fighter squadron began operating from Woodlark. The Japanese responded with two light bombing raids which did little damage.

Expected to be a major base for attacks on Japanese-held areas on Bougainville and New Britain, Woodlark was of little use to the Allied forces. The movement of men and equipment and the development work around Guasopa were later thought valuable only because they were practice for other, larger amphibious landings. War had come to the Muruans as suddenly as mining, but it had been on a massively larger scale, was less explicable in terms known to the Muruans and it was briefer. After the war the foreign population declined to about eight: men concerned with planting, saw-milling, trading and prospecting. And sometimes scholars, missionaries and government officers lived on Woodlark. The one obvious legacy of the war was the Guasopa airstrip.

In the post-war years the Muruans changed their way of life in one major respect: they listened to the teachers from the Methodist mission and many became members of the Methodist, later the United, Church. But in 1974 there was still a *waga* pulled up on Guasopa beach. Like others in the area it had been purchased by an exchange of shells and other valuables from the canoe-makers on Gawa Island. The carved prow, the mat sail and the hand-spun rigging were similar to those shown in the earliest photographs taken on Woodlark Island. The houses in Wabununu village on the west side of the Bay are now rectangular boxes with thatched gable roofs; yet people and place seem to be lightly touched by the outside world. A man in Wabununu village still makes a name by acquiring arm-shells in competitive and ceremonial trading with other villages. The old men of Wabununu remember the miners as hard men who took much wealth and gave little. They built fences around their houses in Kulumadau. If you visited a miner's camp and you had fish to sell you were tolerated, but if you went to talk and eat with Papuan labourers you were shouted at and hunted away like a dog. In Muruan eyes the miners were men who did not understand that wealth was something which men could compete for, yet at the same time it could be shared and circulated.

The question for a historian to ask about Woodlark is not what changes the miners brought, but why they brought so few.

More than other Papuans on island goldfields the Muruans suffered a decline in population. In 1900 there were about 2500 Muruans; they had recovered from the effects of the periods of drought and disease in the 1850s. By 1920 there were only 800, and the population remained at about that level until 1940. There seems to have been no one epidemic but many: dysentery, venereal disease, influenza, whooping-cough and German measles. Before the village constable of Dikoias came to Campbell in 1901 to ask for help because many of his people were sick and some had died, he may have been forced to doubt the capacity of Muruans to sustain their own ways; but later he would learn that the Europeans could then do little to stop the dying. The declining population may have made it harder for the Muruans to maintain their independence; it was not sufficient to make them capitulate to foreign science or foreign explanations about the causes of suffering. In 1971 nearly 1200 Muruans lived on Woodlark.

The description of the Suloga quarry is taken from MacGregor to Governor of Queensland, 20 November 1893, and Seligman and Strong 1906. Seligman 1910 and Malinowski 1934 have additional information on the axe industry. Fred Damon showed me an old grindstone near Wabununu village in 1974.

Laracy 1969 and 1970 provides more detailed information about the early Catholic missionaries on Woodlark than is given here. The description of Guasopa Bay and the incident about the baptism of the child are from Verguet 1861. The looting of the Gazelle is from *Sydney Morning Herald*, 14 June 1856.

Tetzlaff's association with the Marquis de Rays's expedition is mentioned in Biskup 1974. The burning of his station is referred to by Corris 1968, Morrell 1960, and Scarr 1967.

Of the 405 men returned to south-east New Guinea by the Victoria only two were from Woodlark; *Queensland Parliamentary Papers*, 1885, Vol. 2, p. 1072.

The sources referred to in earlier chapters were again used to provide information about early contact between foreigners and Muruans, and about the beginning of mining. Two valuable newspaper accounts on mining are *Cooktown Courier*, 22 October 1895 and *Brisbane Courier*, 5 March 1897. Ede's obituary is in *Pacific Islands Monthly*, May 1942. Meek 1913, pp. 73–4 and Monckton 1921, p. 22 give their experiences. The conditions on the goldfield in 1897 can be gauged from Campbell's reports in the Nivani Station Papers and in newspaper accounts; *Brisbane Courier*, 9 April 1897, *Sydney Mail*, 24 April 1897, *Mount Alexander Mail* (Castlemaine) 21 April 1897 and in Musgrave's collection of cuttings. Campbell wrote from Suloga Harbour: 'I have now before me several Northern Queensland papers, but in none of them can I find a word of warning in re the Woodlark Goldfield, although on the other hand there are plenty of, if not absolutely untrue, highly exaggerated, reports of the quantity of gold, the vessels, or miners, carry away with them from this island ... ' He gave an example of a man who arrived in Cooktown and claimed to have 200 ounces; Campbell said he had less than 20. Nivani Station Papers, 20 March 1897.

The account of the strike by white workers is taken from the Nivani Station Papers. Campbell, Nivani Station Journal, comments for September 1900, wrote: 'The men contend, and with truth, that although the rate of wage was £5 they, what with sickness, wet weather, and other causes did not earn more than £3-12-0 per week. The company's pay sheets show an average of £3-13-0 per week per man. The storekeepers are paying £5 per week at present-time.' H.B. Higgins in the Harvester Judgement of 1907 thought 7s. a day was a minimum fair and reasonable wage. Before the Royal Commission of 1906 Fred Weekly said, 'No labouring man on this island gets less than £5 a week. The salary of the Magistrate and Warden hardly comes up to that'. In fact, Francis Gill, the Warden and Assistant Resident Magistrate, received £250 a year.

The statistics related to reef mining are recorded in *Annual Reports* and there are some quarterly returns from the companies in the Nivani Station Papers. Evan Stanley, 'Report on the Geology of Woodlark Island (Murua)', *Annual Report, 1911/12*, pp. 189–208, gives details about the location of leases as well as the nature of the country.

The diary of the Reverend James Walsh is quoted as Appendix D in D.A. Affleck, Murua or Woodlark Island: A study of European Contact to 1942, B.A. hons. thesis, A.N.U., 1971.

The discussion on prohibition is recorded in *Commonwealth Parliamentary Debates* on the Papua Bill, and *Commonwealth Parliamentary Papers*, 'Papua (British New Guinea). Spirituous Liquors — Correspondence Respecting the Suggested Prohibition Against their Introduction, Manufacture and Sale', 1904, Vol. 2. The Chairman of the Royal Commission, Kenneth MacKay, wrote unofficially of his tour in *Across Papua ...* , London, 1909, pp. 65–6.

Events of importance to the European community were noted in the Kulumadau Station Journal and, infrequently, in the *Papuan Times*. Whittens' storekeeper wrote about Woodlark women to Frank Pryke, 23 September and 23 November 1912, Pryke Papers. South-Eastern Division, General Correspondence, C.A.O., C.R.S.G 180, item 15 has a little information on the school and item 55 is a file on the European Armed Constabulary.

There is a tribute to John Taaffe in the Kulumadau Station Journal for 4 June 1920.

Campbell, Nivani Station Papers, 30 December 1897, wrote a long report on the attack on Gallagher and enclosed statements by witnesses. D.H. Osborne has an account of the attack on Gallagher in *Pacific Islands Monthly*, September 1942.

While there are breaks in the government records about Woodlark after the transfer of the government headquarters to Bwagaoia in 1920, some of the reports of H.W. Rogerson, who was stationed at Kulumadau for much of the 1930s, have survived. The Bwagaoia papers include reports of patrols to Woodlark.

In 1905 I.J. Penny died while walking to Kulumadau and it was assumed that he had died of fever, but two years later it was discovered that he had been murdered by Muruans. This second death was ten years after the beginning of alluvial mining.

Dimidau, who established a copra plantation at Muniveo, was another leader who pressed his people into the cash economy.

'Muruans were often able to choose the sort of relationship that they wanted ... ' I do not argue that they made the best choice; just that they chose.

There are two basic studies on the people of Goodenough Island: Young 1971, and Jenness and Ballantyne 1920.

The statement by the leading man of Maiadoma was taken down by Campbell, Samarai Station Journal, 23 August 1903.

The death rates are from *Annual Reports*. I was unable to find any figures for injuries not resulting in death.

Campbell wrote down the Muruan objection to the missionaries in the Nivani Station Journal, September 1900, comments at the end of the month. The Minutes and Journals of the Annual District Synods, held at Dobu, United Church Papers, Box 20, New Guinea Collection, University of Papua New Guinea, have references to Woodlark, 'one of the most trying and perplexing stations in the District'.

The account of the war and Woodlark is based upon Australian War Memorial, C.A.O., 609/7/9, 616/4/1–5, 616/10/1, 519/1/12 and 519/6/71; and ANGAU War Diary. A report by A. Timperley lists the foreign population on Woodlark on 1 January 1942, ANGAU 80/6/4. I am grateful to Jerry Leach who directed me to the records in the War Memorial. Additional information on the war from: *Allied Geographical Section, Southwest Pacific Area, Terrain Study No. 23, D'Entrecasteaux and Trobriand Islands*, October 1942; Dexter 1961; Krueger 1953; Miller 1959; and Morison 1950.

Fred and Nancy Damon, resident in Wabununu village in June 1974, hosted a meeting with the people of the village and translated their talk. I am indebted to them for much general information.

OPENING THE MAINLAND

Tree House, Koiari, after Stone 1880

5

The Laloki

a beautiful country but a failure

In 1871 seventy-five 'spirited young men' paid £10 each to buy a leaky old brig, the *Maria*, and sail her from Sydney to New Guinea where they hoped to find gold. None reached New Guinea and less than half survived the voyage. Off the Queensland coast the *Maria* struck a reef and only about fifty of them were able to get away on boats and rafts. Some reached the settlement at Cardwell, eight drifting further north were kept alive by 'good-natured savages', and about seventeen others landing at different points on the coast were killed by the Aborigines. The first rescue party in a fight for possession of one of the boats from the *Maria* shot eight Aborigines dead and left another eight wounded. A few days later black troopers in a punitive raid caught a party of Aborigines in their camp and acted with 'unrestrained ferocity'. The Police Magistrate at Cardwell, Brinsley Sheridan, thought that without harsh immediate action no boat would be safe on the coast, and settlers 'or even the town itself, [might] be attacked by savages'. In spite of the fate of the *Maria*, the Reverend Wyatt Gill of the London Missionary Society, who landed the first Polynesian teachers on the southern coast of New Guinea in 1872, believed that many men were still eager to join another expedition to find gold on the island. Six years later news that a few specks of gold had been found was sufficient to cause several groups to charter boats for Port Moresby.

In 1877 Jimmy, a New Caledonian who had been on the Queensland goldfields, picked up a piece of quartz containing a few flecks of gold in a river bed about 15 miles inland from Port Moresby. His employer, Andrew Goldie, a naturalist, and the Reverend William Lawes of the London Missionary Society spread news of the find and cautioned against a rush. Lawes wrote in his journal that if the diggers came 'with gold fever on them, the natives [will] be shot down like kangaroos as they have been in Queensland'. In letters printed in Australian newspapers Goldie reported that he had found further signs of gold at every prospect in the black sands of the river which he had given his name; and he warned that the people of the area were intelligent and would be quick to revenge any wrong. It was for this reason, more than any other, that he feared disaster would follow any rush of white miners to New Guinea. Mr and Mrs Lawes and Goldie were the only Europeans with permanent homes in Port Moresby. They lived more than 200 miles from any other Europeans in a land unclaimed by any European power.

Map 7 The Laloki

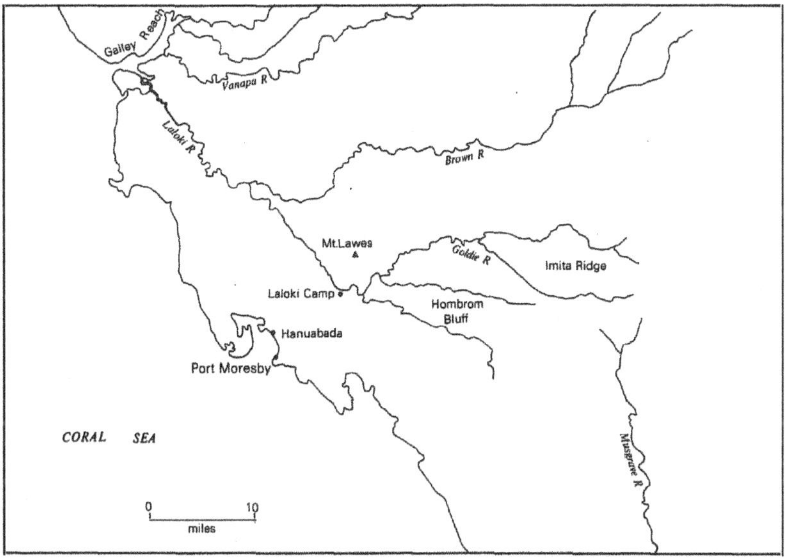

The first diggers arrived in April 1878 at the end of the wet season when the hills around Port Moresby were covered in lank grass. It looked 'a beautiful country' to John Hanran, whom the diggers had elected their warden. Over a thousand Motu and Koitapu people lived close to the harbour 'in nice, clean grass-built houses on the beach, so at high tide the water comes rolling underneath the houses which are built on piles'. 'The women', said Hanran in his report carried by ship to Cooktown and telegraphed to other centres, 'are virtuous, agreeable, and willing to work, and have not the wild appearance of Australian aborigines'. For four years the villagers had been listening to the Polynesian and English representatives of the London Missionary Society and they had been visited by traders, naturalists, fishermen and Her Majesty's seamen. They met the miners with fish and coconuts ready to barter for tobacco, and they cut canoe loads of grass to feed the diggers' horses.

About 100 white miners and two notorious goldfield followers, Annie Smith and Jessie Ormaston, arrived in Port Moresby in 1878. Most of the men crossed the low grass and thinly timbered country behind Port Moresby to a camp on the Laloki; the two women returned destitute to Thursday Island where 'the uncivilized native women' were much shocked by the behaviour of 'their white sisters'. The Queensland Government accepted William Ingham's offer to act as its agent but as he had no police and was outside the colony of Queensland he had little formal power. The diggers decided to try before his peers any man who 'used violence wantonly' against the local people or their property, and they pledged themselves to sacrifice 'their lives and fortunes to prevent Chinamen from landing'.

For the short time he remained in Port Moresby Ingham was not greatly restrained by his lack of formal authority. He paraded eight armed diggers in the coastal village of Hanuabada and handcuffed a villager to a house post until some stolen goods were returned, and he ordered a digger to pay a tomahawk to a man whose dog was shot. But there was little trouble between the miners and the coastal people. Soon the diggers, who had previously travelled only in armed groups, began to move alone and unarmed between the Port and the Laloki camp. Prospecting parties crossed the Laloki and began cutting tracks to the north and west. Eventually a few men reached the headwaters of the Goldie River and crossed the Brown, another tributary of the Laloki.

The land beyond the Laloki belonged to the Koiari. They had seen little of the missionaries at Port Moresby, but they traded with the coastal people, who respected their power as spearmen and sorcerers. The Koiari fought later exploring parties and government patrols, yet their early encounters with the miners were peaceful. Hanran reported camping on the Goldie River close to a village of about 800 people who took the miners on a pig hunt, carried their swags and showed them tracks. To both Ingham and Hanran the Koiari seemed mentally and physically superior to the coastal people. Without government protection and confronted by numerous people in difficult country, most of the diggers realised that they could not prospect if the villagers were determined to resist. It was a fact clearly understood by Patrick Minnis, an illiterate digger who gave evidence against one of the few miners who acted brutally. After seeing a miner attempting to rape a struggling woman while her shrieking child stood by, Minnis told the man that he could not promise to keep quiet about the affair because 'it was a fearful thing endangering men's lives, and he should be cautious'. But within a few months the miners had lost their confidence in the goodwill of the Koiari. A lone miner, surrounded by Koiari panicked, fired his revolver and ran. The Koiari hit him twice with spears before he reached the protection of other miners. Hanran thought that the Koiari had at first just been curious, but the men at the Laloki now believed that the Koiari were massing for an attack and they abandoned the camp while one party was still out prospecting.

Koiari ridge-top village and tree-houses, 1885
PHOTOGRAPH: J W. Lindt

Only five diggers remained in Port Moresby at the end of the year, and they were trying to make money by shooting birds of paradise. Although some diggers had thought that the river valleys looked 'better than even the Palmer', they had stopped prospecting. They had been turned back by the country and malaria, not by the people. On the grasslands south of the Laloki they had been able to travel easily, and the wallabies, scrub hens, ducks and *goura* pigeons made 'beautiful eating'. But once they reached the country around Mount Lawes the rain forest became so dense that they had to abandon their horses and those who attempted to cut tracks suffered from exhaustion and scrub itch. They found it difficult to carry stores, their swags were often wet, and they saw little game to shoot. Most men preferred to stay about the tents and log hut at the Laloki camp. After riding out to see the diggers, Henry Chester, the police magistrate from north Queensland, reported that the men were 'a most respectable class' but 'from a digger's point of view, some who have come to New Guinea would have been better employed in wheeling a perambulator at home …'.

Malaria was a 'fearful reality' for the diggers. Nearly all suffered, a few died, and many return ed to Cooktown believing that they had to leave the Laloki to stay alive. Ruatoka, a Cook Island missionary stationed in Hanuabada, carried two sick diggers back to Port Moresby, and he and his wife nursed many of the sick in their house. This 'truly good Christian' was given a testimonial by the diggers and a 'splendid breechloading fowling piece' by the Queensland Government.

The return of the wet season in December finally forced the diggers to abandon prospecting. The last expedition led by Frank Jones, one of the most active of the prospectors, was caught in floods on the Goldie, horses were drowned and stores and swags swept away. Before they left New Guinea some diggers took schooners along the coast to have a look at the land behind Hula in the east and Yule Island in the west.

After eight months in New Guinea the diggers agreed that 'the New Guinea goldfields may be pronounced a failure for the present'. But many were sure that it was only a matter of time before gold was found in payable quantities; and they had learnt something about the people and the difficulties of travelling in the interior. They had to accept that nearly all prospectors in New Guinea would suffer from fever. At best they could say that most miners survived, that the rate of sickness had been worse on the Palmer, and that the fever might be less severe in the interior. They were able to describe the country around Port Moresby by comparing it with various areas of northern Australia; it was like the Warrego, or 'the back of the settlement at

Somerset'; but they knew of no parallel to the endless series of sharp ridges at the head of the Goldie, each smothered in a tangled mass of growth below the pleasant variegated green of the top canopy. The newspapers in the diggers' home towns had carried reports denying that New Guineans were mere savages. The diggers had seen different communities with different skills and ways of behaving. The New Guineans were clearly unlike the Australian Aborigines; and because they were more numerous, looked more as if they owned their lands, and took a higher place in the scale which white men then thought they could apply to all races, the diggers knew that they could not treat them in the same way.

The horses left behind by the diggers were caught and used five years later by George Morrison, the leader of the *Age* expedition to cross New Guinea. Just beyond the Goldie River Morrison shot and wounded a man who stole a knife. The next day Morrison was speared. Recuperating at Geelong in southern Australia, Morrison wrote a series of articles which, added to others published at the same time, gave Australian readers a greater appreciation of the problems of travelling in New Guinea.

While the diggers were taking the first gold from the Louisiades other north Queensland miners were prospecting to the west of Port Moresby on the Alabule, a river the miners called the St Joseph. By the end of 1888 over twenty diggers had tried the St Joseph, but they had found only colours. On other parts of the coast prospectors who chose to say little about where they had been made trips into the country behind the beaches and the mangroves. One such man was George Sharp who, having left his job as a coalminer in New South Wales, tried goldmining in the Kimberleys of Western Australia and tin mining near Cooktown before making several journeys into the country west of the Fly River between 1887 and 1890. In the Kimberleys he had shot Aborigines who tried to drive him from the water he needed to stay alive, but he did not relish violence and in New Guinea the rifle did not give him the same superiority. Sometimes Sharp, his partner and a few carriers from coastal villages were surrounded by several hundred armed men whom they had to conciliate or evade; he found it difficult to travel among communities who were at war or hostile to each other for if he accepted help from one group, another regarded him as its enemy; and he finally left New Guinea after his partner had been killed and he had a narrow escape. George Sharp had never been to school, but after trying different jobs and looking for gold in Western Australia again, the Northern Territory and Africa, he attempted to write about his experiences. He said that he had learnt that 'there are brainy men in these wild countries, many of them being able to reason with you if you could understand them'; and he thought that they were not to be blamed for 'keeping whites out of their country'. Jolley, a man from Kataw (Mawatta), a village west of the mouth of the Fly, had accompanied him on his New Guinea travels. Without pretension or qualification Sharp wrote of him as his 'best friend'.

Tobacco pipe, Koiari, after Haddon 1894

For many diggers the islands were preliminaries to the big fields which must soon be found on the mainland. The presence of gold on the islands confirmed what they had long believed; that somewhere in those ranges, looking close and accessible on clear mornings and disappearing each afternoon in thick cloud, there must be gold. The prospectors' only tracks to the interior were the rivers. With the opening of the goldfields in the Louisiades they began looking for openings into the country of the east end.

Lewis Lett begins *Papuan Gold*, 1943, with an account of the *Maria*. The quotes in paragraph one are from Moresby 1876, chapter 4. Moresby picked up some of the survivors. The Reverend Wyatt Gill wrote in the *Leisure Hour*, 3 January 1874.

H.J. Gibbney 1972 has written in greater detail about the rush of 1878. The main reason why this chapter is able to add to Gibbney's article is that he lent me his file of newspaper cuttings relating to the rush. Particularly valuable were articles from *Town and Country Journal*. British Parliamentary Papers, 'Further Correspondence Respecting New Guinea', 3617, 1883, includes articles from newspapers and statements from observers who were not government officials.

Marjorie Crocombe 1972 wrote a biographical article on Ruatoka. and Lovett 1903, pp. 132–8, quoted Chalmers on Ruatoka.

George Morrison's travels in New Guinea are described by Pearl 1967, pp. 45–54.

It is difficult to find information about prospecting trips between the rush of 1878 and the discovery of gold on Sudest in 1888. The St Joseph expedition was referred to in the *Sydney Morning Herald*, 4 October and 10 November 1888. The *Cairns Post*, 12 December 1888 said that Bill McCord, a well known prospector, was going to try the St Joseph. It is this sort of entry which indicates that private expeditions were going to New Guinea: but they give little indication of where the prospectors went or what happened to them.

George Sharp wrote an autobiography, Elusive Fortune: the Adventures of George Sharp while Prospecting in the Kimberleys, New Guinea and Coolgardie, 1885–1907. The manuscript has been edited and has an introduction by Gibbney, Australian National University. The manuscript is in his possession.

6

The South-east

a few fine colours and malaria

By opening St Patricks Creek Frenchy and Jimmy the Larrikin had led the Cooktown miners to north Misima; but their experience on the mainland made other prospectors pause before trying the south-eastern end of New Guinea, that line of mountains pointing to islands where gold had already been found. In August 1889 they landed on the south coast near Abau Island. Having purchased beads and tobacco from Thomas Andersen's trading station at Dedele Point, they moved to Domara village and prepared to go up the Bomguina River. For protection they fashioned armour from ships' copper and carried two revolvers, a Winchester rifle and a double-barrelled shotgun. Ino, the son of Tuari the head man of Domara, having failed to persuade them that their plans were foolish, agreed to take them by canoe to Merani, the first inland village. Accompanied by men from Merani, Frenchy and Jimmy left the next day to trace the river back into the foothills. A few miles upstream the Merani led them across a swift-flowing tributary; and, as some villagers held the prospectors' hands to steady them against the current, others axed them. The Merani sent a revolver and some gold (which the prospectors had brought from Misima) to Ino who, knowing there would be trouble, refused to accept them. Ten years before when the crew of one of Her Majesty's ships burnt Dedele village, the coastal people had learnt of the Europeans' determination to look after their own kind.

Two months later MacGregor, his police and several armed Europeans arrived. They found Merani and the adjoining village of Isimari deserted. MacGregor occupied the village while the Merani, camped outside their own stockade, shouted an admission that they had killed the prospectors and hurled spears and stones into the village. After three days of stalemate a government party went out to make contact with the Merani. Having surprised the villagers, the government force took up a position in a grassed area and the Merani 'came on with great spirit, shouting and yelling as only savages can'. Before they came within spear range MacGregor ordered his men to fire: 'In about one minute [the Merani] ceased to yell, then faltered, turned, and fled to the forest.' Two were killed instantly and four died of wounds. MacGregor ordered the destruction of the sapling stockade and the treehouses and fighting platforms built 60 to 80 feet above the village. He named the river the McTier, and a point just below where the prospectors were killed Rochefort Falls. Neither maps nor people now use Jimmy's or Frenchy's family name. Two prisoners were taken to Port Moresby where one died and the other was released when it was found that he was suffering from leprosy.

Map 8 The south-east mainland

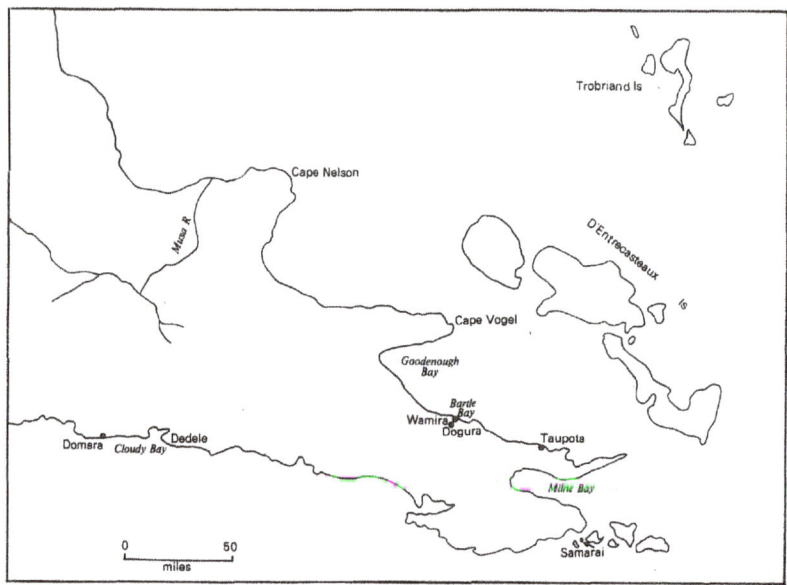

MacGregor summarised his actions for the Governor of Queensland:

> A fierce and powerful tribe has been defeated in open fight, and been thoroughly humiliated and demoralized; their strongholds have been occupied by us and destroyed, and they are for the time completely expelled from their houses and lands, and we occupy their dwellings and live on their food.

As MacGregor saw the situation, either he executed the Queen's warrant or he succumbed to raw savagery. He was confident that his actions carried out with a 'strong hand' would be 'once for all'. He was wrong. The Merani killed Ino, who had helped the government take prisoners; and the Domara, afraid of further reprisals, moved to Dedele. In 1894, after they had returned to Domara, the inlanders attacked them again, setting fire to the houses and spearing the occupants as they fled. Nineteen Domara were killed and more would have suffered but for the arrival of Joe Fiji and his shotgun. Brought from Fiji to serve in MacGregor's police, Joe had married a Motu woman and joined the polyglot population then trading along the coast and, incidentally, giving some coastal communities access to greater wealth and power than that available to more isolated people. Joe had only two cartridges but one blast hit a man in the side, caused alarm among the marauders and turned their triumph into desperate flight. Government officers and a troop of police returned to Domara and the cycle of pursuit and arrest began again. Nor did the 'strong hand' influence other peoples along the southern coast. A year after MacGregor's first expedition against the Merani a prospector, Neil Anderson, was killed at Baibara Island, about 60 miles east of Domara, by villagers who had probably heard the talk about the fighting at Cloudy Bay.

Prospectors hesitating to go inland were further deterred by reports of the sickness suffered by the government party. MacGregor's first expedition only stayed a few days on shore, but twenty-three of the twenty-five were sick with fever before they left the area. The land behind Cloudy Bay, MacGregor reported, was remarkably fertile and unhealthy. Ten years later Cloudy Bay was crowded with schooners and men were trekking inland to the new field at Keveri.

For five years after the deaths of Jimmy and Frenchy few of the miners leaving the declining fields in the Louisiades attempted to prospect the interior. But from 1894 the Queensland miners sent several well-equipped parties to test the north-east coast. Fifteen Cooktown diggers led by James Hurley employed carriers from Bartle Bay to take them south into the ranges. The hill people, who spoke a different language from the coastal villagers, had seen no Europeans cross their lands, but they made no immediate attempt to stop the prospectors and some agreed to carry for them. Having located a patch yielding a little 'scaly' gold, Hurley left the main party to search ahead for a richer deposit. Five other miners and some coastal and hill peoples went with him. As they travelled up the bed of a stream Hurley bent over to help retrieve a panning dish lost in a pool. The hill man who had been carrying Hurley's pack snatched up an axe and killed him with the second blow. The other hill men immediately attacked the rest of the miners, who were able to use their revolvers to kill one man and drive the others off. The Cooktown miners withdrew, and the police who came in pursuit of Hurley's murderers reported shooting another man and taking six captives before the hill people abandoned their homelands for more inaccessible territory. As there was no evidence to show that the six prisoners had anything to do with the attack on Hurley they were convicted of resisting the police and put to work in the Samarai prison gangs. The 7 ounces of gold found by Hurley's party was sufficient to bring more men from the Louisiades and north Queensland to Bartle Bay.

In Cooktown the public raised funds to send a second expedition to New Guinea, and a committee selected twelve men from a list of volunteers ready to make the trip. Three weeks after the Cooktown party under the leadership of Pat Riley left on the *Merrie England* in March 1895, eleven miners of the Cairns Prospecting Association went on board the schooner *Meteor* for Samarai. The best known miner in the Cairns Association was George Clark, about fifty-five years old, a founder of the Charters Towers and Mulgrave Goldfields, and a prospector on the Palmer, Hodgkinson, Herberton and Russell fields. Before leaving for New Guinea he was testing the conglomerate country on the Palmer for a Cairns syndicate.

New to New Guinea, the Cooktown and Cairns miners looked closely at Samarai, their first sight of their neighbouring British colony. Burns Philp's store and Whitten Brothers' boat sheds and copra store were built on the beach. In the centre of the 50-acre island the 'residencies' of Burns Philp's manager and the senior government officer looked out across the crowns of the coconut palms. Gangs of prisoners were taking trolley loads of earth from the side of the one low hill and tipping them into a recently drained swamp. When completed the work would reduce the incidence of malaria and give Samarai a fine cricket ground. The men arrested after the killing of Hurley were pointed out to the Queensland miners. With its profusion of palms, gentle undulation, sheltered anchorage and neighbouring islands, Samarai presented 'a beautiful picture'; but a north Queenslander who had taken a temporary job as storemen with Burns Philp told the miners that a man got lazy and demoralised with the natives doing all the handling.

Anchored off Taupota on their first night out from Samarai on their way to Bartle Bay, the Cooktown men watched women bobbing up and down 'like dancing emus' as the light from burning torches flickered on their bodies and grass skirts while they fished. At Wamira village where they established their first camp the miners thought the people docile and curious. One of the Cooktown party, charmed by the hymn-singing of the villagers, compared it with 'the treble singing of the paid boy choristers at St. Paul's in London'; and R.J. Walsh, travelling with the Cairns Association, thought it 'very comical' to see the missionaries 'riding the niggers over the watercourses'. But he too chose to be carried over the creeks. At Easter next year the Anglicans were to baptise two men, their first Papuan candidates for the assembly of Christians. From their arrival in 1891 at Dogura on the west of Bartle Bay the Anglicans had obtained a strong influence over a few coastal villages. On the beach the Queensland miners were meeting Papuans who had encountered a variety of foreigners over the previous twenty years.

Inland the miners admired the productivity of the hillside gardens, the capacity of the carriers to handle 50-pound loads on the boulder-strewn gullies used as tracks; but they thought the New Guinea spearmen less skilful than the north Queensland blacks, deplored the spread of skin disease in an otherwise handsome people, and ridiculed the village dogs 'so miserably poor and wretched looking that they have to lean against a tree when they want to howl'. Although the Cooktown miners were never in danger of attack Jack Christie, the best shot in the party, gave a 'salutary check' to each new group by demonstrating the power of his repeating rifle. The miners were aware of hostility between different groups in the area south of Wamira for they noticed that their carriers brought spears with them and attempted to interest the miners in raiding the mountain villages for food and women; but from their own observations and talk with the Anglican missionaries they concluded that 'Plenty of preparations and a vast deal of palavering is the upshot of most of the tribal fights'.

The Cooktown miners made three trips inland, one from Wamira and two from Paiwa further north on Goodenough Bay. Most of the time the prospectors suffered a monotonous, enervating routine. Attempting to cut across as many streams as possible, they were forced to scramble up and down steep, broken ridges; they lived on johnny cakes, salt beef and rice, and vegetables and occasional pigs bought from villagers; at times unable to obtain carriers, they attempted to lump their own packs; frequently lashed by rain they camped at night in tents floored with cut saplings to keep stores and packs above the mud; and most experienced the shaking, profuse sweating, cold and delirium that went with malaria. The Cooktown party found no payable gold; and they thought themselves lucky that only one man died. J. Turner shot himself at the Paiwa camp while suffering a 'delirium of fever'.

The Cairns Association split into two groups: one left Bartle Bay and cut across the tracks of Hurley, and the other, led by Clark, recruited carriers at Taupota for an expedition from the north coast over the ranges to Milne Bay. According to Bill Alexander, one of the group chasing Hurley's prospect, the rivers behind Bartle Bay came down at such rate that not even fish could survive in them. The only alluvial they could find was small beaches on protected bends where they washed a few 'fine colours'. The villagers, Alexander said, were 'quite peaceable'. Each community was prepared to travel with the miners, but only within its own territory. To their surprise the prospectors became dependent on the villagers. As Alexander, a man 'used to Queensland bush life for many a day', explained to the reporter from the Cairns *Argus*:

> You cannot find your way from one place to another without having native guides, and the way they dodge about is very curious. We would sometimes be sure our guides were taking us right away from where we wanted to go; but no, they were always right.

The Cairns miners prospecting behind Bartle Bay returned to Samarai when most of them 'got down with the fever'. Jack Martin died before the *Meteor* reached Cairns. Bill Alexander decided to pack his 'traps' and return to the Towalla field in Queensland; but he was sure that if there was gold in New Guinea it would be found soon. With the temporary slump in the copra, pearl and bêche-de-mer industries many of the traders were idle and, Alexander predicted, they would now 'dodge about the coast with a couple of boys, do a little trade with the natives, and try all the places they think likely'. Unknown to other miners one of the trader-prospectors, Charlie Lobb, was already working gold on Woodlark.

Clark's party crossed to Milne Bay without finding payable gold. Having passed miles of almost impenetrable country Clark was convinced that prospectors were most likely to succeed where they could enter gold-bearing country by boat. The Mambare River, recently mapped by government officers, he now thought could be the track to the gold of the mainland. Clark took a canoe from the head of Milne Bay to Samarai where he met Johnny 'Fiji' (Cadigan), owner of the cutter *Seagull*. For shares in the Cairns Prospecting Association Cadigan agreed to take six miners, two Queensland Aborigines and fifteen Taupota carriers, their stores and a whaleboat to the mouth of the Mambare. Before leaving the south-east Clark wrote a note to the Resident Magistrate at Samarai asking him to send a boat to pick them up in about eight weeks' time.

MacGregor's dispatches recording his pursuit of Frenchy and Jimmy's killers are reprinted in the *Annual Report 1889/90*, pp. 26-35. The *Report* includes a map by J.B. Cameron. Dutton 1971 locates villages and language groups along the south coast. Green, letters and Abau station papers, describes events in the area in the 1890s.

Lett 1943 has an imaginative account of the killing of Hurley. Winter's dispatch referring to the incident is in the *Annual Report 1894/95*, pp. 4, 5.

All the members of Riley's expedition are listed in the *Cooktown Courier*, 19 February 1895. Clark's life is outlined in obituaries in the *Cooktown Courier* and *Cairns Argus*, 10 September 1895. These two papers carried long reports spread over several issues on the work of the prospectors.

The *Annual Report 1895/96*, pp. 17, 18, prints an exchange of notes between MacGregor and Moreton on whether or not Moreton had committed the government to provide transport for Clark's expedition from the Mambare to Samarai.

THE NORTHERN RIVERS

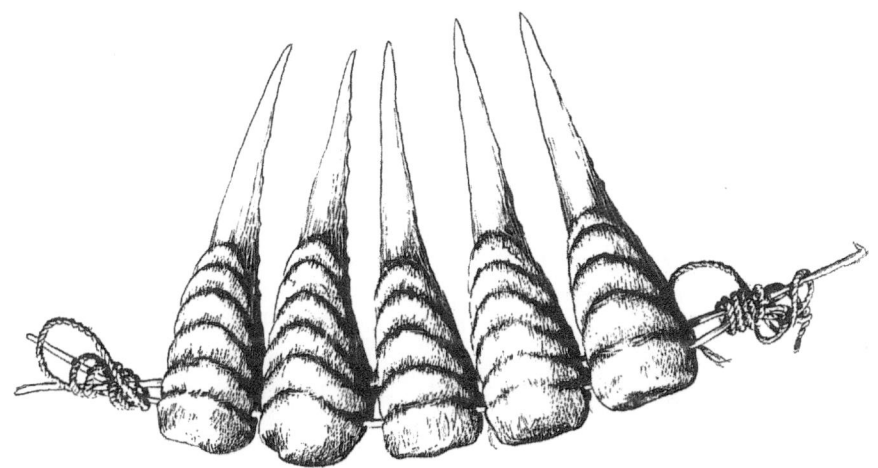

Hornbill headdress, Kumusi River, Australian Museum 1907

7

The Mambare

natives of the fighting variety

Away from the main shipping routes, the people living north of Cape Nelson had seen few Europeans. When Captain John Moresby took the *Basilisk* along the coast in 1874 he was conscious of his privilege in being directed to chart one of the last areas outside the polar regions still marked on world maps by guess-work and dotted lines. Having been able to land and mix freely with the peoples of the southern coast and the islands of the south-east, Moresby attempted to take a boat into shore at a point where about a hundred men had gathered. But the men, decorated with paint, bird-of-paradise head-dresses and shell necklaces, danced on the beach and waded out into the water, holding aloft their spears, stone clubs and shields. The warriors ignored gestures of peace, seized presents, offered nothing in return, and attempted to drag the boat inshore. Moresby abandoned the attempt to land and named the place Caution Point. The next day, further along the coast, Moresby fired into the shield of a warrior to protect a shore party collecting wood. He named Ambush Point and Traitors Bay to commemorate the events near the mouth of the Mambare, a river he called the Clyde.

Inspecting the limits of his sovereignty near the border with German New Guinea in 1890, MacGregor found no evidence that missionaries, traders or fishermen had been there before him. The people he met south of the Mambare mouth knew nothing of iron or tobacco. When MacGregor tried to obtain spears and stone clubs by barter 'They laughed at the idea of giving one of their weapons for anything we could offer them in exchange'. For the information of other travellers MacGregor marked on his published map that the area was 'Occupied by a powerful and friendly tribe'. Moresby, comparing them with other New Guineans, had called them a 'fiercer race of savages'.

In 1894 MacGregor, the first foreigner to ascend the northern rivers, made short trips up the Gira and Opi, and took a steam launch as far as possible up the Mambare and Kumusi. At the end of his tour MacGregor wrote that the district was 'without exception the most

Map 9 The northern rivers

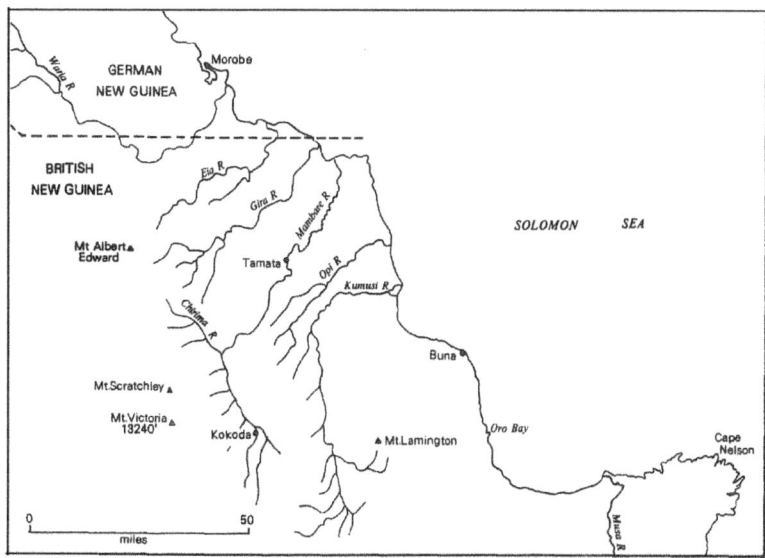

attractive one I have seen in New Guinea'. Between the mountains and the north coast was a fertile plain about 40 miles wide cut by broad rivers. It supported a large population. Several times MacGregor confronted armed parties, and he described their dancing in terms from his homeland (a 'strathspey') and his learning (a 'pas seul'); but he was uncertain whether they danced to intimidate, propitiate or welcome. After their initial suspicion had been allayed, most of the river villagers were prepared to trade; and MacGregor decided they would be 'easy to gain over'. There were land and people on the northern plain, MacGregor decided, to welcome the trader and the planter; but of more immediate consequence, he reported seeing gold in the wash of the upper Mambare, and two of his officers found colours when they tested ground away from the river.

On the Mambare the people had shouted the greeting, 'orokaiva', and the foreigners had assumed it was a password for those who came in peace. Later travellers used it throughout the district and within three years they had applied it to the people as a group name. Away from the Mambare some 'Orokaiva' first heard the term when it was shouted at them by approaching foreigners.

The Orokaiva, loosely related peoples sharing a basic common culture, occupied the land bounded by Oro Bay, Mount Lamington and the headwaters of the Kumusi in the east and north; the hills leading to the great peaks of Mount Victoria, Mount Scratchley and Mount Albert Edward in the west; and the Gira and Eia Rivers in the north. Most of their lands were covered in dense rain forest broken by patches of grassland and swamp. In the wet season from October to April the rivers flooded, becoming wild torrents on the northern slopes of the ranges and spreading deep and still over thousands of acres of low country. The Orokaiva were constantly clearing forest for new gardens, the abandoned gardens soon disappearing under secondary growth. Their basic crop was taro, but they also cultivated yams, sweet potato, sugar cane and bananas, and harvested the tree crops, coconuts, okari nuts, breadfruit, sago and betel nut. By hunting and fishing they added to their food supply, the most successful hunting taking place at the end of the dry season when the villagers gathered to burn-off patches of grassland and drive the fleeing game into lines of nets held by men armed with spears and clubs. Policemen and labourers from other districts coming to the northern plains for the first time marvelled at the piles of betel nut in the villages, the numbers of pigs, and the gardens often extending for over 100 acres of planted ground. Among New Guinea peoples Orokaiva were rich, numerous and aggressive.

They were constantly forming alliances, fighting, celebrating a victory or gathering a force to fight again. They recognised different classes of warfare: the ceremonial in which opposing lines of spearmen advanced and retreated, challenged and posed, while their leaders engaged in shrill oratory; fighting between clans prepared to wound and kill but where one clan would not attempt to annihilate the other and the victors would not eat the enemy dead; and *gitopo itoro* in which an alliance of clans attacked people from another language group, sometimes attempting to kill all of them, eat their dead, destroy their gardens and occupy their lands. Raiding parties made surprise attacks on enemy villages, but the Orokaiva also met in open warfare with the lines of spearmen clashing, and behind them were the drummers and conchshell blowers, the sorcerers and strategists, and the women screaming encouragement and bringing up more spears. Young men were trained to use weapons and all listened to the clan histories of attack, defence and alliance. At maturity the men were formally presented with their fighting weapons: the long spear, the stone club and the pointed wooden shield wrapped in woven cane. A man made a name by his prowess in war, and the prestige of his group depended on its capacity to fight. If a member of a clan were killed, all the men felt an absolute obligation to seek vengeance. The widows and children of the slain were a constant reminder to the warriors of their obligations; a specific obligation to those who had suffered and a general obligation to the unborn, the living and the dead to maintain and extend the power of the group. A man caught in a hopeless position by a raiding party would call his own name, proclaim his past victories over his attackers and die shouting defiantly of the terrible retaliation that his clan would inflict.

The Binandere, the most northern of the Orokaiva peoples, occupied the central and lower Mambare, the lower Gira and central Eia Rivers. Less than 100 years before the arrival of the Europeans, the Binandere had left the Kumusi and from Eraga, a settlement on the Mambare above Tamata Creek, they killed, dispersed and absorbed the Dogi, another immigrant peoples, and the Girida, who had previously possessed the area. One of the last decisive battles was at Tai Hill on the Eia. Kewotai of the Yema and Waie of the Binandere combined their forces to attack the Girida. The Girida saw smoke from the raiders' fires and a man shouted, 'Who are you?' Waie replied, 'Have you fucked your wife?' implying that he had better for it was his last chance. By watching the pigs enter the stockade in the evening the Binandere and Yema found a way into the village and attacked just before dawn. Some Girida escaped to the tree-houses and fighting platforms, but the raiders piled wrecked houses at their base, set fire to them, and forced the Girida to jump. The Binandere and Yema feasted on the slain.

By the time the miners reached the Gira, the Yema too were being absorbed by the Binandere, who were continuing their northward expansion. Fighting parties travelled almost to Salamaua, and Binandere clans had formed alliances with the Suena at the mouth of the Waria for a series of raids against the Zia and Mawai peoples higher up the river. In the south the Binandere were still engaged in sporadic conflict with other Orokaiva peoples. The Binandere took their canoes from the river villages to travel over 100 miles south-east to round Cape Nelson, but they clashed more frequently with their Aiga and Taian Dawari neighbours.

Moresby had named the place where he had fired a shot into a warrior's shield 'Traitors Bay'; to the Binandere it was Totoadari, the place where Totoa, the head man of Girida, had been killed. It was close to Taian Yabari where the Binandere had counted the Taian Dawari dead to make sure that they had killed at least as many as they had lost when the Taian Dawari had attacked them in an earlier raid; and to Dawari Odari, the site of a long-remembered battle. Europeans who entered the Mambare did not merely encounter a 'fiercer race of savages'; they met a warrior people who had occupied the area by conquest. The land everywhere reminded them of other triumphs and defeats. They were bound by the system of 'payback', the obligation to exchange the spirits of the slain; and each clan felt that its survival depended on its ability to maintain delicate alliances.

Sometime before the arrival of the Cairns prospectors the Pure clan had been the strongest on the Mambare. They raided and looted with little fear of counter-attack. Ribe village, the centre of their strength on the middle Mambare, was said to be so big that the people at one end would only learn of an attack at the other because they heard the blowing of the conchshells. But Dandata, a war leader from higher up the river, secretly gathered a strong force by uniting Binandere clans from above and below Ribe, and calling in allies from the Gira and Waria on the west, and from the Aiga in the east. The Pure of Ribe village, whose heavy spears had for so long knocked aside the shields of their enemies, were overwhelmed and their village destroyed. The Pure survived as sections of other Binandere villages. When the foreigners entered the Mambare many clans along the river were suspicious that the Pure might attempt to recover their position by forming an alliance with the outsiders; or that other groups would exploit the new forces to enrich themselves and destroy their neighbours.

Stone clubs, Papuan South Coast, after Stone 1880

At the end of June 1895, Clark's party entered the north-west mouth of the Mambare and anchored about 2 miles upstream. The villagers near the river mouth were willing to trade, even agreeing to sell the prospectors the canoes that they needed to take some of the load from their one whaleboat. After five or six days the miners had reached the more densely populated parts of the middle river. Large numbers of men, some in canoes and some on shore, began to follow the miners upstream. The Binandere robbed one of the expedition's canoes, but when Clark appealed to Bousimai, recognised by the miners as 'a big chief of the tribe thereabout', the goods were returned. The miners noticed that the crowd accompanying them was increasing and included no women and children. On about 11 July the Binandere threw stones into the camp and Clark ordered them to keep back, supporting his command by firing a shot at a canoe. The expedition encountering rapids at noon the next day, four miners took the end of a rope to haul the whaleboat into clearer water; Clark stayed on board to hold the steering oar. Binandere came forward as they had on previous days to haul on the tow-line; but unnoticed by the miners one warrior cut the rope allowing the whaleboat to rush back into the following canoes. Jumping into the boat, the Binandere threw the rifles and shotguns into deep water and began looting the trade and stores. Clark fired his revolver into the Binandere then jumped overboard using the side of the boat to protect himself. Tom Drislane, swimming to Clark's assistance, saw Clark hit over the head with a paddle, speared and disappear. The Binandere withdrew without attempting to press their attack on the miners, who were now armed only with their revolvers and two rifles. The two Queensland Aborigines, Tommy and Milori, deserted and were not seen again.

The miners decided to return to the coast. In the villages they heard women crying for their dead, and jeering groups of warriors, gathered on headlands, threw spears at the miners who cleared their path with rifle fire. Believing they were the only foreigners within 100 miles of the Mambare, the diggers suddenly encountered the cutter *Mayflower* 20 miles from the river mouth with seven members of the Ivanhoe prospecting party on board. The two parties combined under the leadership of William Simpson to go back up the river. In the villages of the men thought to have been responsible for killing Clark, the miners burnt houses, broke canoes and made 'free use' of pigs and poultry. Beyond the highest Binandere villages the diggers prospected 'likely looking' country to the south-west, then, when most of them became ill with malaria, they rafted back to their highest camp on the river.

At the mouth of the Mambare the miners divided, one group establishing a camp and the other leaving on the *Mayflower* to report the death of Clark and pick up stores in Samarai. A month after the attack, Europeans in southeast New Guinea learnt of Clark's death, and in another month the news reached Cooktown. Ignoring the fact that for several days the Binandere had obviously been gathering their strength and testing the foreigners' power, and that Clark had fired on the Binandere the previous night, Australian commentators saw the killing of Clark as yet another case of the generous and trusting white man murdered by treacherous natives. It was to be quoted by men who wished to show that 'a fatal mistake is made by those who place any reliance on the apparent friendliness of the native races'.

The miners on the *Mayflower* met MacGregor on the north coast: he sailed immediately for the Mambare. On the lower river he found that the Binandere had visited the miners' camp and they were prepared to trade and talk with the government; but higher up the river they were suspicious and hostile. In deserted villages the government party saw goods taken from Clark's boat, and signs that men had been making new weapons. At Eruwatutu armed men pressed forward as MacGregor and the police boarded their boat. Seeing two men about to throw spears MacGregor called to his secretary, John Green, standing on the bank. Green wrote to his family:

> I turned round like a flash and saw a native just in the act of lifting his spear at me. I shot him in the side with my shot gun, and he dropped his spear like a hot iron & off into the scrub. I ran after another man who had a spear and shot him in the seat of honour.

Always determined to show that the government never retreated, MacGregor landed again at Eruwatutu and at another village higher up the river; but having only six police he made no attempt to arrest anyone.

The miners from the camp near the coast left the Mambare with MacGregor, and while the *Merrie England* went on to collect more police and carriers MacGregor took the miners up the Musa by steam launch. The government officers met people anxious to trade for the steel axes and knives which they had heard of but never seen; the miners prospected beyond the highest navigable point but found no gold. On their return to the mouth of the Musa, John Green watched

> between thirty and forty large canoes, each full of armed natives, about four hundred fine big men … all decked out in war gear. In the centre and largest canoe were two men evidently chiefs, each holding a banner made with white feathers fixed on to spears. I called the Governor and we prepared for a fight. But they passed us as if we were not worth looking at. The swish of about 300 paddles in the water and the weird and savage appearance of such a number of men … was a sight I shall never forget.

After collecting the additional police from the *Merrie England* MacGregor took a force back up the Musa where they saw gardens devastated by the raiders and in their canoes abandoned against the river bank were 'roast legs, arms, ribs, heads, backbones etc.; some partly eaten'. The government party determined to shoot the raiders 'like pigs'. Having destroyed the canoes, the five white officers and twenty-five black police confronted the warriors. They shot three dead immediately and wounded others as they pursued them through the scrub. Remembering the body of a young girl found in a canoe, her skull smashed and her body prepared for cooking, Green felt no mercy as he fired.

The six men from the Cairns and Ivanhoe prospecting parties travelling with MacGregor must have thought now that even the most lurid stories of brutality in New Guinea were told with restraint; and that the strengthened government force would exact a savage penalty from the communities involved in the killing of George Clark.

But MacGregor did not attempt to crush the Binandere. While he led a party for seven days' walk beyond the point where the prospectors were attacked, the Honorable Matthew Moreton, the Resident Magistrate of the Eastern Division, Archibald Butterworth, the Commandant of Police, Tom Drislane and a troop of police built a camp at the junction of the Mambare and Tamata Creek just above the highest Binandere village. Gradually the number of people coming to the camp to trade increased. They were truculent and bold, selling food to the Commandant, stealing it, and selling it again. Drislane having pointed out some men involved in the attack on Clark, Butterworth ordered their arrest. In a short violent struggle six or seven men were shot dead and another six taken prisoner; they were the first Binandere to wear the government chains. As the steam launch carried them past the upper villages men onshore wept and uttered a 'plaintive wailing shout', which was answered by the prisoners on the deck. MacGregor reported the conflict at the Tamata camp as another example of the one 'thorough and complete' defeat that 'never fails to put an end to fighting in a district'. But the Binandere tell stories that the men who were shot and arrested were from a clan which took no part in the attack on Clark. The fight at Tamata , they say, helped unite the people of the upper Mambare against the foreigners.

MacGregor appointed John Green Government Agent at Tamata to protect the miners moving back up the river and bring the Binandere under government influence. Green was thirty years old. His home was a farm near Healesville in Victoria and in letters to his family he remembered the tennis, football and rifle clubs, playing the organ in the Presbyterian church, concerts at the Mechanics' Institute and 'our blacks' at the Coranderrk Aboriginal station. The family sent copies of the Healesville *Guardian*, the Lilydale *Express* and Saturday's *Age* with their frequent letters; yet more than any of MacGregor's other officers Green was at home in New Guinea. Having arrived there in 1892 hoping to establish a plantation near Kabadi, Green had immediately taken an interest in the men and women who chattered boisterously as they unloaded his goods from a Thursday Island schooner. Sometimes judging his own countrymen by puritanical standards, he was tolerant and curious of New Guineans. When they burnt the grass at Kabadi he joined the men in 'scenes ... wild beyond description' as they tried to net and spear pigs and cassowaries. He was a good bushman, able to endure the climate and willing to eat local foods, but most of all his capacity to talk to New Guineans gave him knowledge and ties unavailable to other white men. Soon after he began work at Kabadi he discovered a talent for mastering new languages, and within three years he could converse in three of the languages of the south coast. On the Mambare he saw that his first task was to acquire the Binandere language.

Map 10 The Mambare 1895

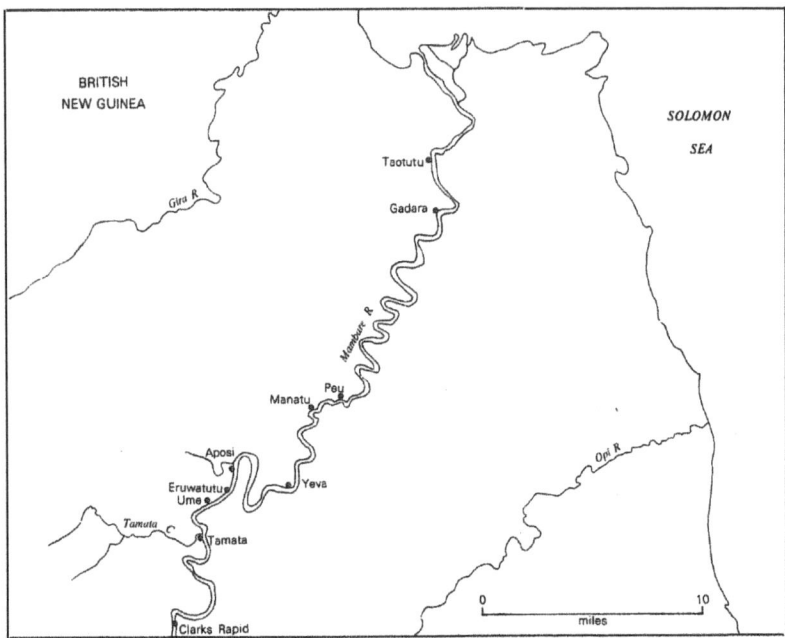

In October 1895 Green, his personal servant, Gemaruya of Fergusson Island, Corporal Sedu of Kiwai and nine constables occupied the four tents enclosed by a log stockade on the point formed by the meeting of the Mambare and Tamata Creek. From the camp Green believed that with his Martini-Henry rifle he could command the 200-yard breadth of the river. The police, who spoke the languages of Kiwai, Mailu, Orokolo and Taupota, could only talk to Green and Gemaruya in Police Motu and a few phrases of English. Later, complying with MacGregor's general instruction, Green insisted on English at all times, fining in tobacco those who used 'broken dialects'. Green, not expecting further stores or instructions until January, began cutting timber for a permanent station with an axe and cross-cut saw, clearing land and planting a food garden, drilling the police, evading the flood waters which crept slowly into the camp after storms on the Mambare headwaters, and trying to build up his contact with the Binandere. All the men at Tamata camp were held together by the knowledge that they were dependent on each other to survive.

Having deserted the camp after the shooting and arrest of the men thought to have taken part in the attack on Clark, the Binandere gradually returned to trade. An old man, Gaina, came in regularly; women brought food for sale; and eventually the young men again began to visit. At least for a while it was to be *orokaiva*. Green set values in beads, fish-hooks, plane-blades, cloth and axes for food and building materials. At the end of 1895 the Binandere completed the building of their first canoe made with steel tools. Gaina was introduced to *raisi, bulamakau, kuku, kisi kisi* and *si* (rice, tinned meat, tobacco, biscuits and tea). But Green found it difficult to gain influence among the Binandere. After a month he had a list of only forty Binandere words. Knowing that at least twelve men had been shot and six gaoled since Clark's death, Green thought that the Binandere had suffered sufficiently. And he worried that more of them were likely to be killed. They knew that he had shot the two men at Eruwatutu, and nearly every night Green and the police heard the laments in the villages, keeping alive the memory of the dead, and making it impossible for any Binandere men to feel at ease in the presence of the foreigners.

By February 1896 Green was more confident. He believed that the Binandere accepted that the foreigners were going to stay and that he was 'master of the Mambare'. The police went unarmed to some of the villages. But in April Alex Clunas of the Ivanhoe party reported that on his return up river he had been threatened by a large group of armed men at Peu village. The next day Green and the police took the government whaleboat down to Peu. As they approached the village they could see a large crowd of men and 'in handy positions' were 'scores of spears' stuck in the ground close to the bank. Green instructed six police to arrest three men while he and the other police held the rest of the villagers at bay. As soon as the boat touched the bank the police seized three men and, using their rifles as clubs, helped Green drive the rest of the surprised warriors back through the village. Green fired the only shot; he wounded Bousimai in the leg with gun-shot. The police gathered over 400 spears, shields and clubs and burnt them in the centre of the village. A fortnight later when Green returned with the released prisoners, the villagers welcomed their kinsmen and showed their resentment against the foreigners. Green decided to confront the sullen villagers by landing and cooking lunch. As the police prepared to take the boat, away from the bank the Peu fighting men crowded forward. At a warning shout from the police, Green shot a man about to throw a spear. It was a similar incident to the shooting at Erwuatutu twelve months earlier. Again the police occupied the village and burnt all weapons; and this time they stayed on to force the Peu to spend the night in the bush.

Early in August Butterworth and Ross Johnson took charge of Tamata while Green accompanied MacGregor beyond the headwaters of the Mambare to Mount Scratchley and Mount Victoria. Green then returned down river, and took the *Merrie England* to Port Moresby to meet MacGregor who came overland down the southward flowing Vanapa. Butterworth found that the Binandere were as inclined 'to beat their drums, etc., and to show fight' as they had been a year before. A man from Ume killed a labourer working for the miners, and a large group of warriors attempted to ambush government carriers. Butterworth trapped the Binandere by sending an apparently defenceless canoe-load of carriers up river. When a combined force from Ume and Aposi attacked the carriers, two miners concealed in the canoe protected them by firing over the heads of the Binandere. Then, knowing the home villages of the aggressors, Butterworth landed and destroyed ornaments and weapons, shot pigs and shattered canoes with dynamite.

On his return to Tamata in December Green, the police and a group of prisoners from Samarai gaol began shifting to a new site on Tamata Creek away from the frequent Mambare floods. To obtain materials and labour, and increase contact with the local villagers, Green wanted the Binandere to help build the new station. He could be hopeful that they would. A large group had gathered at the old station on his return protesting friendship with the government; he could now speak a little of their language; and one of his policemen, Dumai, came from the upper Mambare. Arrested at Tamata by Butterworth after the killing of Clark, Dumai had become a prison warder and then a constable. But Dumai had not abandoned his own people when he put on the police *rami*. According to Binandere stories he wept when he saw the children and widows of the men killed when he was arrested; and he told the fight leaders that after seeing the white man's settlement in New Guinea he knew that there were few white men, and that their strength lay in their *epidi* (rifles).

Because of the constant rain at the beginning of the wet, the increase in malaria, and the need to return labourers at the end of their contracts, most of the miners working on the creeks on the upper Mambare took the track down to Tamata. In January 1897 Fry, Haylor and six labourers left by canoe and raft for Mambare beach. A few days later Davies, Steele and Olsen and twelve labourers followed on two rafts. Soon after leaving Tamata they passed jeering, threatening crowds. From one of Fry and Haylor's labourers they learnt the reason for the excitement along the river. While Haylor and Fry had been ashore at Peu, the Binandere had seized their rifles. In desperate flight, the miners and labourers rushed back to the river where Fry and two labourers were clubbed to death on the raft; Haylor and the other labourers escaped in the canoe. The Peu pursued Haylor who was 'bad with fever' and killed him on the beach. The rescued labourer believed that he was the only one of his party still alive. Having kept the 'yelling mobs of cannibals' on the river at a distance, Davies' party camped on the coast to wait for a boat going to Samarai.

Perhaps unaware of the violence down the river Green continued work on the new station. The police suspected that an attack was likely, but Green, who knew that the people from the nearby villages would not work on the station alongside armed men, insisted that the police put their rifles aside. On 14 January the Binandere suddenly attacked. Speaking in Binandere Green called to Dumai: 'I have helped you to sleep comfortably and I have given you good food, and taught you good things for your benefit; but you are not loyal to me, and are here with your people to kill me.' But Dumai did not attempt to stop the warriors. All those working at the new station were killed: Green, Corporal Sedu, three constables, Kess Kess (Green's cook), and three prisoners. The five police remaining at the old camp secured their defence before they could be taken in a surprise attack. Petari, one of the men who took part in the attack on Tamata station, told his grandson that the Binandere laid the bodies in a line and felt triumphant when they saw that the foreigners had not shot as many of their kinsmen as they had killed on that day.

The reports of the violence on the Mambare, which reached Samarai in mid-February and Australia ten days later, were shrill cries of a 'terrible massacre in New Guinea'. Six Europeans, nine police and thirty labourers were said to have been killed. The fate of another three miners and their labourers still on the headwaters of the Mambare was unknown; perhaps they were already murdered. Again there were calls for a 'more efficient lesson' to be handed out and MacGregor, contrary to fact, was condemned for making the country a reserve for the missionaries while keeping it closed to 'enterprise and civilization'. The *Cooktown Independent* took consolation from the thought that the massacre would only mean a delay; the miners were at the front of 'the advancing race before which the native and receding race must eventually disappear'. In Healesville the death of a son in the 'far-off land of savages' cast a 'palpable gloom' over the community; here was further proof of 'proverbial' native treachery. But during the following days the apparent magnitude of the tragedy declined. Three of the assumed dead, Davies, Steele and Olsen, had been in Sydney for a week before news reached Australia of Green's death. When the Binandere had attacked their camp at the mouth of the Mambare, the miners and their labourers put to sea on two rafts. The miners and three labourers on one raft drifted helplessly for several days before landing in German New Guinea, purchasing canoes and reaching a Lutheran mission station. Taken to Sydney by the German cruiser, *Falke*, they were unaware that Green had been killed while they were at their camp on Mambare beach. Nor did they know what had happened to the nine labourers on the other raft which had become separated at sea. But two at least survived: they identified themselves after hearing their home language being spoken by carriers working for Matt Crowe on the Waria in 1909. Three of Haylor and Fry's labourers were picked up from canoes 80 miles along the coast from the mouth of the Mambare. The three miners on the upper Mambare, a week's walk beyond the highest Binandere village, learnt about the attack on Tamata station from one of their carriers bringing up stores. They were never in danger of attack, but it was another five months before anyone from the lower Mambare knew that they had survived. In all, during January 1897, the Binandere had killed three Europeans (Green, Fry and Haylor), four police, three prisoners and about fifteen labourers.

Again the government was slow to gather its forces. MacGregor was at Boigu in Torres Strait when he heard that Green had been killed, and it was April before he arrived on the Mambare. Moreton had already been to Tamata, but had made no attempt to arrest or punish. The Binandere, MacGregor now conceded, were 'more warlike, pugnacious and cunning' than any other peoples he had encountered. They were also better armed. If the Binandere had retained all the weapons from their victories over the miners and the government, they had up to fourteen rifles and hundreds of rounds of ammunition; and Dumai was trained in their use. The government officers knew that they had kept some rifles for men at Peu and Tamata had fired on Moreton's party. But in spite of the involvement of people from the beach to Tamata in attacks on the foreigners and wide distribution of loot, the Binandere were still divided. On the Gira and at the small villages of Manatu and Yeva on the Mambare the people were prepared to tell the government the names of those who had taken part in the fighting and to accompany government forces on patrols.

After ten weeks on the Mambare and the deployment of fifty police, several white officers and two temporarily enlisted miners, MacGregor could report little success. The acting Commandant of Police, G.H. Livesey, had been most ineffective. On his return from pursuing Binandere warriors across the divide between the Mambare and Gira, Livesey had been unable to say where he had been; he and twenty police had withdrawn in the face of jeering spearmen; two of his four prisoners had escaped; in separate incidents his police had shot and killed an old woman from the upper Mambare and a man from Gadara on the Mambare; and he had suffered frequent attacks of malaria. MacGregor was pleased to accept his resignation. Apart from an occasional shot fired from a distance, the Binandere had not used their rifles: they had frustrated the government, not confronted it.

They always knew in advance when the government men would try to surround villages or trap groups of warriors. Only once the Binandere misjudged the government's strength. A group of warriors taking refuge on the Gira fought the police and lost six dead. MacGregor saw the clash as decisive: they had been 'completely humiliated in the eyes of the other tribes' and it would only be a matter of time before the people along the Mambare were 'pacified'. MacGregor left the area with Dumai, Bousimai, the leader of the people on the lower river, and Amburo Apie, who had gathered the fighting men of the upper villages, still free. To enforce peace and arrest those involved in the January fighting MacGregor re-opened Tamata station and posted a troop of police at the mouth of the Mambare.

Head-dress with bird-of-paradise plumes from Kumusi River presented to the Australian Museum, 1907
PHOTOGRAPH: G. MILLEN, THE AUSTRALIAN MUSEUM

In the two years after the attack on Tamata station many more Binandere died from spear and club than from rifle fire. To the foreigners the dominant encounter was between them and the 'natives'. To the Binandere the clash with the foreigners overlay and complicated older wars and alliances. What mattered most to the Binandere was the strength of one clan or alliance of clans relative to others; and the foreigners were most important because of the ways they changed, and could be induced to change, the strengths of different groups.

Between the attacks on Clark and Tamata station there had been fighting between Gira and lower Mambare peoples, and between the upper Binandere villages and the Orokaiva of the Opi and Kumusi. In 1898, while the government forces made further fruitless attempts to arrest Binandere leaders, about 200 men from surrounding villages attacked Yeva killing thirteen, capturing ten women, and looting the abandoned houses. Evading the government and hostile clans, some Mambare peoples sought refuge on the Opi, but after a month the Opi turned on the Binandere and killed some ten to fifteen. Allied to the refugees who had been attacked, the Binandere of Onombatutu on the Upper Gira crossed the Mambare, raided along the Opi, and left the exchange of the spirits of the dead in favour of the Binandere. The Manatu and Yeva, having suffered great losses and fearful of further attacks, built a temporary village close to Tamata station. Government officers assumed that the marauding Binandere were determined to massacre the Manatu and Yeva because of the assistance they had given the government. That was partly true, but it did not explain why the Yeva and Manatu had been prepared to help the government in the first place. The Manatu, it seems, were looking desperately for allies before the killing of Green. MacGregor on his first trip up the Mambare in 1897 had noted that 'The Manatu tribe was friendly as usual', and three weeks later on the Gira he had found it 'very well understood' that the Manatu were friends of the government. But as allies the government had failed the Manatu: they had not protected them, nor had they 'paid-back'.

Eighteen months after Green had died the remnants of the Manatu and Yeva at Tamata were the only Binandere living on the Mambare. The police, dominant on the river, had harried the villages on the banks, seized an occasional prisoner (usually a woman or child), and shot a few people; but they had been unable to capture Dumai or Amburo. Bousimai and eight others from the lower Mambare, arrested in mid-1897 and imprisoned at Port Moresby, had escaped. Put to work on a road leading north from Port Moresby, the Binandere prisoners had learnt that they were cutting a track to take the miners overland to the Mambare; they left to walk home. Having crossed the Owen Stanley Ranges they looked down on the headwaters of the Mamba which they followed to their villages near the coast. Two men died of starvation and exposure on the track. A year later Henry Stuart-Russell led the first government patrol to cross the 'Gap', and reported that it would be possible to build a road from Port Moresby to Tamata, a distance of about 140 miles. Another forty-three years later thousands of Australian and Japanese troops followed the same route across the Owen Stanleys; they found neither a road nor a gap, but an endless series of ridges, the same ridges seen by the miners cutting their way up the Goldie in 1878. An Australian journalist in 1942 called the line scarred by soldiers' boots and carriers' feet the Kokoda Trail.

MacGregor saw the futility of his officers' policy of sporadic harassment. In May 1898 he directed them to tell the Binandere that while the government would never make peace with Dumai and the men directly responsible for killing Green, Fry and Haylor, all others could return freely to their homelands. Although Michael Shanahan, who was then directing the government forces on the Mambare, thought that clemency would lead to disaster, most of the Binandere were keen to return and rebuild. Debera was able to convey the government's message. One of the men arrested in 1895 after the killing of Clark, Debera had, like Dumai, joined the constabulary. Having completed the normal two-year period of service for recruits, he now returned with MacGregor to his home in Ume village. When Debera told his people that the government wanted *orokaiva*, an old woman embraced Butterworth and told him that they had no homes, no gardens and they feared attacks from the Kumusi people who had already killed some of them; they needed to be friends with the government. With the people from the lower river constantly on guard against further raids from the Gira, and the Yeva and Manatu sheltering near Tamata, all the Mambare Binandere were scattered and unable to invest the wealth of their gardens in alliances; they were losing strength relative to their neighbours. In transmitting MacGregor's offer of peace Butterworth probably knew that he was making an alliance with the Mambare villagers consistent with agreements they had known in the past; some other officers realised that the Binandere sought peace out of self-interest; and some thought that they had changed their ways because they were becoming 'civilised'.

Dumai, Amburo, Bousimai and the other Port Moresby escapees, and two men from Peu and one from Ume said to have separately delivered the death blows to Corporal Sedu, Fry and Haylor were arrested before the end of 1898. Binandere who had been to Port Moresby or villagers who had entered agreements with Butterworth helped persuade the wanted men to surrender to the police.

Accompanied by fellow villagers, they then met the white officers at Tamata or the beach. Bousimai was released almost immediately as there was little evidence against him; Amburo was returned home in January 1899, taking his betel nut on the platform of his house without any sign of emotion while his people celebrated his release. Dumai and two other men found guilty of manslaughter were sentenced to five years in prison; they suffered the harshest penalty imposed on any Binandere by the formal processes of law.

At first receiving food and seed plants from the government, by mid-1899 the Mambare villagers were selling taro and other garden crops to the miners. New villages were built at Duvira at the mouth of the Mambare, Mowata, Aposi, Ume, Umbogi and Beya. Peu, its inhabitants scattered on the Gira, Mambare and Opi, was the most outstanding of the old villages not rebuilt. At Duvira men were contracting to carry miners by canoe to Tamata at a standard price of an axe per passenger. The Binandere would not carry the miners' stores overland from Tamata to the goldfields; but they strengthened their new alliance by working for the government. Following the outward movement of the miners, the government forces were then coming into collision with peoples on the Kumusi, Opi and Upper Gira. By volunteering to carry for patrols the Binandere associated themselves with government victories, and were sometimes able to loot deserted villages or influence whether the government would have peace or war with the peoples encountered. The very presence of particular Binandere clans with a patrol helped neighbouring groups decide whether they should welcome or fight the 'government'. But those Binandere who took the *dabua* (uniform) and *epidi*, who joined the police, were better able to demonstrate their alliance with the government and direct the government's power.

Taller and blacker than the people of the south-east, the Orokaiva 'looked like fighting men' to Europeans. In 1906 Colonel Kenneth MacKay, recently commander of the New South Wales 6th Imperial Bushmen's Contingent in South Africa, believed they would serve their Commonwealth and Empire with distinction. His review of Australia's resources in its new Territory included the statement: 'what splendid material we have for soldiers in Papua'. Within a few years of the attack on Tamata station most of the police serving in the Northern Division had been recruited locally, and in many villages the leading men were ex-policemen. In 1908 John Higginson, who had served at Tamata station, wrote that the 'local natives' formed the 'cream of the admirable native constabulary'. Over the next thirty years government officers who worked alongside them repeated his judgment.

Related men joined the police: Bia and Barigi, two of the most praised N.C.O.s; Bakeke and Tamanabae, brothers of the man who killed Clark; Poruta and Oia, Bousimai's sons. Of 'a high order of intelligence', Bousimai exploited his close ties with the government. In 1900 he obtained the assistance of three policemen from the camp on Mambare beach to raid the Gira peoples where they killed one man and looted gardens. Arrested and gaoled at Tamata, Bousimai persuaded his guards to set him free. Poruta, newly recruited into the police force, negotiated his surrender. Impressed by the bearing of the 'powerfully made man' and aware of his predominance in Duvira, Lieutenant-Governor George Le Hunte, educated at Eton and Cambridge, extended the privileges of high birth to a 'chief'. Bousimai was not to be treated like 'a common criminal', but he and his wife were to stay for a period at Cape Nelson station where his two sons were serving in the police. Monckton, the Resident Magistrate on the station, watched Bousimai assume a position of influence:

> on his first day at the station, [Bousimai] began by sitting on the steps of my house; on the second day, he had oiled himself into my office, where he sat upon the floor, whilst I did my work or heard native cases, throwing in a little advice at intervals; on the third day, he had made up his bed in my room; and on the fourth day, he had picked up the largest axe on the Station, and was acting as general overseer and adviser.

On patrols Bousimai spoke for the government. In 1901 he was said to have given 'good fatherly advice' to Opi and Kumusi peoples who appeared to hold him in great respect. On the same patrol the police shot some villagers and made friends with others. But Archibald Walker, the Resident Magistrate, who wrote about the 'instructive object lesson' his patrol had given, could not speak the language of the people he was appointed to rule. He could understand only the hoots, derisive laughter and slapping of bare backsides as signs of hostility; and nose-rubbing, boisterous welcomes or pathetic submission as evidence of peace. He could not know the details of the shouted exchanges between patrol members and villagers, nor could he have known that all Bousimai had said was good and fatherly.

At Duvira Bousimai served the government as village constable, briefly commanded the police camp on the Mambare (although the government Secretary doubted that he was fit to hold such office), and after the police were withdrawn he was responsible for the government store near the beach. But he took more from the government than some of its officers thought just. He persuaded the government to protect Duvira. Walker issued an extra thirty rounds of ammunition to each man at the police camp when Bousimai, Poruta and Tein (another leading man of Duvira) explained that the people on the Opi were planning to attack them. In 1904 he was gaoled for three months for possessing a police uniform and leg irons; presumably the resident magistrate was more concerned about the use of government property than just ownership. Five years later Bousimai killed Anjiga of Bongata village on the Mambare and escaped to Buna where his son, now Corporal Oia, could intercede for him. Having convinced government officers that he had suffered great provocation, Bousimai again avoided imprisonment. But he lost formal power; Poruta succeeded him as village constable. A warrior when Moresby had sailed along the north coast, a leader on the lower Mambare during the conflicts with the foreigners, Bousimai had been able to use the new institutions to pursue old aims and spread his name throughout the Division.

Warfare had ceased on the Mambare by 1899; but over 2000 Binandere on 40 miles of the lower Gira continued to celebrate and suffer after raid and counter-raid. As the miners were not interested in the swampy lands of the lower Gira and they reached the gold-bearing country of the upper river by crossing overland from Tamata, the Gira villagers saw little of the foreigners. Still uncertain of the power of the police, they sent a challenge to the police camp after the Mambare Binandere had decided it was better to join the police than fight them. In 1901, with the assistance of some Mambare warriors who had old debts to pay, the Gira raided the villages of the Zia on the Waria. In spite of losing fifteen killed, the raiders looted villages and brought back canoe loads of Zia dead. This was *gitopo itoro*; but it was now given another name, *kiawa itoro*, a whiteman's war, for the Binandere had used rifles. John Waiko collected and translated the story of the fight from the grandson of one of the raiders:

> the men with the rifles waded into the water [of the Waria] until it reached their armpits. They stood in the water with rifles on their shoulders. The other fighters stood on the bank of the river and they beat their drums and sounded their conchshells to match the tune of their war songs.
>
> These attracted the Jia tribesmen to come across like ants in order to kill the Binandere men. But as they came the Binandere men killed them all with the rifles.

A year later ex-constable Ade reported to Tamata that a woman and child had been killed in fighting on the Gira. He was given a Snider rifle and twenty rounds of ammunition to secure peace. But Ade had told less than the truth. Six Waria men returning home from Tamata had been killed at Umuta on the Gira. In retaliation the Waria surrounded Umuta at night, set fire to the houses, killed twenty-six and wounded others. Ade aimed to lead an attack against the Waria, but news of the extent of the fighting at Umuta reached Tamata before he could act.

In 1904 Higginson reported that people on the Gira 'seemed to have settled to civilisation'. The village stockades had been taken down and men from German New Guinea and the Mambare were visiting to trade. The changes on the Gira were not imposed by force of government arms, but by villagers, knowing what had happened to their neighbours, deciding that the old days with their ideals and savagery of warfare had gone.

The Binandere were not a defeated people. Foreigners noticed that they laughed a lot and were generous to each other. They passed cigarettes from man to man, and if one obtained a tin of *bulmakau* (meat) he took a share and handed it on. H. R. Maguire, who surveyed dredging leases on the Gira and Mambare in 1901, said that the Binandere looked you straight in the face and you knew that you were among 'men of the highest calibre — as far as physical perfection, courage, and savage nobility is concerned'.

Believing that the resettling and civilising of the Binandere would be accomplished more successfully with the assistance of teachers of Christianity, MacGregor invited the Anglicans to open a mission station on the river! MacGregor's warning that he could not prevent another mission entering the area if the Anglicans failed to respond may have helped the Anglicans discern 'the guiding of God's hand' directing them to a new field. The Anglicans, still expanding slowly from Dogura in the southeast, had to pass 100 miles of pagan coast to establish St Andrews on the Mamba in 1899. Short of staff and money, the Anglicans found it difficult to maintain an isolated station. The Reverend Copland King stayed from 1900 to 1903 and produced the first texts in the Binandere language, but most of the European staff left after a few months, debilitated by fever. Ten years after the founding of St Andrews the Anglicans could claim only one 'hero of the Mamba', David Tatoo, a teacher recruited from the Melanesian islanders taken to work in Queensland as indentured labourers. Arriving at St Andrews soon after the first mission party, Tatoo did not take leave for five years, and often he was the only missionary on the river. The Anglicans spoke of him with condescension and admiration. He was 'not the most intelligent of our coloured helpers', nor was he as enterprising 'as a white man would be'. His triumph was personal. He had demonstrated the 'grace of perseverance'; he was 'a splendid example of the loyal, self-sacrificing Christian that a South Sea Islander can become by the Grace of God'. About twelve students attended his school, and on Sundays he preached at six places to small congregations. After ten years' work he had made no converts: the Binandere still thought all his talk was 'gammon', and the few children adopted by the mission and educated at Dogura were without influence at home. The Binandere were to accept the teachings of the Anglicans but in 1910 their celebrations were still much concerned with the warfare that had ended.

Moresby 1876 recounted his voyage along the north coast and MacGregor's dispatches outlining his explorations were printed in the *Annual Reports* of 1890/91 and 1893/94. Both reports included maps.

The comments on Orokaiva history and society were taken from Williams 1928, 1930, Waiko 1970, 1972, Wilson 1969, and Chinnery and Beaver 1917. I am also indebted to Richmond Tamanabae, Joe Saruva and other Northern District peoples with whom I talked either in Port Moresby or in their home areas.

The activities of Clark's party were described by T. Linedale and W. Day (two members) to the *Cairns Argus*, 10 September 1895; A. Symonds, purser on the *Merrie England* to the *Cooktown Courier*, 10 September 1895; and T. Drislane and S. McClelland to M. Jones, Commander of the *Merrie England*, *Annual Report 1895/96*, pp. 15-17. There is an interesting difference in the sequence of events in the various accounts. In the *Cairns Argus* the Binandere began looting, Clark fired on them and then they attacked him. In the *Annual Report* and the *Cooktown Courier* (both probably from the same source) Clark was attacked and then he fired. If the *Cairns Argus* report is correct then loot, and not Clark, was the first concern of the Binandere.

Green wrote of the incidents at Eruwatutu and on the Musa in his letters. MacGregor's dispatches about the same events are in the *Annual Report 1895/96*. Two Binandere have written about the killing of Green, Barereba (Tago) 1964 and Waiko 1970, 1972. An obituary of Green was published in the *Healesville Guardian*, 5 March 1897. Green's letters to his family are detailed and give an insight into his relations with the police, and relations between the men at Tamata and the people in the villages. The escape of Davies, Steele and Olsen is reported in Reichskolonialamt Records (information supplied by Dr Stewart Firth, A.N.U.). Their arrival in Sydney was noted in *Sydney Morning Herald*, 19 February 1897. In *Annual Report 1896/97* MacGregor listed all those who he thought had died on the Mambare. At that time, 28 April, he still included the men who had reached Sydney on 18 February.

There are two well-known accounts of the killing of Green: Lett 1943 and Monckton 1921. Lett explains the attack almost exclusively in terms of Green's 'theory of appeasement': he treated the Binandere too leniently. But Lett has many major and minor errors in his account of what happened on the Mambare. He says that three miners, Patterson, Davis and Steele, left Tamata on 5 January, passed hostile peoples on the river, and put to sea after being attacked on the beach. Green, Lett says, heard a rumour of events down river but did nothing. Fry and Haylor then left Tamata. To this point Lett has made several errors. The three miners were Olsen, Davies (sometimes spelt Davis) and Steele. From the statements the miners made to the press in Sydney and from the reports of the Germans who saw them on the Mambare beach on 14 January it is clear that they followed Haylor and Fry down the river, and that they were still on the beach when, unknown to them, Green was killed. Lett describes the killing of Fry and Haylor, and claims that Green visited 'the scenes of the crimes', but made no arrests. This seems unlikely. Lett says Haylor was killed on the beach, and other sources agree on this point. Had Green gone down to the beach, a journey there and back of about four or five days, he would have seen Olsen, Davies and Steele; but there is no indication that Green and the three miners met on the beach. Also it seems that Fry and Haylor left Tamata on 7 January and Olsen, Davies and Steele followed a few days later. If the dates are accurate, then there would scarcely be time for Green to have gone down to the coast and been back at work on the new station on 14 January. Green may have heard rumours about the events down the river but he probably had no specific news, nor is it likely that he visited the 'scenes of the crimes'. Given Green's experience in New Guinea, he would have been much more cautious and much less naive than Lett suggests had he known what had happened to the miners. After all Green had shot men on the Musa and had twice used his gun against the Binandere; he did not, as Lett claims, think that the Binandere were 'innocent and harmless children'. Lett concludes his account by stating that Green's 'confidence in native integrity' resulted in six white men losing their lives. In fact Lett only names three Europeans who were killed; and strangely enough that is the correct total.

Monckton is also astray in presenting the sequence of events. He says that the Binandere killed Green and then attacked the miners who 'fled like curs' for the coast; 'five of them were accounted for as being butchered on the way to the coast, but probably others were killed'. All contemporary evidence indicates that the miners were attacked before Green, and the few miners still in the area did not flee. Only two miners were killed on their way to the coast, not five. Later, p. 193, Monckton says that 'Bushimai' had 'killed my brother magistrate'. Bousimai was probably involved in the attacks on the miners, but may not have taken part in the attack on Tamata station. The government officers who investigated the killings did not find evidence that he had killed Green.

Apart from confusing the order of events both Monckton and Lett have many minor errors of detail and fictitious moments of melodrama.

The actions of the government forces after 1897 are given in *Annual Reports* and in some incomplete records from Tamata station. Green's call to Dumai is quoted from Barereba 1964, and Monckton 1921 described Bousimai's increasing influence at Cape Nelson. Two books concerned with the exploits of the Papuan police are Hides 1938 and Lett 1935.

Dabua is Police Motu for 'clothes', and *epidi* is 'rifle'.

The comments on the Anglican mission are based on Chignell 1913, Tomlin 1951, White 1929 and records and pamphlets of the Anglican mission kept in the library of the U.P.N.G.

8

New Ground

all golden country but very poor

Three months after Clark was killed, William Simpson, eight other miners and twenty-two Taupota carriers returned to prospect the 'likely looking' country they had seen on the upper Mambare. From 'Clark's Fort', a log hut about 12 miles upstream from Tamata, the miners cut a track south to a point high above the western bank of the Mambare where they built 'Simpson's Store'. Using the store as a base, the miners spent five months cutting tracks and testing the creeks feeding the Mambare. They prospected the Chirima, the main tributary coming in from the west, and followed the Mambare beyond the Chirima junction into the lower Yodda Valley. The track from Clark's Fort to the Chirima junction crossed no land much above 2000 feet, but it was a hard walk with many creek crossings, thick undergrowth and cliffs which had to be climbed with the aid of vines and makeshift ladders. John Green took only thirteen days to travel from Tamata to the junction and back, but at the end he threw away the new boots he had put on at the start. After MacGregor had seen the prospectors' tracks he wrote that they had carried out 'by far the most arduous undertaking ever performed by any private exploring party in the colony'.

When the miners came down the river in January 1896 they had 46 ounces of gold. The upper Mambare, Simpson reported, was 'all golden country, but very poor'. Beyond the Chirima junction they had prospected one creek which, they thought, might be profitable. Although their gold would not pay the costs of their expedition, they had found enough to bring them and others back to the Mambare. John Green wrote home suggesting that the young men of Healesville should consider trying their luck in the area.

In March the returning miners and their Taupota carriers began relaying canoe loads of stores from Mambare beach to Tamata. Among the first group to arrive with Simpson were Clunas and MacLaughlin of the old Ivanhoe prospectors and McClelland from Clark's party. They found that MacLaughlins Creek, flowing from the spurs of Mount Scratchley, was worth working. By August twelve miners and over eighty labourers had obtained 600 ounces to send away on the *Merrie England*. Some of the Taupota men were constantly carrying stores up and returning for another load, the round trip of about 150 miles taking fifteen days. Already news of the strike on MacLaughlins Creek had reached Samarai, and when Australian papers reported the find they added to the interest already aroused by the talk of the new alluvial field on Woodlark Island. There was general agreement that after the 'wet' there was 'bound to be a big rush' to New Guinea.

Although Australians had been mining in the islands for ten years, most of those about to leave for the mainland knew little about the country they hoped to work in. The newspapers sometimes informed and cautioned them; and sometimes misled them. The diggers were not told they were going to a foreign country. The eastern Australian colonies provided £15,000 a year to pay for Sir William MacGregor's administration; British New Guinea was another frontier for Australians to develop. Settlement ought to be no more difficult than any other area north of Brisbane. Just as development in the eastern colonies and Western Australia had been stimulated by goldrushes, now it was to be New Guinea's turn. And again the 'Munchausens' who located minute reefs in remote places would be able to milk the 'British capitalistic cow'. Most of those general beliefs accepted by Australians were proved false eventually; within a few months the diggers knew that a lot of the details they had read were wrong. Several early reports said that the gold-bearing country was closer to the southern coast than the north, and the *Cooktown Independent* went so far as to announce that a 45-mile track fit for mules and horses could be cut from Port Moresby to MacLaughlins Creek. Much of the route, said the *Independent*, passed over 'well-grassed and pleasant tableland'; it avoided 'collisions with hostile natives' along the Mambare; and there was no 'miasma which creates fever on flats'. MacGregor, who had walked from MacLaughlins Creek to the south coast in 1896, diplomatically turned aside requests to cut tracks for pack animals and to revoke the regulations against importing horses from north Queensland.

Just before the diggers left Australian ports Davies, Steele and Olsen reported the deaths of Haylor and Fry, their own narrow escape, and the fact that the most successful of them had only 10 ounces of gold. Eight days later the papers announced the 'massacre' at Tamata. Men who had been to the Mambare spoke to make sense of the contrasting pictures of 'treacherous arrow and sneaking fever', and pack-horses winding over peaceful up land meadows. Charlie Lobb on his first trip south of Townsville since his arrival in the southern hemisphere and

William Simpson in Sydney after prospecting on the Mambare and the Musa gave long interviews. Both ridiculed the idea of using horses, and warned that the only way to work the new field was to employ large numbers of carriers to transport rations from Tamata. To meet expenses and to cover the fact that there was no work for the unsuccessful digger to fall back on, they thought that no man should leave Australia without £100. If he was to work on the goldfield he would have 'to say goodbye to the coast for at least six months', and Simpson said, if he was unsuccessful, he was probably saying goodbye forever. Lobb warned that even the crews of boats anchored off the Mambare mouth for a few days suffered from fever. He predicted that if a rush occurred at least half the men would die. For himself, he would return to New Guinea, but to the islands: 'Let the madmen go to the mainland'. The best course for the present, he suggested, was to let north Queensland men who were most inured to the climate continue their work of defining the location and value of the new field. From Port Moresby MacGregor wrote an official warning against a rush to the islands or the mainland. He was supported by Walter Gors, Burns Philp's manager in Port Moresby, who said that the 'patch' on the Mambare was worked out, further payable gold was yet to be found, in the event of a 'big rush' many would die; and he repeated the false advice that the only practical route was from the south.

At the request of the Government Secretary, Shanahan prepared a shopping list for those still prepared to go. He recommended '2 pairs moles or dungarees, 2 pairs flannels', boots, blankets, hammock, oilcloth, tent, twine, scissors, needle, towel and straps to make up a 40-pound swag; cooking utensils and stores; shovel, pick, dish and two tomahawks; 1 bottle sulphate of quinine pellets, 1 bottle Dover's powders pellets, 1 bottle anti-febrine pellets and antibilious pills; and a revolver and a shotgun. Shanahan thought that for most men the shotgun was a better weapon than a rifle, and while he conceded that the bow and arrow had a better range than the shotgun, he pointed out that all the people in the auriferous country were spearmen. Shanahan believed that his 200-pound pack would support one miner for a month, it would require six carriers to transport it, and the carriers would need additional stores. Captain John Strachan, who had survived dramatic encounters with the people of New Guinea, advised all miners to wear a broad flannel belt to prevent cholera.

Nearly 1000 diggers sailed to New Guinea in the first half of 1897. About half went to Samarai and on to Woodlark or the Mambare, and about 400 landed in Port Moresby. The men on the south coast tried four starting points to take them across the Owen Stanleys: Rigo, Port Moresby, the Vanapa and the Alabule Rivers. One prospector wrote to the *Ingham Planter* that he and five others had attempted to go inland from Port Moresby. At Sogeri they realised that they 'could do no good with horses': some animals had rolled 'packs and all' for 100 feet down the first range rising from the coastal hills. They could not get carriers; and

> As regards the country, I never dreamt there was such rough and broken country in the world, the only thing I can compare it to is country that has been boiled up into huge boulders ranging from 300 ft. to 12,000 ft. high, with gorges hundreds of feet deep, and covered with dense scrub, just like the Johnstone scrub.

But most of the miners who landed in Port Moresby went north to try the Vanapa Valley where a government party was said to be marking the route and erecting rest houses. They found no easy stages. Most of the 130 or so miners who reached the Vanapa gave up after two days on the track. The few who struggled on were led into difficult country inhabited by aggressive peoples with little knowledge of Europeans. Instead of following MacGregor's tracks through Gosisi village and over the eastern flank of Mount Scratchley, the large and struggling government party had followed the Vanapa north towards Woitapi. One miner, Martin Dabney, was killed on the upper Vanapa, and another group led by George Wriford, an ex-government officer, was trapped in camp until its besiegers were routed by MacGregor's police. Some miners had shot pigs and looted gardens, increasing the hostility of inland peoples. The coastal villagers would not carry into the lands of their enemies across tracks at over 8000 feet where men wept as the cold rain swept in each afternoon. And the miners could not travel without carriers. A digger returned to Rockhampton said, 'So far from being able to carry your own swag, you do very well if you can carry your own carcass'. For the independent digger proud of humping a 100-pound pack to the Western Australian fields, this was an admission of defeat and a recognition that New Guinea was another country. When a miner opened a bag of flour, the basic food of the Australian bushmen, 'you [could] smell it fifty yards away'. The New Guinea prospector had to use rice, a lesson learnt earlier in the islands. About six miners died on the track, another six died in the temporary hospital in Port Moresby and others died at sea or in Cooktown. The extent of the rush and the tragedy had been less than many had predicted.

By June 1897 only two parties had reached the Mambare from the south coast, and both included men accustomed to travel in New Guinea. G. O'Brien reached a miner's camp after his two companions had been drowned when their raft capsized on the upper Mambare. Without stores, equipment or carriers, O'Brien was unable to mine and he was given a temporary job at Tamata station. W. Nettle and

W. Kelly reached MacLaughlins Creek after two and a half months on the track. Dependent on the people of one village to move them on to the next, Nettle and Kelly had often been delayed, but they had been generously supplied with food, and only near Woitapi had they been in immediate danger of attack. They too arrived with their stores almost exhausted, and could not replace them without leaving the field. But MacGregor with his normal indifference to the physical hardship suffered by himself and others still asserted that 'the journey across' ought to be 'an easy one for, say, fifteen days'.

Three men, Schmitt, Ryan and Burns, had remained on MacLaughlins Creek during the wet season of 1896–7. Isolated from the rest of the European community for five months after the attack on Tamata station, they survived because the people of Neneba village were willing to supply them with food. Schmitt lived in the village for six weeks, recovering from an injured foot.

A small community of about 100, the Neneba occupied a cluster of leafy huts on Asiba Creek, a tributary of the Chirima. They had no spears, shields or beheading knives; their only weapons were a few stone clubs and small, weak bows blackened by the smoke of cooking fires. They seemed 'peaceful and amiable', dependent on their isolation to protect them from their aggressive neighbours. Originally a group of Mountain Koiari who had been pushed north, the Neneba had no contact with the Binandere, and infrequent meetings with the Orokaiva and Koiari peoples to the south and west; but they had formed an association with the Fuyuge villagers higher up the Chirima and by the 1890s they were beginning to adopt their language. Through the people of the upper Chirima they met other Fuyuge speakers living south of the ranges in the Woitapi Valley. In their gardens on the slopes of Mount Momoa the Neneba grew sugar cane, bananas, taro, yams, sweet potato, tobacco, and maize, a crop which had spread recently from the south coast in a series of exchanges by neighbouring communities. The Neneba had approached Simpson's party in peace in 1896; Goiye, the village leader, had returned with Green to Tamata to meet the Binandere who made the smoke visible far down the Mambare Valley; and they had been generous hosts to MacGregor's overland expedition, supplying him with four pigs and many vegetables. Tolerant of foreigners and keen to trade, the Neneba would sometimes carry for miners travelling in their area, but they were not inclined to work on the goldfield or to go away as indentured labourers.

At the end of October 1897 there were only about twelve miners on MacLaughlins Creek. The two most successful diggers, Gilbert Hudson and Moses MacClelland, had each taken over 600 ounces from the cold water and shifting boulders of the Mambare creeks. But now the

Sketch of Neneba village and man from Neneba made by a member of William MacGregor's patrol, 1896
BRITISH NEW GUINEA ANNUAL REPORT. 1896/97

rich patches had been worked out and the miners were struggling to find payable ground. In search of a new strike Robert Elliott and Alex Clunas prospected along the Chirima to the west and Elliott, Clunas and MacClelland traced the Mambare east beyond MacLaughlins Creek into the Yodda Valley. The Yodda was gold-bearing, they reported, but it was too far from Tamata to be worked by men dependent on carriers. Even on MacLaughlins Creek the miners found that their carriers had eaten most of the stores by the time they reached the camps. And the prospectors had twice been attacked by spearmen. In the second encounter they had shot three men 'in self-defence'. Clunas and Elliott decided that it was worth attempting to cut a shorter track overland to the Yodda, but changed their plans when they heard that Shanahan had found 'good colours' on the Gira.

Near Shanahan's camp, three days' walk west of Tamata, Clunas quickly obtained 20 ounces from a creek, but it was not until others entered the area early in 1898 that the miners learnt about the extensive auriferous country on the upper Gira. Twenty-five miners took 1200 ounces from one creek, and after a pause while men fossicked and re-worked old ground, a rush to a new gully yielded another 2000 ounces. On Shanahan's recommendation the Gira was proclaimed a goldfield in November, and regulation was confirmed a fact the following year when the miners washed 6000 ounces from the head of Tamata Creek. Working in shallow, narrow gullies the early miners quickly exhausted the richest areas, and again there was a dull period.

During 1898 and 1899 an average of 150 miners worked on the Gira and Mambare, most of them always arriving too late to peg rich claims. Many of the men suffered from malaria and dysentery. The death rate, higher than on Woodlark, reached one-third; it was, as the warden modestly claimed, 'appalling proof of the almost pestilential character of the district'. As on Woodlark some men were as keen to rob the dead as they were to mine. The sick who struggled back from the mining areas lay in the bush-material settlements at Tamata and Mambare beach waiting for a chance to leave. A temporary hospital opened in December 1898 was closed four months later. The government was not prepared to pay all its costs, the successful miners left, and many of those who needed the hospital had no money. The man most responsible for opening the goldfields on the northern rivers, William Simpson, died at Tamata in September 1897. His real name was William James Shearing and why he chose to have another name is now unknown. Shanahan died at Mambare beach on the way to Samarai. His brother officers made a coffin from two sheets of galvanised iron and buried him on the same day for his body had already begun to putrify. The next three officers appointed to take charge of Tamata also died there or on the way to a healthier climate.

MacGregor wrote repeated requests for a doctor to be appointed to Tamata, but could not persuade his superiors to act. Finally he stated that disaster was likely and asked to be exonerated from blame. After MacGregor left the colony at the end of 1898, Doctor C.A. Brough arrived. A man of 'advanced middle age, who had been leading a sedentary life', he stayed in Moresby a few days and left without going to the Mambare. But when Joseph Blayney, the Resident Magistrate for the Central District and the only doctor in British New Guinea, visited Tamata in April 1899 he reported that much of the sickness was 'brought on by the reckless mode of living'. Most of the miners, he said, expected to work as hard as they did in Australia, exposed their heads and necks to the sun, ate poor food, used polluted water, 'drank heavily of alcoholic drinks', and lingered in the area after they became ill. Blayney's comments were least comfort to the six men who died in the week before he arrived. Doubtless the miners lived and worked in unsanitary conditions, but the deaths and the 'absence of a really healthy face' noted by the Acting Administrator Francis Winter, were a result of malaria, not a reckless disregard for the rules governing the care of a white constitution in the tropics.

The Gira revived briefly in 1900 when runners from the Tamata stores took word around the camps that Robert Elliott had found gold on the Aikora, the south-western branch of the Gira. About sixty miners, many without carriers, left Tamata together on the six-day walk to the new strike. Only the first camp was enlivened by the effects of rum and whisky taken from the centre of carefully rolled swags. From then on the miners settled into a rain-washed camp each evening and lit fires in a clear dawn to dry their tents so they would be lighter to carry. Elliott's find was high on the slopes of Mount Albert Edward. Frequent rain storms had scoured nearly all gravel and alluvial from the creek beds, but crevices and rock bars had formed natural traps holding rich, easily worked deposits. Further down the Aikora at Campions Beach the miners worked a more extensive area of alluvial. Some gold-bearing ground was still undisturbed when news arrived that payable gold had been found on the Yodda.

Once they knew that there was a limited amount of gold on the Gira, men had returned to prospect the Yodda. In 1898 Clunas, Clark, Nelsson, Close, an escort of police loaned by MacGregor, and seventy carriers set out from Tamata to cut a new and shorter track to the Yodda. After several false starts they marked a track to the Opi, south along the Kumusi, and then west over a low range onto the upper Mambare or Yodda. They followed the Yodda valley to the north-west, meeting old tracks cut by prospectors coming upstream from MacLaughlins Creek. Although they avoided a major fight, the miners saw large numbers of peoples on the Opi, Kumusi and Mambare who seemed ready to attack and loot. 'As far as prospecting is concerned', Alex Clunas wrote to MacGregor, 'nothing was got except fine colours throughout the whole trip'. Other miners attempting to reach the Yodda were harassed by the Opi and Kumusi Orokaiva, and some were forced to turn back.

In mid-1899 Henry Stuart-Russell, surveying the 'road' from Port Moresby to the north coast, reported that 'Colours of gold are obtainable anywhere' along the streams flowing into the Yodda Valley. The 'tribes' of the Yodda and beyond to the north, he said, were 'numerous, warlike, and treacherous'. Stuart-Russell, who had served at Tamata after the death of Shanahan, was surprised at the boisterous confidence of the warriors who crowded to greet his expedition. The expected attack took place as the police and carriers prepared to cross back over the Mambare. Believing that their shields would protect them, the spearmen advanced boldly to be cut down by rifle fire; 'and, though they came on again and again with the usual bravery of all natives belonging to that district they were repulsed everytime with loss, and eventually drew off, not a man in my party having been injured'. Foreigners had now approached the upper Mambare and Kumusi from MacLaughlins Creek, overland from Tamata, and across the ranges from Port Moresby; all had encountered what they thought were bold and aggressive peoples. At the end of 1899 miners on the Gira, the Mambare and in the south-east began moving up the Kumusi track; they had heard that Matt Crowe and others had opened a new field on the Yodda.

By 1900 the general location of gold-bearing ground on the northern rivers was known. On the Mambare it extended from MacLaughlins Creek for about 30 miles along the Yodda Valley, and to the west of the Mambare there was gold on the headwaters of Tamata Creek, the Gira and the Aikora. It was a vast auriferous area, but apart from a few patches it was poor. The miners standard comment was, 'There's a lot of gold in New Guinea, but there's a lot of New Guinea mixed with it'.

The main sources used were Green's letters. *Annual Reports*, Tamata station papers and Australian newspapers. A. Musgrave made a useful collection of papers, many concerned with the goldfields, and they are now in the Mitchell Library. Dutton 1971 and C.F. Jackson, patrol report, July-August 1914, Appendix A, Kokoda station papers, outline the history of the Neneba. MacGregor's visit to Neneba is reported in detail in the *Annual Report 1896/97*. Sketches of the people and the village are included, D.H. Osborne wrote his memories of the Gira in *Pacific Islands Monthly*, January 1943.

9

The Yodda, Gira and Waria

unavoidable mishaps which constantly recur in warfare

In 1899 about 150 white miners were working on the Yodda and Gira. A few came and more left, but for the next ten years about 100 miners and 600 labourers washed gold in streams on the Yodda and Gira, and for a time crossed to try the torrents in the broad bed of the Waria. The storekeepers told the wardens that the men were obtaining somewhere between 10,000 and 12,000 ounces a year; but as some diggers did not pass all their gold across the store counters the exact total was never known. The gold was taken by hard work and violence. The early signs that the Orokaiva would fight the *ijiji-avujo*, the puzzling foreigners who travelled without purpose across their lands, first in one direction and then in another, were fulfilled. Their attacks were ferocious, persistent and futile.

In 1900 William Armit, the Resident Magistrate of the Northern Division, left Tamata station to make the first government patrol to the upper Yodda. On the Kumusi he attempted to contact the people who had stolen the stores of a mining party two years earlier, but the villages were deserted. 'Just as I was on the point of leaving', he wrote in his journal,

> two villainous-looking individuals, with blackened faces and wearing war-plumes, marched defiantly into the village. To seize these gentlemen, tear off their plumes and wash some of the black pigment from their faces was the work of about one minute. Then I clapped two heavy swags on their backs and sent them ahead. They did not like it at all.

After a fortnight Armit reached Papaki where he was met by 'quite 250 people'. He wrote:

> I ordered them to put away their arms, but they laughed at me, and one big man, taking two or three rapid strides forward, deliberately poised his spear at me. He was instantly shot dead. A fight commenced, but only lasted some few minutes … These [people

had earlier clashed with a group of miners] and being a very powerful and aggressive tribe, it became imperative to teach them a salutary lesson.

Thirteen men were killed. As Armit crossed from the Kumusi to the Mambare there was a series of clashes. Three days after leaving Papaki Armit and his police shot seventeen people when they followed a retreating group of warriors close to their village. Included in the seventeen were two women whose deaths Armit explained as 'one of those unavoidable mishaps which constantly recur in warfare'. Before the end of the patrol another twenty-four men had been killed and an unknown number wounded. No police or carriers were killed. Le Hunte received Armit's report with 'great uneasiness'; and he wrote to Armit stating that while he did not doubt the necessity for the action he must have more precise information about who was responsible for the shooting and how many people were wounded. Armit replied by listing the total killed at various places, claiming that he had seen no wounded and omitting any further information about responsibility for particular deaths. Armit was unrepentant when he made his annual report in the middle of the year. The people of the Kokoda area were 'treacherous, truculent, aggressive, cruel and cunning'. He hoped to prevent them from re-occupying their villages for another two years. While regretting having to fight them, it was

> incumbent on myself to uphold the prestige of the Government, and secure the safety of the miners who, I knew, were following in my wake.
>
> Again, it was preposterous and intolerable to even dream of permitting a horde of savages to browbeat and intimidate a Government expedition with impunity, and as a consequence of their ill-advised action they lost a number of their warriors.

Six months later Armit died at Tamata station: the prestige of the government had to be upheld by others. It took them another five years to 'pacify' the peoples on the Kumusi.

Because of the death or delinquency of earlier officers and the reluctance of men to serve in the Northern Division, Alexander Elliott, who came to the Mambare to mine, was appointed Assistant Resident Magistrate at Tamata. He opened Bogi station to protect miners and carriers moving up the Kumusi track to the Yodda. In January 1901, a few months after Elliott's arrival at Bogi, Sam McClelland reported that his two prospecting partners, Tom Campion and John King, had been killed higher up the Kumusi. Before the attack the neighbouring villagers had appeared friendly, entering the camp, watching the work, and trading with the miners. McClelland first knew he was in danger when his two labourers yelled a warning that his shotgun and rifle had been stolen. An hour later another two labourers who had been testing a creek with Campion and King rushed into camp with the news that the two miners had been killed. McClelland and the four labourers were attacked frequently, but by cutting through the bush, staying close together and McClelland keeping 'his revolver going all the time', they reached Bogi.

After a wait of nine days for more police to be sent up from Tamata, Elliott and McClelland left for the upper Kumusi. On the first day out they shot four spearmen who rushed them at a creek crossing. During the next two days Elliott's force killed another thirty-six men and left seventeen with their legs broken; Elliott thought that 'There must have been a few more wounded of those who got away'. The last and bloodiest clash had not been a case of a patrol shooting when attacked. Elliott deployed his police so that they could kill, not merely drive the chanting warriors from a stronghold on the edge of their own gardens:

> I wanted the police to get round behind them before I started the fight.
>
> They howled and hooted at me to their hearts' content, and also once fired a revolver at me. It was two and a-half hours before I heard the first shot. This was followed by a volley, and then I started in earnest at 250 yards. I did not waste a shot, as I was firing low — mostly for the legs.

McClelland, whose presence with the patrol was justified because Elliott needed him to act as a guide and to identify the people involved, 'opened up with his rifle' when spearmen rushed towards his position. Later, miners and government officers believed that Campion and King were captured alive, tortured and eaten; but Elliott and McClelland had probably not heard that story when they set out on their savage punitive patrol. Sam McClelland, a member of George Clark's expedition of 1895, died of 'fever' soon after he returned to Bogi. One of Campion and King's labourers, 'merely a lad', was picked up two months later on the coast east of the Kumusi mouth; government officers never learnt what happened to the other labourers.

At Papaki on the upper Kumusi old villagers remember stories of miners who came into their area and began prospecting along Homa Creek, a tributary of the Kumusi. After a while the villagers made contact with the miners and presented them with a pig to demonstrate that they wanted peace. But later Hara, the man who had owned the pig, became angry. The pig had been named after his mother and in allowing it to be killed he felt that he had dishonoured her. He blew the conchshell of war and the villagers attacked the miners. The decision had been taken quickly. Pipiri, a Papaki man working in the creek with the miners, did not know what was going to happen until he saw the approach of men decorated for war.

The Papaki also talk of clashes with the patrols which followed the killing of the miners, and of one incident in particular. A government force entered the village, ordered the villagers to line up, forced a piglet to squeal, and while the people's attention was diverted, opened fire. At least two people, who were children at the time, were still alive in 1972. It is now impossible to tell whether the accounts of the conflicts between villagers and foreigners differ because poor memories and loose talk over seventy years have distorted events, or because men chose not to write down what they knew had happened.

Spread over 40 miles of garden land and rain forest between Bogi and the head of the Kumusi, about 5000 Orokaiva lived in scattered settlements. In defiance or in ignorance of the power of the government patrols they continued to attack the carriers on the Yodda track. Accurate information did not pass quickly from one community to another, and different communities had conflicting stories to tell. Many villagers were shot, but others intimidated the miners, forcing them to flee, or they found the carriers easy victims. In 1901 another large government patrol made a slow irresistible progress through their lands teaching many more people about the power of the rifle.

Table 7
Gira Goldfield production and population

Year	White Miners	Papuan labourers	Gold in ounces
1899/1900	90	450	7000
1900/01	30	150	2400
1901/02	50	250	5500
1902/03	50	250	6000
1903/04	55	275	6000
1904/05	52	260	6000
1905/06	55	330	6000
1906/07	42	300	5000
1907/08	37[1]	459	5000[2]
1908/09	40	298	4500
1909/10	3[3]	40	NA

1. Includes twenty-one men on the Waria River.
2. About 3000 ounces of this came from the Waria. The next two years also include production from the Waria.
3. Most miners had left for the Lakekamu field.

In February 1901 Le Hunte appointed two new officers to the Northern Division, Archibald Walker and the Honourable Richard de Moleyns. Walker, the son of an Australian Senator and a director of the Bank of New South Wales, went to New Guinea for adventure and gold. He had already made trips to the Aikora and other parts of the goldfields before entering the government service at the end of 1900. De Moleyns, son of the Baron Ventry of Kerry, Ireland, had also been 'visiting different parts of the Possession'. He volunteered for a government post while waiting for a response to his application, supported by Le Hunte, for 200,000 acres of land near Mullins Harbour. Walker and de Moleyns left Bogi with twenty-five police and about seventy carriers recruited from the Mambare and lower Kumusi. Travelling up the west bank of the Kumusi they passed the 'lookout tree'. From its branches over 100 feet from the ground the Orokaiva maintained a constant watch on the Yodda track: it was manned as the patrol passed. At Memekowari, 10 miles south-east of Bogi, the houses were deserted when the patrol arrived, but the people gradually came in bringing food and indicating that they wanted peace. They also persuaded Walker that higher up the Kumusi were warriors from a 'big bullying tribe' who had killed some of them and were boasting that they would soon wipe out the government.

Table 8
Yodda Goldfield production and population

Year	White Miners	Papuan Labourers	Gold in ounces
1899/1900	90	450	7000
1900/01	120	600	10,000
1901/02	70	350	6000
1902/03	70	350	6000
1903/04	45	225	5400
1904/05	48	180	5000
1905/06	45	225	6000
1906/07	61	305	5000
1907/08	39[1]	NA[2]	3600
1908/09	24	437	3400
1909/10	5[3]	60	NA

1. Twenty-one miners were on the Waria River.
2. The figures for miners and labourers are the totals on 30 June. Although this figure is not available for 1907/08 other information for the year is given in the *Annual Report*: a total of 632 labourers were employed by miners and storekeepers; 312 were employed in mining, 188 carrying, 102 mining and carrying, and 30 general. This sort of distribution would have applied in other years.
3. Most miners had left the Lakekamu field.

The gold was worth about £3 12s. 6d. an ounce.

Accompanied by ten men from Memekowari, the patrol passed through hundreds of acres of gardens before approaching the first settlements in mid-afternoon. The constant sound of drumming was broken by yells and the first shower of spears fell. On the rush into the village the police shot two men and the carriers axed two others. Fighting was then 'not of a desultory character but continuous and determined' until dark. Spearmen made attack after attack, one group screaming defiance while another group rushed into the village from a different direction. By dark the villagers were crossing their own dead on every track. The Orokaiva made two attempts to break into their village during the night, and in the morning de Moleyns, attempting to leave, was ambushed within a few yards. Sporadic fighting continued until midday when the people whom Walker called the 'enemy' withdrew. Close to the area where the patrol had been besieged were twenty-one villages, each of ten to twenty houses. In his official report Walker said that it was impossible to tell how many men had been killed, but he knew of twenty dead and thought that twice as many had been wounded. The police had fired 200 rounds 'all at close quarters'; their Martini-Enfield rifles, Walker said, had 'proved their serviceableness'. No member of the government party was injured. Charitable to the defeated, Walker wrote: 'These natives are the most aggressively hostile and the most determined and pluckiest I have met, Mr. de Moleyns and the police concurring.'

Before returning to Bogi the patrol was involved in minor skirmishes with local villagers close to where Campion and King had been killed. Other groups did not fight. Walker believed that the communities higher up the Kumusi retreated or tendered food as gestures of peace because they had heard of the defeat of other groups. The Government Secretary conveyed to Walker the Lieutenant-Governor's appreciation for 'the stand made against the natives, & the efforts they made to secure peace'.

Six months after his patrol Walker wrote that the people at the head of the Kumusi and Mambare were still in a 'constant state of turbulence and revolution'. They had fought all government patrols entering their lands and 'attacked the miners' and the storekeepers' carriers practically every week for the last eighteen months'. They had suffered 'crushing defeats', been harassed and driven from their gardens. Yet, said Walker, they would not desist or make peace:

Table 9

The Northern Division		
October	1899	Northern Division created from area called north-east coast and Mambare district.
April	1909	Northern Division divided into Mambare and Kumusi Divisions.
May	1920	Mambare and Kumusi Divisions amalgamated to form Northern Division.

'There is no doubt a remarkable strain of courage and pertinacity running through these people.' When seventy white miners and 200 carriers left the Yodda for the new rush at Cloudy Bay:

> they seemed to think they had achieved their one purpose and aim in life namely that the stranger they loathed was evacuating.
>
> They became most jubilant, hordes of natives from the other side of the Kumusi crossed over and joined in the jubilation.
>
> They harassed carriers and whites for quite 40 miles of the 60 miles to the goldfield, hooting and yelling from daylight till dark.

Fearing an attack, the miners remaining on the Yodda moved close to the stores or to a central camp further down the valley on Finnegans Creek.

During the early years of mining on the Yodda the Neneba alone continued to keep peace with diggers, carriers and government officers. Some time before 1901 they had shifted to Beda, closer to the Chirima River; and it was by this name that they were now known to the miners on the Yodda and MacLaughlins Creek. Labourers from the camps at the northern end of the field visited them to buy food, communicating with them in a mixture of English, Motu and Dobu. The Beda had fixed prices: a tomahawk would buy three bags of potatoes; a plane blade, one bag; they no longer accepted payment in beads or calico. To provide a surplus beyond their own needs they had increased the area of their gardens. In 1901 Robert Hislop on his first patrol as a government officer estimated that they had 1500 acres under cultivation. With food purchased from Beda, miners on MacLaughlins Creek and the northern Yodda could afford to work poorer ground than men dependent on imported rations carried from distant stores.

To give greater protection to the men on the Bogi-Yodda track the government opened a new station at Papaki. According to Walker, de Moleyns was a 'decided triumph' at Papaki. His ascendancy was either imaginary or brief. After de Moleyns, emaciated by malaria, left for Australia, Allen Walsh, Assistant Resident Magistrate at Papaki, found that the Orokaiva were still ready to challenge the 'government'. Although he reported no major fight his patrols were a routine of pursuit and skirmish. In December 1902, attempting to find the people who had killed Baiwa, the village constable of Koropata, and many of his people, Walsh was jeered at and threatened on his second day out from Papaki. The police shot two men in a brief attack and later in the day they shot another three and a carrier was wounded by a spear. In two incidents the next day six more were shot and another carrier was speared. On the fourth and fifth days a further four were shot. A year later his patrol diary was still a flat record of sporadic violence:

> Aug 22nd A very wet night. Started from camp with 6 police & 20 carriers to look at the surrounding country, leaving Lance Corporal Waibua in charge of camp with 3 police & the balance of the carriers. Travelled SSE through old gardens passing through several villages 2 or 3 houses in each & got sight of mountain at back of PAPAKI 221°, Mt. Lamington 163°. Travelled then by very winding tracks seeing some natives on the way, one man being killed resisting capture, to a garden above a deep creek. Passed through 3 villages in one of which were 2 fresh skulls. Natives on the opposite ridge called out to us that they were WASETA men & wanted to fight us. Crossed the creek, name unknown & climbed a steep bank to the village where the men had been but found both it & another village close to it deserted. Picked a site for camp having travelled about 6 miles & keeping 3 police & 3 carriers with me sent the balance back to bring on the camp. They had not long gone when armed men appeared at both ends of the village. They retired on 5 shots being fired. No damage was done. Soon after the natives were shouting all round us & the men I had sent to camp returned saying the natives were mustering in force. Started for camp & the natives at once attacked us but retired on one man being wounded. In the meantime the natives were mustering strongly across the creek, where the track led up a very steep track about 100 ft high. One carrier was speared in the arm climbing the bank while a spear passing within a foot of me nearly hit the man behind me. It was a wonder we got off as lightly as a number of spears were thrown and the natives had a very strong position in our front & were in considerable numbers in our rear. They here lost 3 men killed & I fancy 2 more were wounded. About $1\frac{1}{2}$ miles further on they were so numerous & close on our flanks & rear that I halted outside a village & drove them back with a loss of 5 men. They soon however came close on to us again so as I had a very nasty gully to cross 1 mile further on I attacked them & drove them back. They fled losing 3 men & after that contented themselves with hooting & shouting at a respectable distance leaving us altogether some time before we reached camp. Lance-Corporal WAIBUA reported all had been quiet during my absence. Shortly after my return to camp a carrier said that the natives were coming along the track but they did not come in sight of the camp. All the country travelled through was populated, very thickly just where I turned back & I consider that there were at least 200 armed men following us on our homeward journey. The soil is rich & there are good gardens. In two villages

there were signs of recent cannibalism. 2 skulls in one village, one of which showed the mark of a tomahawk, & in the other a thigh bone & pelvis.

A total of twelve Orokaiva were killed on 22 August and five were shot soon after the patrol moved off the next morning.

Faced with persistent attacks and behaviour which they thought insolent, government officers found it difficult to avoid violence. In 1903 a large expedition including the Acting Administrator, Christopher Robinson, William Bruce, the Commandant of Police, and Monckton, went up the Bariji River attempting to reach the Yodda by passing to the south and west of Mount Lamington. Two days after leaving the coast the police at the head of the patrol shot one man and captured two others. Later in the same day Corporal Bia shot a man who threw spears at the expedition from a treehouse, and fired at another who escaped into the bush. In the evening Robinson told the police that he

> desired as little bloodshed as possible and enjoined them not to kill unnecessarily, but to endeavour to make captives from whom they might be able to obtain some information, and not to shoot native scouts if it could be avoided.

The following day the police shot two men and Constable Maioni was wounded by a spear. Frequent fights between police and spearmen took place, and Robinson revised his instructions: Now 'every native scout if armed and apparently hostile [was to] be shot'. By the time the patrol reached Papaki, Robinson was

> convinced of the fact that as a general rule before it is possible to pacify and maintain friendly relations with Papuans who are disposed to be insolent and hostile, it is necessary to inflict a short sharp punishment. This is what some of the natives hereabouts need as they have been treated too pacifically in the past ...

The station journals from Papaki and Bogi for early 1904 continued to report inflicting numerous short sharp punishments. On 22 March Elliott's police shot five men. Three days later they shot four more. Elliott, finding nearly all the villages deserted and people constantly aggressive, was unable to find anyone who would listen to the government's plans for their improvement. When he finally captured two men his message was brief: if they brought the goods stolen from the white men back there would be no more trouble — but if they didn't plenty more fight would come up.

Early miners on MacLaughlins Creek had said that a shorter overland route could be cut from the coast to the Yodda. After his patrol through the Northern Division Robinson won the gratitude of the miners by agreeing to survey and clear a track from Buna to the upper Mambare. The £1000 set aside for the construction of the road was then the largest single allocation the government had made to open the interior. Again the prospect of a field supplied by mule trains was held out to the miners. Crossing the closely settled country between the Kumusi and the sea, the Buna track exposed new communities to direct contact with the foreigners, but unlike the peoples living close to the goldfields they were to have their first prolonged and dramatic encounters with government patrols and construction gangs.

The first patrols from Papaki and Bogi to the coast had the same experience as others which crossed the water courses, bush and gardens of the plains. In some places they found deserted villages but heard distant hooting and drumming; a few communities tendered gifts of food; and some groups fought the patrols, the spearmen making desperate rushes and constantly looking for unguarded points along the flanks. But the period of violence ended more quickly than on the Bogi-Yodda track. Two reasons why peace came quickly were the greater frequency of government patrols and the conscription of villagers to work on the road.

The government abandoned the stations at Bogi and Papaki, selected to guard the old track up the Kumusi, and took up a new site, one day's walk from the Yodda on the Buna road. Within eighteen months of its foundation in 1904 Kokoda was a show-place of the Papuan field service. Built on a short plateau jutting from the main range, Kokoda overlooked the flat, steep-sided Yodda Valley to the north-west and Oivi Ridge, the plains and Mount Lamington to the east. Immediately behind the station the Owen Stanleys rose in massive blue peaks. As the mist left their high ridges in the morning the sun struck white patches of water tumbling across rocks. The Mambare, normally a series of swift channels dividing and meeting along a strip of boulders and gravel where it passed just to the north of Kokoda, increased in volume as it picked up the creeks draining the Yodda Valley. From 1905 there was regular overland contact between Kokoda and Port Moresby. Sometimes each village constable along the track was responsible for seeing that the mail was handed on to someone in the next village, but normally members of the Armed Constabulary took the mail from Buna and Ioma to Kokoda and fresh men carried it over the 'Gap' to Sogeri and Port Moresby.

The first government officers appointed to Kokoda expected to be healthy for they found no *Anopheles* mosquitoes there. It was a basis for optimism unknown when other stations had been opened in the Northern Division. While the death rate was at its height on the Gira and at Tamata, Ronald Ross in India had written to Patrick Mason in London telling him that 'the mosquito theory is a fact'. A year later in 1899 he wrote Memoir One of the Liverpool School of Tropical Medicine, *Instructions for the Prevention of Malarial Fever, for the Use of Residents of Malarious Places*. By 1902 the pamphlet, revised and expanded, had reached its ninth edition. One of the earliest administrators to apply Ross's teachings was the new Governor of Lagos, Sir William MacGregor.

All equipment, building material not available locally, and stores for the officers, police and prisoners at Kokoda had to be carried over the partly made track from the coast. Government officers also frequently needed carriers to support patrols or service road gangs. Villagers could not avoid the education of the 70-mile walk from Buna to Kokoda. When the people near Buna ran away rather than carry, the Assistant Resident Magistrate, Henry Griffin, said, 'I did what I had threatened to do and went to the villages & shot 3 pigs, then I went to their gardens & took enough taro for one day'. Armed Constable Donabai, recruited from the area, told the people that Griffin would continue to feed himself and the police on their pigs and taro until the men agreed to carry. At Kokoda government officers obtained carriers by instructing the village constables that either they sent men in or they spent some time in irons; or a police troop went out and brought in a line of handcuffed men. In January 1905 Rayner Bellamy wrote:

> I found a SISERETA native at the top of a big tree busily engaged in chopping out some sort of animal. I told him to come down. He refused. I repeated the order. He persisted in remaining up the tree. I then told him I would chop the tree down & when the tree arrived he would come too. I pretended to be about to carry this plan out with tomahawks. He came down & joined the carriers. I gave him the preliminary fee of one stick of kuku [tobacco] & he carried in to camp with the rest.

Still believing in his right to choose whether or not he worked for the government, he deserted during the night.

The road gangs were also conscripted. Having cleared the bush, the gangs built up the low sections with corduroy or with stone laid between log borders. Between Buna and Samboga they bridged forty-five creeks, up to 100 men being needed to drag the logs used on broad spans. At the Kumusi they slung a cable from bank to bank and travellers crossed by pulling themselves over on a platform suspended from a pulley. The 'wire-rope' had been carried up from the coast looped like a giant snake across the shoulders of an extended line of carriers. While the police often had to force reluctant communities to work, most men probably found the communal labour close to their own lands more congenial than carrying. They had the compensation of using steel tools, participating in a new mastery of their environment, and continuing communal rivalry in a novel form. Groups demonstrated their strength by appearing in large numbers and working hard and flamboyantly. After a few days they might still begin work at seven with a show of energy, but by mid-morning only a few were still singing, and by eleven men were stopping to ask for matches, and perhaps a deputation would approach the overseer to say that a man had died at their village and they must join the mourners. They would be given ten minutes to smear themselves with mud and return to their work. The labour gangs were given three meals a day, and one gang was always sent ahead to make temporary shelters and sleeping platforms. At the end of each day the men 'lined' to receive a stick of tobacco and a piece of paper; at the end of six or seven days when one group ceased work and another community took over each man was paid three sticks of tobacco and a box of matches. Although most of the work was done by peoples living close to the road, men from as far away as the Mambare were directed to work on sections of the track.

While supervising Dobuduru men building a swamp crossing, Bellamy received some knotted pieces of grass which accompanied the spoken message that Poumbari, a leader from near Bogi, had eaten three men and now he was hungry again. He was unafraid of the 'government' and if it came let it bring plenty of police because he would like to eat one. Bellamy did not respond to the challenge.

By the time the road was completed in 1905, it was also safe for travellers. Forcing men to build the road had probably been more important in changing old ways than the rifles carried on infrequent patrols or the lectures about the road being a sanctuary for all men. In his 1904-5 report Monckton wrote in self-congratulation that in the Kokoda district no European had made a complaint against any villager while in the previous year 'hardly a day passed without its story of outrage and robbery'. Some Orokaiva agreed to carry for the Yodda storekeepers, and others living along the track sold food to the carriers, sometimes being paid in bottles which they smashed and used in trade with more isolated villagers. In January 1905 the peoples within a few days' walk of the station demonstrated their new relationships with the government and each other by accepting invitations to dance on the cut grass of the Kokoda parade ground. They celebrated from early one morning until noon the following day with breaks for divertissements from an alien culture. The government officers organised races for different age groups and a greasy pole-climbing contest; only four men were able to wrench the tomahawk from the top of the pole. Before the arrival of Christian missionaries the 'Christmas' celebrations at Kokoda were an annual event.

The building of Kokoda station brought the government closer to the Mountain Koiari, a people then known to miners and officials as the Biagi or Isurava. From their small stockaded villages they kept a constant guard against Orokaiva raiding parties. In 1899 before he went down to the Mambare Stuart-Russell had been shown the remains of six men on burial platforms at Iuoro village, the result of an Orokaiva raid. Themselves a people with a warrior tradition, the Koiari used their lookouts to survey developments on the Yodda; and from 1904 they attacked carriers and stole from deserted camps. They took firearms and ammunition whenever they could. Less flamboyant in fighting manner, not using the charge of massed warriors carrying spears and shields, and away from the main carrier tracks, the Koiari were not involved in such bloody conflicts as those which took place on the plains. But neither did they quickly submit to direction by the foreigners at Kokoda station. In 1906 they killed two labourers employed by the naturalist and sometime miner, A.S. Meek, who was, he said, forced from the area by the 'ferocity of the natives'. Reluctant to visit the station except to bring in the mail or join the Christmas dances, the Koiari seemed reserved and independent to the government officers. Even in 1909, on the eve of the exodus of the alluvial miners, Laurence Henderson, the Assistant Resident Magistrate, investigating the attempted spearing of a miner's labourer, found that all the men had left the villages and Village Constable Babila could not persuade them to come and talk to the *gavamani*.

After about five years of peaceful trading the Beda too were drawn into the violence on the Yodda. Unknown to any European officer until long after the event, the Papaki police in 1902 killed a Beda man, intimidated others by handcuffing them, seized two women, smashed a house and stole valuables. To retain the friendship and trade of the Beda one of the miners compensated them for the damage caused by the police. But the Beda were also harassed by raiding parties from the Kokoda area and by miners' labourers. Some of the labourers, having run away from their employers, lived by plundering gardens; others, sent to obtain food, found life in Beda more attractive than mining, and loafed about the village. In 1903 the Beda killed a labourer caught stealing from their garden, and in 1904 Elliott went to investigate reports that in another clash the Beda had killed four labourers.

Delayed by flooded creeks and the rough track along the Yodda, Elliott eventually arrived at the gardens on the slopes leading up to Beda village. Corporal Bakeke took a troop of police ahead while the rest of the patrol followed. Elliott heard three shots before climbing out of a gorge to see the stockaded village full of people and in front of them four bowmen advancing towards the police. He ordered the police to fire, fix bayonets, keep firing and charge. Caught in a 'blind gully' Elliott did not see any fighting, but by the time he entered the stockade, the police had occupied the village, captured a woman and two children, and they showed Elliott the bodies of three villagers who had been shot. The patrol camped in the village for three nights. They ate pigs and vegetables, destroyed spears and arrows, and confiscated all objects obtained by trade or theft from mining camps: tomahawks, knives, billy cans, frying pans, pannikins, cloth, singlets and one home-made flannel. Unable to speak to the woman, Elliott released her and the two children. Police patrols attempted to capture one of the men, but while they did not use their rifles and narrowly escaped being speared, they took no prisoners. At night the Beda gathered outside their village and Elliott (his prose this time not corrected for publication) ordered 'a volley to be fire where we Heard them there Must Have been a large Crowed & a good few Must Have been wounded as I found the tracks of Blood the Next Morning'. During their last day in the village Elliott again instructed the police to use their rifles to clear a ridge of spear throwers. Another three Beda were killed. After a last attempt to capture some of the men, the government party left. Elliott regretted that 'these Natives Have turned out bad', but he thought that they had 'seen that it is useless for them to fight the Government'.

Government officers from Kokoda gradually re-established contact with the Beda. In November 1905 Koiari mailmen reported that some weeks earlier the Kokoda Orokaiva had killed two Beda men. The two Kokoda village constables were gaoled until their people brought in compensation to pay the Beda. Once sufficient payment was held at Kokoda the police asked the Koiari to bring the Beda to the station. When he learnt of events at Kokoda, Musgrave, the Government Secretary, was disturbed that people should be allowed to buy immunity from punishment; it was, he said, contrary to law and to practice elsewhere in British New Guinea. But the Beda left Kokoda apparently satisfied that justice had been done; and government officers acted in the same way to bring peace to other areas of the Northern Division.

In 1907 the Beda again visited Kokoda station accompanied by Koiari villagers. They presented a pig to the Assistant Resident Magistrate and told him that they still sold food to miners on MacLaughlins and Finnegans Creek. Ten years later when there were only two or three miners in the area the Beda continued to sell to the foreigners. Sometime between 1915 and 1917 they combined with another group of people who had broken away from the Koiari to form a new village, Nairoda. Known as the Karukaru people they numbered just under 100 in 1919, but by 1930 Nairoda was deserted. Most of the Karukaru had been absorbed into villages higher up the Chirima. The transfer of the people's cultural allegiance from Mountain Koiari to Fuyuge, probably under way when they first encountered Europeans in the 1890s, was complete.

Drum, Orokaiva, after Williams 1930

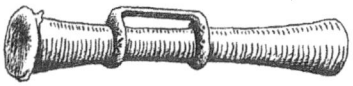

The violence along the Kumusi disturbed briefly the uneasy peace on the Mambare. According to Elliott, whose opinion was forthright but whose knowledge may have been slight, the Binandere had strongly resented Walker's gaoling of the twenty-five men who took part in the raid on the Waria. Having been to the Kumusi and seen the clans there robbing carriers and hurling abuse at the foreigners, some 'restless spirits' among the Binandere attacked Tamata station in 1902. No European officer was present at the time and the police repelled the raiders with five shots. Elliott gave his 'word of honour' to the Government Secretary that if they came again 'half of them will stop just where they may have the luck to fall — gaol is no good to them at present ...'. The attack may have been an expression of Binandere bitterness and a test of the police defences; but it was scarcely a serious attempt to destroy the station. For the officers at Tamata the violence between miners and villagers on the upper Gira was more serious and more difficult to control.

The early miners on the head of Tamata Creek and the Gira saw no people in the area but during 1902 the men further west on the Aikora were constantly skirmishing with a community which they called the Seragi or Red Creek tribe. Driven out of their homeland on the Waria, the Seragi had moved to Red Creek, a tributary of the Aikora. A small group speaking a different language from their distant neighbours, the Binandere and Chirima, the Seragi lived in temporary shelters, planted little, and hunted over a wide area. In October 1902 the Seragi killed James Blackenbury and James Jassiack (Jimmy the Austrian). Both men had claims close to other miners who arrived too late to save them from sudden attacks. Jimmy, a 'poor old man 65 years of age and as harmless as a child', was too battered to move and his body was burnt; Blackenbury was buried by building a stone wall around him and covering his body with earth. Earlier in the year James Delaney, Fred May and two carriers had been wounded by spears, and camps had been robbed. Among the miners 'all hands' were crying for 'blood' and the government officers began organising a patrol which they called a 'punitive expedition'. After attending divine service at Tamata three government officers, Halkett Parke, Walsh and Hislop, twenty-one police and a large crowd of miners and carriers left for the Aikora. On the fifth day they saw people who, by dress and physical appearance, the miners recognised as the marauders. So as not to alarm the Seragi, the police removed their uniforms, and two men who came into camp were held captive to prevent them from warning the rest of the community. Instructed to take captives and to fire only if attacked, the police attempted to surround a settlement. When seen by a woman, who uttered a piercing scream, the police and carriers rushed forward. Amidst the spearing and shooting four Seragi were killed and three women and some children were taken prisoner. Police on independent patrols during the next few days shot another ten men: 'No rougher country could be imagined, & no prisoners were taken', Parke reported. Although police, carriers and labourers could 'muster some 18 or more languages' no one could communicate with the prisoners. But having found saucepans, billy cans, tools, two rifles and other goods in the houses, some of them taken from Jimmy the Austrian's camp, the government officers and miners were confident that they were punishing the right people. Unable to find sufficient food in the Seragi gardens, most of the expedition withdrew after a week on the Aikora.

Three years later the miners on the Aikora were again appealing for protection, and six of them signed a letter threatening to 'take matters into their own hands' unless the Seragi were stopped from robbing their camps. Government officers patrolled the area and police were stationed at some of the outer camps. But both miners and government employees found it difficult to stop the looting of camps and attacks on labourers. Men from the upper Chirima and Kambesi were also visiting the mining areas to trade and pick up anything of use from deserted camps. The Seragi were not constantly hostile: they maintained friendly relations with one miner while robbing and harassing another. Using prisoners taken on earlier patrols, Bell assembled the Seragi and told them that they must stop robbing the camps or be punished severely. There were only about 100 people to hear the government's message. But no peace was made.

In 1908 Joe Sloane told Arthur Lyons, the Resident Magistrate at Ioma, that he and his partner Charlie Ericksen had been living with two women from the Waria: the Seragi had looted their camp and abducted the two women. Sloane admitted that in an attempt to arrest some of the Seragi, he and Ericksen had killed two men. From camps in the area the Seragi had stolen four rifles, a shotgun, a revolver, ammunition, and other goods, and killed a labourer signed-on to John Butler. The Seragi may have had a specific reason for killing the labourer: Butler had been feeding his men on their gardens without paying them, an action which angered the other miners who feared that it would stop all trading and lead to violence. Lyons sent a rifle and ammunition to Ned Ryan, who was left without protection, and instructed Corporal Bokina and five police constables to work with a few Seragi men who remained friendly with the miners to recover arms and arrest the murderers. Patrol Officer James Keelan followed the police to the Aikora and while he waited about the mining camps Bokina captured ten Seragi men, and Domata, recognised as the leader of those Seragi still working and trading with the miners, brought in two men supposed to have killed Butler's labourer. In great 'jubilation' Keelan and the miners contributed tins of meat and jam to a feast for the police and Domata's men; and to demonstrate further 'how good boys are treated' he rewarded Domata with trade goods and put the prisoners in irons for two days. Lyons was disturbed by Keelan's graphic history of events at Aikora. He pointed out that the government punished men only after they had been found guilty at a trial: it was Keelan's duty to make friends with the Seragi, not to intimidate them.

During 1909 the Seragi decided that they would be government men. Domata was appointed their first village constable. On a site above the Aikora about fifty people began building a permanent village, their houses, a government officer noticed, were the same design as the huts built by the miners on their claims. The Seragi fulfilled their first obligation to the government by using their skills to build vine and bush-timber bridges across the Aikora and Gira Rivers. Domata held office for only four months. He was murdered by a released prisoner he was escorting from Tamata to Sloane's camp on the Aikora. Lyons hoped that the Seragi would develop into 'useful people for road work in and around these localities'. It was a lowly station which they did not have to accept, for after ten years of skirmishing with the miners the Seragi had changed their ways just as most miners left the area. The Seragi too abandoned the Aikora, perhaps because the gardening land was poor, and settled on the headwaters of the Eia River where optimistic government officers attempted to transform them into cash farmers by issuing them with rubber tree seedlings. The trees grew but the Seragi saw little cash.

Fighting spear, Orokaiva, after Williams 1930

Miners and government officers believed that the Seragi would not have continued fighting and thieving for so long had the lawless communities on the Waria been punished. It was true that the Waria killed seven labourers during 1908–9 and 'punctured' others, and it is possible that the Seragi knew about the actions of peoples with whom they shared a language; but some of the Waria communities had been 'punished'.

By as early as 1903 miners had crossed from the Aikora to prospect along the Waria. They found gold, but not in payable amounts. Then in 1906 Matt Crowe and Arthur Darling, with a strong party of thirty-five labourers carrying over twenty rifles and guns, left Finnegans Creek on the Yodda and passed through the Aikora to come down the south-western tributary of a large river. Through their labourers, who were able to persuade some local villagers to talk to them, the prospectors learnt that they were on 'the Wariah for certain'. During July they travelled slowly upstream sometimes keeping close to the river and at other times crossing the spurs and creeks on the flanks. After a hurried trip to the 'head' of the Waria the prospectors came back downstream to Gamundu. They had not been attacked by any of the Waria people and they had frequently been able to barter for food. At Gamundu, which could be reached by canoe from the sea, the people had met other white men. Nearly four months after leaving the Yodda the prospectors arrived at Tamata.

Crowe said little to the miners eager for news of a new strike, but Darling admitted finding over 40 miles of shallow river beaches returning a few grains to the dish and 'in places up to a quarter of a weight'. Looking for evidence to help them decide what to do the miners noticed that the prospectors did not apply for a reward claim, but this could be explained by the fact that Crowe and Darling 'were rowing the whole time, and bust up as soon as they reached Tamata'. And both planned to return to the Waria. While the prospectors had been away some miners had recruited fresh teams so that they would be able to spend the full period of their labourers' contracts on any new field. Now about twenty miners decided that although it was 'nearly sure to be a tough affair' they would make the long overland trek north-west to the Waria. One group of seven miners, 100 labourers, a government officer, twelve police and carriers left from Tamata, and another group started from Waterfall Creek, a tributary of the Gira. The miners quickly spread over many miles, racing up the valley of the Waria to work the richest patches and try the creeks coming in from the west. Men beyond Garaina were more than ten miles inside the German New Guinea border and nine days' walk away from those men working downstream where the Waria looped southwards into British New Guinea.

Back in Tamata after a year on the Waria the Pryke brothers, Frank and Jim, had 300 ounces. Fred Kruger, they thought, had more, but many had less. After a spell in Australia the Prykes spent another year on the Waria in 1908–9 and again won just on 300 ounces. But they decided they would not return to the Waria; as a goldfield 'she [was] done' said Frank. In the year of greatest production 1907–8, the miners had taken about 3000 ounces from the Waria, but their expenses had been high. Carriers took fourteen days to make the round trip from Tamata to the Waria, and the highest camps were too far away to be supplied by porters. Even on the middle Waria a miner needed to find a prospect giving more than half an ounce a day before setting to work. And after the government employed over 200 men to improve the Tamata-Waria track and Whittens put in a store at Jijingari, isolated miners could still work only if they traded with local villagers.

Villagers from the lower Waria had visited Tamata before 1906 and during the early months of the rush the first village constable was appointed at Gamundu; further downstream the village officials were appointed by Germans. The people of the lower Waria were 'spear-men' speaking a language related to that of the Binandere whom they met in trade and war. But on the upper Waria the people had developed a different culture, they spoke other languages and their main fighting weapon was the bow and arrow. Neither those living south of the river in British New Guinea nor those beyond Biawaria through to the Ono in German New Guinea had seen the foreigners from Tamata and the sea. Frank Pryke, one of the first miners to travel to the upper Waria, reported that another party 'had some trouble' and he was hampered by people cutting bridges, but never attacked. The vine bridges, 'cobwebs' Billy Ivory called them, nearly 100 yards long and swaying high above the water, were often the only means of crossing the turbulent Waria. The miners used them but had to carry their dogs across.

Needing to trade with the villagers, the miners and labourers were tolerant of men who were at first aggressive or who came to their camps out of curiosity. Frank Pryke wrote to his brother Dan in Australia:

> There are a lot of villages and a big population of niggers two days up from here and I have established friendly relations with them. I have been up along the river to there on both sides; and I think we will be able to get plenty native tucker from there.

In his diary for 1908 he recorded frequent visits to his camp of 'boys & gins' bringing 'a lot of tucker'. Local men carried for him when he shifted camp, and he made an entry: 'One boy (Agunomi) started to work for me.' The labourers added to the rations by shooting birds and game. But because of the divisions between the different groups on the Waria miners could trade with one community and fight another. In fact once they formed a close relationship with one village they were more likely to be drawn into local feuding, either by a friendly village inducing them to fight for them, or by another village assuming that they had joined their enemies.

On 20 October Frank Pryke noted in his diary that two of Coleman's labourers had disappeared, 'supposed to be eaten'. Two days later he wrote, Driscoll came up today — one of his boys killed by the Wakaia boys'. The Pryke brothers, Edward Driscoll, twenty-one labourers from south-eastern Papua and fifteen Waria men set out for Wakaia. Pryke's diary, always brief, gives no indication of their actions at Wakaia, but the miners had clearly set out to punish. Describing another incident to his brother four months later, Frank Pryke was more explicit. One of Arthur Darling's labourers was killed when he left camp to get water. Frank explained:

> It was too late that night to do anything, but next morning Darling and I were among them just at daylight and gave them a bit of a shock, but I think by the way they got to cover they are used to being surprised or else they train for it ... . Anyhow they suffered heavily in pigs and would also have to build fresh houses.

He described the people they had fought, probably from Guswei near the junction of the Ono and the Waria:

> They are a rather unsociable lot and are armed with the bow and arrow or skewer as Darling calls it. These weapons are much better than the spear as a native can send them over a hundred yards on level ground, and in that open grass country they must be able to send them long distances down the sides of the steep hills. Of course there was no chance of the nigs making a stand against us in a fair go as we were well armed, I had a Lee Enfield, Automatic Winchester, and two ordinary winchesters and a shot gun and Darling was even better fitted out, but there are places about there where a large rock rolled with a bit of judgement, could wipe out an army.

At the time Frank wrote there were only seven miners still on the Waria. Soon after the Prykes left to recruit fresh labourers and look for another field: 'we have', Frank said, 'kept our noses in front of Bill Whitten's books so that is alright'. A government officer in mid-1909 thought that the 'take per man' over the past year had been about 200 ounces valued at £3.15s. an ounce.

When the miners first worked on the Waria they had been uncertain of the whereabouts of the eighth parallel of latitude which on maps clearly marked the border between British and German New Guinea; and they did not know what German officials would do to Australian miners found in German territory. In shifting backwards and forwards across the border the miners were violating the customs regulations of both colonies; they contravened the Native Labour Ordinance by taking their teams outside British New Guinea, and their Miner's Rights purchased for 10s. at Ioma or Samarai gave them no authority to take gold from German New Guinea. The Germans responded to the border incursions and stories that their black citizens were involved in fighting by establishing a station at Morobe, just to the north of the Waria mouth.

In spite of rumours among the miners that the Germans would not let them reach the upper Waria by entering the river mouth, Governor Hahl on a visit to the area in 1909 told them that they were free to use the Waria for transport where it passed through German territory. German officials did not attempt to impose licence fees or collect customs dues. And they did undertake to punish the Wakaia. One member of the Papuan Armed Native Constabulary helped the Germans to capture four Waria men said to have killed miners' labourers and then accompanied the German police and their prisoners to Tamata. There Lyons was perplexed by the whole operation: he could not read the letters in German carried by the police and he had neither evidence nor witnesses to use against the Waria prisoners. By the time most miners were leaving the Waria the German officials were displaying a lot of energy by proxy'; that is, they were compelling local villagers to cut a track from Morobe inland along the Waria Valley.

From their meetings on the border the Germans decided that the Australian miners were guilty of wanton violence; the Australian officials thought that the German officials were too ready to punish by shooting. The Reverend Percy Shaw of the Anglican mission on the Mambare complained that the German police were 'Kidnapping women for immoral purposes', and using their arms to intimidate and extort. Murray, after talking with Governor Hahl, wrote a general condemnation of his imperial rivals in his diary: 'The Germans take any land they want without payment, also if the Government wants labour they simply take it, likewise women and food.' But the two governments combined to place two teams of surveyors in the field to mark the boundary. Malaria, sago swamps near the coast and hail-swept mountains inland forced the Anglo-German boundary commission to abandon its plan to mark all the border where it followed the eighth parallel. After much hard work and nearly a year in the field the two teams had erected only a few posts and cairns at significant points. Several times the police and government officers with the Australian surveyors had fought off people who objected to the foreigners crossing their lands. They reported that they had fired mainly to frighten and had killed no one. By the time the surveyors decided that most of the bed of the Waria known to contain gold-bearing wash lay in Australian territory there were only four alluvial miners left to rejoice at the news. Men who held dredging leases may have thought it an important decision; but they soon found that their leases were worthless.

When Australian troops replaced German officials in 1914 the people on the upper Waria were still beyond the area administered from Morobe station. They maintained their stockaded villages, fought each other, and they fought the patrols directed by Australians which entered their valleys in the 1920s. By that time the Lutheran missionaries, who had come early to the coast, were well established on the middle river and had walked beyond the headwaters of the Waria and on to the Bulolo.

The lives of most peoples of Papua and New Guinea who lived in mining areas were profoundly changed, and even after the alluvial miners moved on they immediately faced other agents of change. But the people of the upper Waria had had only a brief encounter with the miners. They had fought them, traded with them, and perhaps a few men had worked for them and a few women had slept with them. It was more than ten years before foreigners again lingered in their area. The coming of the miners had been more of an interruption than a revolution.

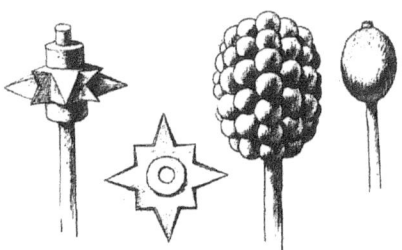

As on the Mambare in the 1890s, the fighting on the Kumusi and the Yodda was not simply a clash between villagers and foreigners. The Kokoda Orokaiva fought the Mountain Koiari and the Beda, but the most violent encounters were between alliances of Orokaiva clans. On his pioneering patrol to the Yodda in 1900 Armit counted 102 houses burnt at Sisireta; the gardens were devastated and everything of value taken. In 1901 Walker, travelling with a strong patrol near the Kumusi mouth, passed through numerous deserted villages and pillaged gardens. He saw many people suffering from spear wounds and he was told that thirty-three people had just been killed on Mangrove Island.

The first patrols to mark the track from Kokoda to Buna crossed country which was, in Monckton's colourful prose, the 'theatre of constant warfare and scenes of cannibalism. Groups of cocoanuts and desolate stretches of fertile country, which were once the site of villages, were to be seen everywhere, whose inhabitants had long since passed into the cooking pot.' The sight of plundered villages, the dead, and the bodies gutted and prepared for feasting gave miners and government officials a ready excuse for harsh actions: the people they were dealing with were savages, they bowed to force but never to reason or compassion, and they were killing one another at such a rate that lives would be saved if the government quickly imposed control. And again as on the Mambare, clans attempted to form alliances with the government, for protection and to direct the government's power against their enemies.

The desperate grasp by clan leaders for government help can be seen in the journals of officers who may not have known what was happening. Walsh at Papaki wrote in May 1902: 'The only natives near here that I can really depend on are the KOROPATI tribe. They bring in far more food than any other tribe and if I tell BAIWA (V.C.) to do anything I find it is nearly always done at once. ' In October Baiwa told Walsh that Koropata was likely to be attacked, but Walsh decided that no immediate attack was being planned. In December Baiwa and several of his followers were killed, Koropata destroyed and the surviving people scattered. Accepting his obligation to punish, Walsh led a patrol which shot at least fifteen villagers, but when he reached Baseta, thought to be the home of the raiders, he found it already destroyed by another warring group. It is possible that the intrusion of foreigners, upsetting old balances, and the increased movement that followed their arrival, had increased violence on the northern plains.

In brief asides government officers referred to 'unofficial' violence. Armit reported that Thomas Davitt, a Yodda miner, and his labourers were 'visiting villages, shooting, looting and destroying'. After the killing of Campion and King the miners gathered strong punitive forces, but whether they punished anyone is unrecorded. In 1904 when attacks on camps were frequent and the Orokaiva wounded two miners with shotgun pellets, the government officers could not dissuade the men assembled on the Yodda from attempting to recover stolen guns and ammunition. Later Elliott reported that

> they Had better Have stayed at Home it cost them 8 oz of Gold & they done nothing whilst they were out only frighten the Natives all over the Place they say they Had a good feed of Potatoes & Pig & got a lot of catridges & fly Calico Billy cans but that was all they told me.

The miners were not always so accidentally benign. On a patrol to the Koiari villages in 1906 Griffin noted the old site of Iworo which, he said, had been destroyed by the miners. He was probably referring to the village which Stuart-Russell had called Iuoro in 1899; it was then the 'principal village' of the communities living in the ranges over-looking the Yodda and inhabited by 'Friends'. As late as 1909 Matt Crowe and James Lawrence, prospecting on the upper Kumusi, fought the Wawonga, a general name for several communities speaking Koiari languages. Lawrence 'in the course of conversation' said he thought that two spearmen were wounded. Later the Assistant Resident Magistrate at Kokoda heard that five men were killed and three others wounded. Although no one thought it proper to advertise the extent to which the miners used their guns, it is clear that for fifteen years after Clark fired to keep Binandere canoes at a distance, miners shot villagers on the Waria, Gira, Mambare and Kumusi. Sometimes they supported government forces and sometimes they acted independently or in defiance of officials.

The most numerous strangers encountered and killed by the Orokaiva were miners' labourers. Between 1900 and 1910 there were always from 400 to 1000 Papuans from other areas working in the Northern Division. About half of them were carriers shifting rations and equipment to the mining claims. Some were employed by individual miners but most of the men lumping cargo worked for the storekeepers, Clunas and Clark and Whitten Brothers. In the 1890s Tamata was the start of the tracks: to MacLaughlins, the Gira and Aikora, and in 1899 and 1900 to the Yodda. Crossings, camps, swamps, steep climbs and outstanding physical features became well known to miners and carriers: the Calico, Double Crossing, the Four Mile, the Sisters and the Lookout. The Tamata-Yodda track was the worst for loaded carriers. Men struggled for hours in waist-deep swamps searching with their feet for the logs and roots which kept them above the mud. When floods made creeks impassable the carriers gathered in temporary camps until the water-level fell. Thirty years after the event Frank Pryke remembered how he had:

> Tested all the reaches
> Out to Finnegans and back
> And fattened up the leeches
> Away along the Bida track

By 1901 most men and rations reached the Yodda by going by launch up the Kumusi to Bogi where the stores stood on one point, the government station on another, and the boats anchored in the backwater between the two. The carriers walked for ten days from Bogi up the Kumusi, across the low divide and then down the Mambare to the workings on the Yodda. Most of the Bogi-Yodda track was flat, but some carriers lost their lives when the earth gave way on the narrow benches cut into the banks of the swift-flowing Kumusi. If the carriers went beyond the Yodda stores past Finnegans Creek towards Prospect and MacLaughlins the track became very rough with sharp ridges and boulder-strewn creeks.

The Bogi-Yodda track was the most dangerous used by the carriers. Even those manning canoes bringing stores up from Kumusi beach to Bogi were likely to be attacked or intimidated so that their packs could be looted. For the four years that the Bogi track was the main route to the field the carriers were never free from abuse and violence. When the track was first opened and all travellers were jeered at and threatened by large numbers of armed and decorated men, Clunas and Clark paid the government £10 a month to meet the costs of a troop of police to travel with their carriers. But after the escort was withdrawn attacks continued in spite of the efforts of carriers, miners and irregular government patrols from Bogi and Papaki. In August 1903 Elliott near Kokoda suddenly came upon thirty men chasing a carrier; the leading warrior had his axe raised ready to strike. Elliott fired: 'I caught Him in the small of the back. He dropped at once.' The police also opened fire, killing or wounding others. How many carriers were not saved by the opportune arrival of quick-firing government officers is unknown; their numbers were lost among the many who deserted or died of disease and malnutrition.

The Buna-Kokoda-Yodda track opened in 1905 was safer and easier walking. Many of the indentured labourers travelled alongside groups of Orokaiva who had agreed to carry. In July 1909 Beaver estimated that in the previous year over 1200 'local natives' had carried on the Buna road for miners and storekeepers. With the rush to the Waria another long and difficult track was opened; and even after gangs had spent months cutting and forming sections between Tamata and Jijingari many Binandere, Waria and indentured men found the track to the upper Waria beyond their endurance.

The average size of miners' teams increased from four or five in the early years on the Yodda and Gira to ten by 1909. The most energetic and successful miners such as the Prykes or others who 'worked mates' employed up to thirty men. On the claims the labourers looked after the sluice boxes, cleared away overburden and forked large stones from the alluvial. Once the most accessible deposits had been exhausted the miners began to work higher ground and stream beds. To bring water onto higher claims the labourers dug races, sometimes over a mile in length, and ensured a sufficient and constant flow by connecting creeks and building dams. In some places the stream could be brought across the gold bearing area to scour away the overburden and then the deposit could be 'ground-sluiced', a method which meant that the labourers did not have to carry the silts to the sluice boxes. The digging of races and exposing of new faces was 'dead work' to the miner; he might have to pay his labourers and buy stores for several weeks before recovering any gold. The labourers worked some creek and river beds by building 'wing dams' to divert the stream from successive areas. Where it was difficult to build dams the labourers tried 'blind stabbing', shovelling the alluvial from the flowing water while supported by vines against the current and boulders which moved underfoot and crashed against shins. On the Aikora and Waria the labourers were often forced out of working areas by floods which destroyed dams, races and sluices; on the Yodda dry spells were more likely to cause delays and the labourers had to dig longer races to draw on higher streams. In steep country heavy rains could cause landslides with hundreds of tons of mud and boulders sprawling across a claim, forcing the miner to abandon a working face. Where nests of boulders obstructed mining the labourers would have to 'shoot the stones'. Holes to take explosives were cut with a hand-held drill and a heavy hammer, the gelignite and detonator put in place, and the fuse attached and fired. Sometimes big rocks could be shattered by 'plastering': the charges were laid on the surface and packed down with clay. By law a labourer without a permit could not handle explosives but some did set and fire charges when no white miner was at the mine site.

Early recruits on the Yodda and Gira were often new to mining work, but later many of the 'leading boys' and men in the teams were skilled miners. They knew how to set up temporary camps while prospecting, or build bush material huts and plant gardens when the miner decided to set in on a piece of ground. They adzed logs to make sluice boxes, and hunted, fished and traded to supplement stores. Some labourers from the islands were competent prospectors and went out with the dish and shovel to look for more payable ground. In 1908 Frank Pryke noted in his diary, 'Most of the prospecting on the Waria appears to be done by boys'. There is a touch of disapproval from the son of a man who had mined on the goldfields of Victoria and New South Wales in the 1850s and 1860s.

The death rate among labourers, appalling in the early years on the Gira, declined slowly, but even in 1909 twenty-nine of the 729 labourers who spent at least part of the year on the Gira or Waria did not live to complete their contracts. On the Yodda where 806 men worked, eighteen died. Expressed as a percentage of the average number of labourers in the Northern Division the death rate varied from about 30 per cent in 1898-99, to 10 per cent in 1903-4, and even in later years was probably never less than 5 per cent. The Yodda was normally more healthy than the Gira. In 1905-6 the difference was most marked; the death rate on the Gira was 17.7 per cent against 5.6 per cent on the Yodda. Until at least 1907 the death rate was almost equally high among white miners. But 'bosses' and 'boys' died of different causes. Of the nine European miners who died on the Yodda and Gira in 1903-4 one committed suicide, one was killed in an accident and seven succumbed to malaria. The labourers had some immunity to malaria and while they may sometimes have suffered from fever, especially when they came in contact with new strains of the disease, it did not kill them. They died of dysentery, beriberi and respiratory ailments. The wide fluctuations in the death rate were largely a reflection of the prevalence of those diseases in the mining camps.

In 1896 Bill Whitten brought Lihoiya of Taupota and Tabe of Koira on the Opi to Tamata station. Lihoiya explained that he and another labourer, Yarumeku, had deserted while working for Simpson on the upper Mambare. By keeping in the bush they had passed Tamata and stolen a canoe from Ume, the first village downstream. At the mouth of the Mambare they left their canoe and began walking south along the beach. The Koira people, especially Tabe and his father, greeted the two strangers in a friendly way and allowed them to sleep in the village. But just south of the Kumusi they were attacked: Yarumeku escaped but Lihoiya, the stouter and slower, was caught and injured with club and knife blows. He was saved by a man from a village near Koira who washed his wounds and accompanied him back to Tabe's house where he stayed until he was picked up by Whittens' schooner.

Green set out to return Tabe to his home and look for Yarumeku. He found him living at Oreya village, south of the Musa and over 100 miles from the mouth of the Mambare. Yarumeku had continued walking along the beach, swimming rivers and avoiding villages. He again outran a hostile group near Oro Bay, and when he met Taniava of Oreya he was making his way homeward by canoe. Taniava, who had married into the Oreya community, knew a language spoken by Yarumeku. He acted as Yarumeku's guardian at Oreya where most people suspected that he had taken part in the killing of Oreya warriors on the Musa.

Lihoiya and Yarumeku were two of the first labourers to throw aside their loads and begin odysseys which enabled a few of them to reach their home villages. Men who knew that they had to cross over 100 miles of country inhabited by hostile and warlike peoples, and travel a further 100 or 200 miles by land and sea to reach their kinsmen, chose to make the attempt. They decided that a tough journey and a slight chance of survival were preferable to the life they were leading. Wio of Rossel Island when asked, 'What you do along Mambare?' had replied, 'Me been run away long bush; plenty boy dead'. Fifty-four men had gone from Rossel to the Northern Division, and Wio was one of about twenty to survive. Eight men from Cloudy Bay agreed to go to Samarai with Andersen, the trader from Dedele. Five signed-on there to work for Clunas and Clark and three for James Swanson, a miner. Boie, one of the men who worked in the Northern Division for Swanson, said that his two companions died and he buried them. Swanson had left him in Samarai at the end of his contract and after serving another year on Woodlark he had found his own way home along the southern coast. The other five men deserted on their first trip to the Gira, taking their loads with them. They were not seen again.

When Monckton opened the first government station at Cape Nelson on the north coast the police and nearby villagers frequently brought in men who had already completed the most dangerous part of their journey home. On 12 May 1900 he wrote in the station journal: 'A runaway carrier from the Mambare arrived at daylight looking to be in the last stages of death by starvation and exposure he reported that seven companions had been killed on the road to here.' Later in the month he reported arresting six deserters who were travelling in comfort; they had stolen a cutter and stores at the Kumusi. To encourage the people of the area to bring runaways to the station Monckton paid an axe for each man brought in. One group handed over to Monckton had stolen arms which they had used to shoot a village man and woman before they were captured. The practice of paying bounties for deserters was used on the Mambare too, but there some Binandere were guilty of extracting from the deserter any stolen goods he carried with him and putting him to work in the gardens before taking him to Tamata.

After the Kokoda-Port Moresby track was well marked, deserters from the upper Mambare hoping to reach the south coast could take a much shorter and safer route. Immediately Griffin learnt that the Gulf and the Western Division labourers knew about the road to Port Moresby he sent a message to the Koiari villagers informing them that they would get a 'present' for any deserter brought back to Kokoda. But a few days later a police mailman brought news that six men from Orokolo in the Gulf of Papua had already walked from the Yodda to Port Moresby.

Groups of deserters lived in the bush near the mining camps on the Yodda and the Gira. Perhaps they realised that it was almost impossible for them to reach home alive, or they knew that if they reached home, the *gavamani* would arrest them, make them work in prison and then send them back to complete their contracts. They lived by robbing gardens and camps and planting their own crops. A constant problem for the Beda, the deserters were an irritant to both sides in the general conflict between villagers and foreigners. Miners whose camps were robbed by runaway labourers blamed villagers, and peaceful trading relations between miners and villagers were disrupted by deserters pillaging gardens.

One measure of the rate of desertion is the number of men convicted at government stations in the Northern Division. It is an imperfect measure for it takes no account of those who escaped, were killed, or were caught and charged elsewhere, or not charged at all. And the figures are not available for the early years when the desertion rate was probably highest. Even so the number of men who chose to flee is high. From 1905-6 to 1908-9 about 200 men a year were convicted of desertion under the Native Labour Ordinance. In 1907-8 nearly 17 per cent of all men who worked for at least part of the year on the Gira and Waria were found guilty of desertion. Given the rates of death and desertion, there could have been few cases in which the men who signed-on were still together when they were paid-off, and there must have been many cases when at least half the team was missing.

Miners and government officers frequently said that the men threw down their packs and bolted for no reason. But given the knowledge that some labourers had of the foreign world and wage-labour, they had reason enough. Eni swam ashore from the *Merrie England* when he saw preparations for a feast and feared he might be eaten. In fact the officers and crew were preparing to celebrate Christmas, and Eni, his contract time having ended, was being taken home. Two Good-enough Islanders employed as carriers from Tamata to the Gira gave as their reason for deserting, 'the road was too long'. Perhaps that alone was a sufficient reason, but it was more the answer of men with few English words to express their dissatisfaction than evidence that men absconded for frivolous reasons. And many men did have specific grievances: they were beaten, ill-fed or sick. A labourer, released from Ioma prison where he had served a term for desertion, was thrashed by his employer in front of the policeman who had escorted him back to the camp. Another man reported to Ioma with 'ulcerated testicles in a very advanced stage of suppuration'; he said that he had received no medical treatment from his employer. Others worked on, their shoulders galled by loads and pack straps. Parke in February 1903 noticed that many of the carriers on the Bogi-Yodda track were 'walking skeletons'. Two months later Elliott inspected Whittens' carriers and, finding several too sick to work, helped the storekeeper hold them and dose them with medicine. Bellamy, on his way to the Northern Division goldfields, first encountered captured deserters at Cape Nelson. He was curious about the men working in leg irons, the few survivors of one or two months in the bush. Asked who they had been working for they replied 'Bobstore' (Whittens') or 'Alecstore' (Clunas and Clark's). After seeing the carriers at work Bellamy said that he now understood why some men deserted, were returned, and deserted again; gaol was preferable to carrying.

Bellamy suggested that there was a need for a government inquiry into the treatment of labourers in the Northern Division, but as he wrote for the *Grey River Argus*, a New Zealand provincial paper, neither his revelation nor his opinion had any influence. The few other independent observers who saw the labourers at work tended to accept that they deserted for trivial reasons. The Reverend Copland King wrote for the *Sydney Morning Herald*: 'they dump their swag and bolt on the least provocation'. He added a salutary story: 'One boy asked the magistrate to gaol him; he did so, and I saw the boy doing his morning's work, carrying a bag of rice of 50 lb weight round and round the compound.' Yet there was always some evidence that the many who chose to run were not irrational: the rate of desertion was highest where the tracks were toughest and sickness most common. Men who went to the goldfields ignorant of what work they would have to do or with false expectations were also most likely to desert. Wio and the other Rossel Islanders who went to the Northern Division in 1898 had originally agreed to go to Sudest where other Rossel men had worked for miners and storekeepers. At Sudest they were shipped to Samarai and from there to the Mambare. Ten years later the terror experienced by the men on the Mambare was well-remembered on Rossel. Owen Turner, Acting Resident Magistrate at Samarai, spoke to a man from Milne Bay who had just signed-on to work at Buna Bay; he was certain that he had not agreed to go near the Kumusi. Since the new road to the Yodda had been opened the recruiters had another name free of old associations to offer village men. Or they lied, telling the recruits that they were going to plant coconuts near Milne Bay and then inducing them to sign-on to go to the goldfields when they reached Samarai. In strange surroundings and seeing the recruiter as the only man who knew their home and how to get them there, they agreed to any suggestion. After talking to the crewmen on Edward Auerback's boat and questioning many of his recruits, Campbell decided that Auerback had given his crew arms to impress and intimidate the villagers and instructed them to deceive the recruits about where they would have to work. Campbell thought it a 'moral certainty' that the men recruited by Auerback would desert immediately they reached the Northern Division. He fined Auerback and sentenced him to one month's gaol.

The miners' dependence on indentured labourers and the reluctance of men to go to the Northern Division was a situation equally open to exploitation by knowledgeable villagers. Nicolas of Fergusson Island schooled men to take the ten sticks of tobacco offered as an inducement to go to the Mambare and then swim ashore. He took five sticks from each man for managing the deception. After investigating violations of the Native Labour Ordinance, Campbell decided that he was dealing with 'many scoundrels amongst the natives' and a 'few very blackguardly alleged white men'.

Each boat bringing government mail from the Northern Division to Samarai brought 'as usual a sheaf of "Notices of Death"' of indentured labourers. Officers on patrol paid the next of kin trade goods equal in value to the wages due to the dead men. The payments helped communities tolerate the high death rate; they could release the recruiter from responsibility for the labourer's death in the same way that a clan member guilty of injuring or killing a man could sometimes avoid retaliation against his group by paying compensation. But government officers were frequently asked for news of men who had left the village and not returned; and the officers could find no mention of the men in the labour records. The gaps in the records were a sign of the loose supervision of recruiting during the early years on the Yodda and Gira.

Recruiters' agents were in danger when they entered the villages of men who had died while away. The people on the south coast of Fergusson Island killed a Suau man working for a recruiter in 1904. In their defence the Fergusson islanders said that three men had left their village for the goldfields, and had not returned, and no more would go. Two years later the people of a neighbouring village killed Inade who was working for James Swanson. Inade came ashore with Wai-iupa in a dinghy and joined a group of men talking near the beach. After a while Inade asked if anyone wanted to go mining. According to a returned labourer one of the group then jumped up and shouted, 'no boy he go work Kumusi two boy belong me he go work Kumusi long time now, he no come back I no been get pay now I make you pay'. Rushing forward he tomahawked Inade. Wai-iupa fled as the group attacked the fallen Inade. The body was cooked and pieces distributed so that others would share the victory and unite against any attempt to inflict punishment.

Yet even when the death and desertion rates were high most men went willingly to the Northern Division and some went with a clear knowledge of the work and the conditions. By at least 1903 men from both the Eastern and Western Divisions were signing-on to go to the Mambare for the second time. There were even cases of men, rejected as being physically unfit, stowing-away so that they could go with their age-mates. Miners who recruited their own teams in the villages told them where they would be going and what they would be doing, paid them off in Samarai, made sure that they got value for their money in the stores, and returned with them to the villages, had little trouble persuading other men to try another year on the field. Often they could use their ex-labourers as unofficial agents.

In answer to the question, 'What all something you been get along work for six months?' Mai of Rossel Island listed his goods: one tomahawk, one axe, one sixteen-inch knife, one ten-inch knife, one blanket, one dozen matches, one small mirror, one wooden pipe, two pounds of tobacco, one shirt and two strings of beads. He had lost a few articles because the recruiter had returned him only to Sudest and there he had to pay men to take him home by canoe. The Rossel Islanders decided that they would never again go to the Mambare, but other communities decided that the rewards justified the risks. Villagers had come to depend on some manufactured goods, but they knew of few ways to get them: individual young men went away to acquire wealth, demonstrate their manhood, and see distant places. Men died on the goldfields; but some recruits left villages where men were still killed in warfare and all left villages which suffered periodic epidemics.

Early recruits for the Northern Division came from near the Fly River estuary, the eastern mainland and the islands. In 1904 and 1905 men from the Gulf of Papua agreed to go to the goldfields, but they quickly decided that carrying and labouring for the miners was not the way of life they desired; they broke their contracts and were inclined to be aggressive when instructed to be more diligent. The communities which supplied most of the men to work underground at Woodlark and Misima were the ones most ready to send men to the mainland fields. In 1903-4 over 600 men from the Eastern Division signed-on to go to the Northern Division, and most of those men were picked up by schooners moving about the bays and straits of the D'Entrecasteaux Islands.

While the labourers were the most numerous 'foreigners' on the goldfields and the miners and government officers were the nominal commanders, to the Orokaiva the police may have been the most distinguished, feared and comprehensible. In 1900 there were about forty men in the Northern Division wearing the blue serge jumper and red-braided *rami* of the Armed Native Constabulary: by 1908 fifty-five were attached to the stations at Buna, Kokoda and Ioma. When they had first gone to the Mambare most had carried Snider rifles, but in 1901 they began to be issued with the new Martini-Enfield and bayonet. Men were keen to join the Armed Constabulary. When the Anglican Bishop, Montague Stone-Wigg, accompanied Monckton to the Yodda in 1906, a man from Gona was prominent among the carriers. Monckton suggested that he might like to join the police. Excited at the prospect but appalled at Monckton's obtuseness the man replied, 'I have carried for you up the track and down the track. Now you wake up, and offer me the clothes!' To collect their dark blue uniforms, recruits were sent to Port Moresby where they did their preliminary training. Having come before the Commandant of Police as the 'rawest of savages' they did a brief course in discipline, weapon training and patrol work before being sent to the out-stations. When the force was being built up quickly some recruits were given only one month's training at headquarters, and they left Port Moresby 'far from finished'. Many did not stay in the force beyond the contract period of two years. Praised for their physical endurance, bravery and loyalty, the police were also keen to shoot and loot. Under officers who could not, or would not, restrain them, they could be a ruthless force.

The nature of the country, the work the police were asked to do and the knowledge they possessed frequently put them beyond the supervision of white officers. The police alone manned the base at Mambare beach and in the absence of white officers they maintained other stations for long periods. On Armit's patrol to the Yodda in 1900 the police often acted independently. On 7 February Armit wrote, 'Today, not feeling over well, I sent Sergeant Tomu and six police to pay a few domiciliary visits. They returned at noon bringing in two flies, one long-handle shovel, and an adze.' When the patrol left the Kumusi and crossed to the west of the Mambare Armit 'sent the police out ... to beat up the quarters of the natives who attacked us so wantonly yesterday. They came into collision in a gorge 1,000 feet above the camp, and six of the Twidians were hurt'. (Armit later admitted that 'hurt' meant 'killed'.) Two days later, 'The police while patrolling towards the gap [near Kokoda] were suddenly attacked by six natives four of whom came to grief. Armit could not stop the police from looting. 'They glory in it' he wrote, 'and it is astonishing what awful rubbish they treasure'. Armit's assurance to Le Hunte, 'I always lead my men', was either figurative or untrue.

The surviving monthly journals from Tamata, Bogi, Papaki and Kokoda show that over the next five years the police continued to act independently on short patrols, bringing up stores, supervising the cutting and improving of tracks, and taking messages to government officers, miners and villagers. In 1905 Monckton showed that there was a limit to the use of unsupervised police. At Kokoda he found that a troop of police had been sent on a night raid to a village to recover stolen firearms. Night operations, said Monckton, should only be made by picked members of the constabulary under the supervision of white officers.

Stone-Wigg was worried about the morality of the police. He told Robinson that married policemen should take their wives to government stations and 'it should be laid upon [single policemen] to keep pure'. When Robinson said that he approved of 'temporary liaisons' the Bishop replied that 'Self-control was the great lesson of life'. Apparently it was a difficult lesson to teach. John Stuart-Russell reported that he could not stop his police from procuring women, and he attempted to strengthen self-control by flogging one time-expired constable. As Chief Judicial Officer, Robinson could not approve all 'liaisons'. On his first visit to Papaki he gaoled three constables for two years for rape, but he dismissed most of the charges of procuring women and terrorising the villagers. When the Papaki people were assembled (presumably by the police) they told Robinson they had no complaints. Some officers were even more tolerant. In 1905 Monckton found that the police at the Kumusi crossing had been given permission 'to obtain and keep women there'. He thought this merely an 'error of judgement' by the officer.

It was easy for the police to abuse their powers. On Robinson's first patrol through the Northern Division, Bruce left a troop of police at Papaki to strengthen the local detachment. When Robinson returned to the Division five months later he heard charges against four of them for murder. Corporal Kaio confessed that the Papaki people had offered him a wife so he had agreed to assist them in an attack on the Wasida Orokaiva who lived on the northern slopes of Mount Lamington. The raiders killed fifteen Wasida, whose bodies were eaten by the Papaki people. Corporal Kaio was sentenced to seven years' imprisonment and the three constables were imprisoned for two and three years. In another case a month later Corporal Emanboga was supervising the movement of stores from the mouth of the Kumusi to Bogi. Corporal Bakeke reported that Emanboga had taken food from the stores and failed to pay his crew all the tobacco due to them. Later Elliott learnt that Emanboga had intimidated the villagers along the Kumusi either by waving a piece of paper which he said gave him powers from the government or by threatening to shoot people. At Batow he demanded a woman. When he did not get one, he asked for a pig, and finally he settled for head-dresses, betel and coconuts.

Elliott decided there was no use putting 'this Boy in gaol as he [had] been in to [sic] often'. He had in fact worked in the prison gang sent to Sudest in 1897 to build a road from the coast to the mine on Mount Adelaide Reef.

While police from other divisions could use their position for their personal advantage, once most of the police were recruited locally they had other motives. By 1906 forty-four of the fifty-three constables and six of the nine non-commissioned officers came from the Northern and North-Eastern Divisions. Their knowledge of the languages spoken by villagers and of local alliances and enmities gave them opportunities to manipulate government power to the advantage of particular groups. On the Mambare the men who joined the police were personally engaged in clan fighting before and after the arrival of the foreigners; in the area from Buna to Yodda most of the police were less immediately involved in local conflicts. But they alone knew the clans, alliances and boundaries; they alone could talk with fight leaders; and they told the government officers what they wanted them to know about the fighting in an area and the tracks patrols should use. Acting as forward scouts they could determine peace or war without the knowledge of the government officer. John Waiko has pointed out that in the clan histories which he has collected wars are remembered as being decided by negotiating and fighting between clans and the police; white officers are rarely mentioned.

By the time the Buna road was being built some Binandere, the first of the Northern Division people to join the police, were non-commissioned officers: they used the police not merely to protect the interests of clans along the Mambare but to re-establish that broader power and prestige of the Binandere lost in the fighting between clans and against the foreigners after the killing of Clark in 1895. At least twice Binandere policemen took the names of men they shot in distant parts of the division and gave them to members of their own families. It was a continuation of an old practice in which men appropriated the names of slain enemies, and an indication of the way the Binandere *polisimani* continued to behave as Binandere warriors.

Five years after the early rush to the Yodda, when there were about 100 white miners in the Northern Division, the diggers had become a stable, if wandering, group. Men shifted frequently from one field to another, left to 'spell' and recruit new teams, but few abandoned the goldfields altogether and there was little excitement to attract new hands to the area. The Yodda, Bellamy wrote, was 'a somewhat prosaic field … . It has been a steady-going field with few surprises, few fortunes, just a steady income for those who worked there …'. The miners knew one another: they met in the stores, in Clunn's, the Cosmopolitan and Billy the Cook's hotels in Samarai, and on recruiting trips around the islands; they waited together for boats and better weather; they drank together, played cards (billiards in Samarai), watched their dogs fight and shot the tops off bottles floating down the river. Many of them were over forty. They had followed alluvial strikes across salt pans and sand in Western Australia, through the rain forests of north Queensland and two at least had looked for a big strike on the Yukon in northern Canada. They came from varied backgrounds. One, MacKay said, 'held a good position in Australia before he took up life in Papua; another was a rector's son …'. Frank Rochfort, said Monckton, was 'a born agitator and trouble maker of the de Valera class'. Barton, in more moderate language, agreed that Rochfort was a trouble maker but added that he was an 'exceptionally well educated man'. Others found the pen more difficult to use than the pick and the pan. John Higginson recalled inspecting a claim and being handed a piece of paper on which a labourer had written, 'Tommy, too much he fight'. The labourer had a means of communicating with the government officer without the knowledge of his master; for 'Tommy' could neither read nor write. On this occasion the labourer's skill, acquired at the Methodist mission station on Dobu, won him no reward and enraged Tommy.

Hubert Murray, who first went to the Mambare as 'Judge Murray' in November 1904, thought that the miners were 'very law-abiding'. Stone-Wigg supported his opinion:

> I must confess myself an admirer of the digger, with all his faults. At first sight he seems nothing but a dirty swearing, drinking, spitting animal — on the higher side he is much enduring, generous-hearted, kindly and as a conversationalist both interesting and instructive. Again I noticed how well their carriers looked, and how kindly was the relation between the two classes. I must acknowledge that in this I was agreeably surprised.

The Bishop's diaries give brief descriptions of the incidents on which he based his assessment of the miners. On 30 June the English prelate, educated at Winchester College and Oxford University, recorded a meeting in the disused store on the Gira field:

> H. Osbourne & J. Ward there. Aft. J. Foley and T. Kelly came over fr. their camps. All very friendly & insisted on giving me gold, as Ivory and O'Brien had already done. After tea had some hymns. Neither Ward nor Ivory knew a single one, either the words or the tune. Finished with Bible reading and prayer. Ward has been drover, shearer, miner. Strong labour man. All quote the 'Bulletin' as tho' it were Bible. Laughed a good deal at night.

Disputed claims were infrequent and gold thieving rare. Higginson, who served as Assistant Resident Magistrate at Tamata, wrote in the *Lone Hand*:

> Isolated as they were, in one of the most God-forgotten spots on earth, the miners always supported and gave unquestioning obedience to an authority that must sometimes have seemed unnecessary and vexatious, an authority which had nothing but the feeling of the community to support it in reality.

In his capacity as warden of the goldfields for 1904, Higginson was called to arbitrate on only one case of claim-jumping; and the diggers imposed their own sentence by forcing the guilty man to leave the field. Away from centres of government administration and living according to a code accepted by all diggers, many men did not register their claims; but they did argue about water, and prudent men secured their access to creeks. The frequency of disputes in dry spells and the passing of 'personal animosities' when the rains fell gave rise to the common Yodda comment, 'The more rain the less law'.

But Murray, Stone-Wigg and Higginson knew that there were scars on their portrait of the humane, generous, uncouth digger. In May 1904 Higginson reported that Harry Edmunds, a miner on the Gira, had come to Tamata to report that he had killed Mabie, a labourer from the Bamu River. After taking evidence from two other miners Higginson charged Edmunds with 'murder under provocation' and released him on bail. A month later three of Edmund's labourers reported that he had shot Hakaia, an Orokolo labourer. While Higginson was preparing to go to the Gira, Edmunds arrived at Tamata where Higginson charged him with 'wilful murder'. Two days later Stone-Wigg spoke to Edmunds at Tamata gaol. Afterwards he visited Edmunds's camp, talked to the miners who had given evidence leading to the first charge, and 'saw graves of the two natives [Edmunds] had shot ... one accidentally, one apparently needlessly'. In November Murray heard the two charges against Edmunds; he found him not guilty on the first charge, guilty of manslaughter on the second, and sentenced him to one year's gaol with hard labour. From his arrival in British New Guinea in 1904 until he gave evidence before the Royal Commission in 1906, Murray heard only two other cases against Europeans on the goldfields. He found Peter Lawson not guilty of shooting John O'Toole with intent to cause grievous bodily harm, and he sentenced Steve Woolf to death for the murder of Manewa, a decision which the Executive Council commuted to three years' imprisonment. Murray had also tried one labourer who shot a miner and another who was charged with putting poison in his master's tea.

These cases were one extreme of relations between miners and labourers on the field. At the other extreme was Joe Faulkner, an old miner working at Prospect Creek on the Yodda. He treated his team with gentlemanly consideration: 'Won't you have another cup of tea?' 'Let me cut you some more bread and butter.' A bemused Hubert Murray thought that Faulkner's men worked well for him. But most labourers learnt a tougher code of relationships. While they were working on the Waria Jim Pryke wrote a note to Arthur Darling: 'I owe you an apology for helping one of your boys to rise with the boot. It was certainly a breach of the ordinance which gives every man the sacred right to "boot his own nigger".' The diggers had added to the unwritten law which they brought from Australia, for in New Guinea the miner was also a master. Rochfort, who mined on Woodlark before shifting to the Northern Division, explained his success as an employer to the Royal Commission of 1906:

> In working boys there are three principal rules I work on —
> 1st. I see that they are well fed, and have a good camp.
> 2nd. I don't nag, am not brutal, nor do I expect them to work like machines.
> 3rd. I allow no familiarities. Let the boys respect you, and when you give an order see that it is carried out.

He claimed that apart from one gang of Orokolos who cleared out in a group no man had deserted from his service. The Commissioners found themselves 'entirely in accord' with Rochfort's views, and decided that if the Papuans were treated according to his three principles they would 'gradually own towards the higher race an affectionate respect'. There were cases of mutual respect, even affection, between miners and their leading hands: on the field miners and labourers worked together to defend and feed themselves, and the miners knew that their chances of securing more recruits depended on returning satisfied men to their homes. For many miners the ideal relationship was strong master and docile boy: it was an ideal tempered by self-interest and shared experience in isolated and dangerous country; it could also condone sinking the boot into a labourer thought to be lazy or cheeky.

The shared experience could be the capture — by strength, bribery, magic or guile — of village women. No miner legally married an Orokaiva woman, and Sloane and Ericksen, who lived with two Waria women for five months, abandoned them when they heard of a new rush on the south coast. But some miners and labourers were eager for quick pleasure. Opportunities were few on the Gira and Aikora but sometimes they were there to be taken on the Mambare and sections of the Yodda track. After Stone-Wigg publicly criticised the miners for 'loose living', King called a meeting of all white miners at Tamata in November 1900 to defend his Bishop and reduce the miners' hostility towards the missionaries. The thirteen men who gathered in the cookhouse of Whittens' store to argue with King and the Reverend Frederick Ramsay made a rare public declaration of their sexual morality.

King told the meeting that on the night before a miner had pursued a prostitute near the mission station, but that she had sworn at him and sent him away. The miners accepted that King's description of events was probably correct; such things happened. When King asked whether they approved of the miner's actions, one man said he did not and all the others thought no harm was done if the miner did not use force. Some added that they did not have sexual intercourse with village women but they did not condemn those who did. In a sharp debate the miners then accused the missionaries of condemning all miners for doing what some scorned to do; declared that the village women saw no wrong in such things; asked why the missionaries did not criticise the government officers who lived with local girls; and, arguing that it was not in man's nature to be pure, challenged the missionaries to name six men who 'had not had connection with women before marriage'. When the miners became more aggressive King said that his colleague would accommodate anyone who wanted to fight. Ramsay, who was known to combine 'calisthenics with religious and secular education', doubted his capacity to handle the miners either in a group or in succession, but the miners chose to battle in words. King countered with the charge that the storekeeper had offered a 'large sum' to break the 'taboos' of the Mambare women, and that if the women did not know right from wrong the woman of the previous night had chosen to spurn the miner. Joe Sloane then admitted that he was the miner involved; he had been misled into believing that getting a woman was 'an easy matter'. Whitten declared that of course the missionaries and miners would never agree about women, and the meeting of frank testimony ended with Sago Bob Harrison making a witty remark which the missionaries chose not to write down. Among the children of Papuan mothers and foreign fathers gathered at Ganuganuana by the Anglican missionaries were several who were the sons or daughters of miners; nearly all the mothers were from the south-east or the islands.

At Papaki where there were no missionaries to question the morality of miners, officials or police, Monckton found 'shameless prostitution' of women. Halkett Parke, arriving unexpectedly at the police barracks in the evening, saw the women 'clearing off into the bush'. At other times the police at Papaki were charged with rape and abduction, and the villagers were accused of pandering. Four Yodda miners took a 'holiday' at Papaki over Christmas 1904, and the next year Monckton ordered the removal of two Papaki women from William Parkes's camp on the Yodda. But from the limited evidence available it seems that the opportunities for sexual adventure were fewer for miners and labourers than for the police.

Miners complained to government officers about the government's failure to protect them, the lack of roads, and the labour laws. William Durietz in 1906 suggested a plan to curb troublesome 'boys' on the Gira:

> There is only one way of doing it, and that is to chase them out of the district — disperse them!
> Meaning?
> Well, I mean they should be fired at, or something like that.

To those miners who passed through the lands of perhaps 10,000 people to reach their claims and were then dependent on local trade to stay in the area, Durietz's policy was neither just, practicable nor desirable. Most other miners demanded that the government be more effective, but apart from pointing to particular officers who were lazy or incompetent they could not say how the government could provide quickly cheap labour, roads and safety.

The miners thought that labourers were too hard to get, too hard to control, too expensive, and on the goldfields too briefly. They wanted the government regulations amended so that recruits could re-engage at the end of their contract without returning to their villages, the time during which a man deserted was not counted as part of his contract time, the contract did not begin when a recruit left the village but when he signed-on, the cost of stolen property was deducted from wages, and employers could cancel the contracts of unsuitable men without too much trouble. To recruit, feed and pay a labourer on the Yodda cost about £60 a year in 1904. As the average miner had at least five labourers and he was paying £1 16s. for a bag of rice, 1s. 6d. for a tin of meat for his own sustenance and £1 for a bottle of brandy for pleasure at the Yodda store, he needed more than 100 ounces to cover his costs. By agreeing to set the minimum wage for miners' labourers at 10s. a month, Robinson effectively reduced wages to that level for all except 'leading boys'. For a miner who previously may have been paying £1 a month for ten men the saving was significant. Many other specific grievances of the miners were met by the Native Labour Ordinance of 1906, but at the same time other regulations were introduced to protect recruits from abuse. Yet even as the legislation stood in 1909 only the most gross cases of physical discomfort were illegal. Before a man could be recruited for mining or carrying in the Northern Division he had to be at least 31 inches around the chest, he could sign-on for eighteen months only (although most men still chose to sign-on for one year), and he had to be issued with two yards of calico, swag slings and four loin cloths every twelve months. He was not to carry more than 50 pounds (plus his food and calico), or if he had to travel more than 12 miles in a day his load was not to exceed 30 pounds; and if he was carrying regularly on the Buna-Yodda or Tamata-Aikora tracks he was to have thirty-six hours rest between trips. The labourers on the claims were not to work more than fifty hours a week. At about a halfpenny an hour, it could still be a hard life.

Writing in 1908 to Atlee Hunt, the Permanent Secretary of the Department of External Affairs, Monckton was pleased to be able to say that the government officers seemed to be neither more efficient nor united after getting rid of the 'effete scions of aristocracy'. Before he retired to New Zealand where he found contentment looking after his hunters, bulldogs, dachshunds and prize pigs, Monckton had been placed by his opponents amongst those officers who were British, pretentious and unpractical. But in his books Monckton presented himself as a man of action impatient with a niggardly bureaucracy and those officers unable to make quick decisions. He assumed the ideals of the English sporting gentleman; he did not like to take advantage of those unable to defend themselves, but there were times when the natives were all the better for a thrashing. He formed closer relationships with New Guineans and he spoke more generously of their abilities than most Europeans of his time. At all times Monckton cast himself as the master with absolute powers; just, firm and terrible in righteous anger. To his police and to the Kaili Kaili and Binandere people he was 'The Man'; they gave him their unquestioning loyalty, they would die for him. While they were surrounded by 'hundreds of stalwart natives' near Kokoda, Monckton told MacKay:

> Nothing can happen to me in this part of New Guinea until these ten men [of Monckton's personal police guard] have been killed, and if they and I were killed, the news would spread fast, and from the German frontier to Cape Vogel the tribes would march to take a bloody vengeance … . That small escort of mine is but the point of many a thousand spears, and the people know it.

Monckton was pleased that his police would interrupt receptions at mission stations and Government House to check that The Man was safe. Several times he reminded his readers that a Kaili Kaili or Binandere attendant slept close to his door. Monckton's picture of himself as the masterful governor and his police as brave, faithful, efficient servants nourished his own sense of importance; but it was inaccurate. While Monckton was active and aggressive, his police were not mere instruments; sometimes they manipulated him (and others) to their advantage.

Griffin, less energetic than Monckton and more inclined to expect preferment by calling on his background of being an old boy of Harrow school, an officer in the British Army and habitué of the best clubs, could more aptly be criticised as an effete Englishman. To his credit he was less violent than Monckton, and he worked patiently to improve relations between some peoples and the government in the Northern Division. To his discredit he was involved in the sale of bird of paradise feathers while still a government officer, and the virtuous tone of his reminiscences was undisturbed by any mention of the children he had fathered in Papua.

Opponents of British' officers praised Australians, who were thought to be tough, practical bushmen. Kenneth MacKay praised John Higginson. 'Young Australians of this type', said MacKay, 'are the men to send to Papua where an ounce of practical experience of how to make the best of things is worth a ton of theory.' Monckton was less fulsome in his praise:

> J.B. Higginson was a man who had served with the Australian forces in the Boer War, with, I believe, distinction, and should have been a very capable officer; but unfortunately he had returned with a fixed idea that any Englishman was firstly a fool, and secondly an awful snob, and generally inferior to Australians. In fact according to his idea, any Australian in six weeks' time would be capable of replacing the admiral in command of the Atlantic fleet, or taking over the See of Canterbury from the Archbishop; and he really believed these things.

In a review and in the discussion which followed the publication of Monckton's first book Higginson had made his public comment on the author: he was an efficient officer but his 'flamboyant posturing [had] made him, in his day, the laughing-stock of his contemporaries in Papua, official and otherwise'.

In fact the distinction between 'Australian' and 'British' officers was forced. It was partly imposed by men concerned with the political change from British New Guinea to Australian Papua, a change long discussed in the Australian federal parliament and formally announced in Port Moresby in 1906. Australian nationalists wanted Australian officials, and they invested them with special qualities and stressed their differences from British officers. The distinction was forced too when applied to particular men. Elliott, an Australian, was not like the literate, practical Higginson. Except when disabled by malaria or squashed testicles after he fell legs-astride a log bridge, Elliott was active enough but he was deeply conscious of his own inadequacies once he entered the office. Unable to spell or mark the beginning and end of sentences, he was also reprimanded for his choice of words. Robinson told him not to use 'sporting parlance' when he wrote about killing and wounding, or to advocate in his official reports the kicking of village constables so that they would be unable to sit down for a week. Illegal acts which other officers glossed over or used to present themselves as forthright men of quick decision, became brutal and thoughtless in Elliott's awkward prose.

Other officers were neither 'British' nor 'Australian'. William Armit was born in Belgium in 1848 and probably trained as a soldier before migrating to Australia. A tough, experienced bushman, writer of 'first class alligator and nigger lies', naturalist and dismissed officer of the Queensland Native Police, he was a man of many talents and wild irresponsibility. After his death Murray heard stories that he had been naked and drunk while he drilled his police and had used crucified captives for target practice. Rayner Bellamy was certainly 'British'. Having left Cambridge and Edinburgh Universities without taking his final examinations in medicine, Bellamy tried acting as a career and then went to New Zealand where he worked as a journalist. After joining the British New Guinea government service in 1904 he was always about to return to England to complete his medical studies, but did not in fact do so until 1917 when he was on leave from service with the Australian Imperial Force in France. Universally liked, Bellamy was praised for reducing the death rate among the labourers on the Yodda. The miners asked him to stay on the Yodda as a doctor. When he said that he was not fully qualified they replied, 'half a medical man was better than none at all'. At the end of six months' service in the Kokoda area Bellamy made a careful assessment of the Orokaiva people and government policy. It was, he said, a mistake to treat the Papuans as though they possessed 'all the cardinal advantages of a

20th century intelligence'. They regarded kindness as weakness, they were to be treated firmly, and they had to learn that all disobedience would inevitably bring punishment. At the same time he wanted to dissociate himself from 'the attitude expressed so forcibly in "bloody nigger"'. Bellamy was humane and paternalistic; he believed he had the knowledge and a duty to chastise, protect and improve Papuans. He chose to stay in the Australian government service, spending much of his time until his retirement in 1937 working to reduce venereal disease in the Trobriand Islands.

The actions of Joe O'Brien caused miners and government officials to declare their beliefs about the proper behaviour of white men on the goldfields, and briefly awoke a wider audience to events in the Northern Division of British New Guinea. Little is known about O'Brien's early life. A Queenslander by birth, he was probably about twenty-five years old in 1905. According to Griffin, who was his gaoler for a month, O'Brien was a 'fine looking man'. For a number of years he worked on the Gira and Yodda Goldfields where he acquired a reputation for his bushmanship and his intense hatred of Papuans. When his fellow miners wanted to praise him, they recalled that he was 'one of the foremost to give his services in the protection of this small community from the depredations of surrounding natives, and assisting to avenge some of the brutal murders that [had] occurred'. And when David Rennie, a miner suffering from *delirium tremens*, was lost in the bush O'Brien found him and brought him back after others had abandoned the search. By 1903 O'Brien's other activities had provoked Hislop, the Resident Magistrate at Tamata, to boast that he 'would make it hot for O'Brien'. Having failed to make it hot for anyone, Hislop was asked to resign. In 1904 O'Brien gave some gold to Bishop Stone-Wigg and later was arrested and charged by Charles Higginson with assaulting Ingala, the village constable of Kuma, who had attempted to stop him seizing a village woman. Three of the witnesses from Kuma also claimed that O'Brien had threatened Ingala with a rifle. Having heard the evidence of four Kuma men, Higginson found O'Brien guilty of assault and fined him £5. O'Brien also appeared in court as a witness for the Crown against Harry Edmunds, who had shot two of his labourers. O'Brien was not permanently attracted to His Majesty's side. Already there were rumours among the miners that O'Brien had captured a deserter from Jim Wallace, appropriated some stolen sovereigns carried by the man, knocked him on the head with a tomahawk, and thrown the body into the Gira. Summonsed for debt by Whittens, O'Brien paid nothing and shifted to the Yodda. There he asked for time and labour, but received neither. Rochfort's Orokolo labourers, briefly under O'Brien's control, deserted at Buna, stole the government whaleboat and were eventually picked up at Cape Nelson. The only debt O'Brien paid was two tomahawks and two pounds of tobacco to the people of Buna; compensation for the pigs he shot when they refused to carry for him. With some assistance from Griffin and Bellamy, O'Brien, barefoot, carried his own swag back to the Yodda where by threatening violence and asking for another chance he tried to evade Whittens' summonses for debt. In April 1905 his attempts to gain time ended.

Shortly after the event, Monckton and Bellamy gave five accounts of their arrest of O'Brien: both described the incident to the Royal Commissioners, Bellamy wrote of it in the Kokoda station journal, and Monckton made a report to Barton and kept his own journal. All are in basic agreement. On 15 April Constable Motu, sent to obtain carriers, returned with five 'weedy' men; all others refused to come because they wished to attend a dance. Corporal Bia and five policemen equipped with handcuffs then went out and brought in enough carriers. The next morning, a Sunday, Monckton, Bellamy, eight police and the carriers left Kokoda station for the Yodda. Most of the miners were out shooting or resting at the stores where they could talk, drink, play cards and listen to the gramophone. Bellamy knew O'Brien and had visited him on the Two Mile only a week earlier. When they arrived at his camp, O'Brien was either reading or asleep and did not see Monckton and Bellamy enter his hut. He offered no resistance as Monckton arrested him for assault; in fact he told Monckton he would have gone to Kokoda on a summons, and pointing to the police he said, 'You need not have brought those black b.........s'. Having heard the general talk that O'Brien intended to shoot the first man who tried to take him in, Bellamy was surprised at the ease of the arrest. They took O'Brien to a camp on the Yodda where Bellamy, under Monckton's instructions, put leg irons on him for the night. It was the act, said O'Brien, of a person who 'had no respect for a white man'. The next day O'Brien was put in Kokoda gaol.

After hearing three cases against O'Brien on 18 April, Monckton found him guilty of assaulting a policeman and a carrier, and dismissed the third charge. Monckton had accepted the testimony of several Papuan witnesses in the face of O'Brien's denials or explanations. O'Brien was sentenced to two months' hard labour for both offences and informed that at the next sitting of the Central Court he would face charges of:

Robbery of gold to the amount of £1,000 [from Whittens' Kumusi store]

Shooting with intent to murder, two charges; Arson, two charges;

Rape;

Unlawfully destroying dogs and pigs, the property of natives, six charges.

When witnesses were brought from distant areas, O'Brien was to be charged with murder and officers were to investigate complaints by the managers of Whittens' stores that O'Brien had been continually threatening to kill them.

In spite of O'Brien's claim that he would die if sent to work in the sun, Monckton instructed Griffin that O'Brien, guarded by two members of the Armed Constabulary, was to clear scrub. But when O'Brien's hands were blistered he worked inside copying out forms where he was 'very useful as his writing [was] neat'. One month after entering Kokoda gaol O'Brien spent the day transcribing labour forms. On the next day, 18 May, Griffin sent him back to cutting scrub under the supervision of one policeman, Griffin's old orderly, Constable Dambia. The rest of the thirteen police on the station were either supervising Orokaiva labourers in another section of the garden, improving tracks, sick or under arrest. Soon after starting work O'Brien hit Dambia several times on the head with a tomahawk, and taking Dambia's rifle but no ammunition, disappeared into the bush. From J. Aitcheson's camp, one of the first on the Yodda road, O'Brien 'took forcible possession' of a revolver and ammunition. That night he visited Rochfort's camp where he made a 'statement':

> I was treated like a dog at Kokoda gaol. They gave me bad taro and dirty unwashed rice to eat. They made me work all day, and put black police over me with rifles.
>
> They allowed the black police to walk around the gaol at night and call me filthy names, and I had to submit to such indignities that I would rather lose my life than stand it any longer. The black policeman told me to work plenty. I struck at him with my hand, my foot happened to slip, and he struck me in the forehead with a tomahawk. Then I knocked him down and cleared with the Government rifle. I would go back, but I believe they have natives retained to swear away my liberty, and that I would have no chance of getting a fair trial. I am fully armed, and will shoot any man who tries to arrest me, but otherwise will injure nobody.

Rochfort said he tried to persuade O'Brien to return but he 'preferred death than suffer the indignities that he had to undergo under the native constabulary'. After Rochfort dressed his head wound, O'Brien 'departed in the night'.

Dazed and bleeding profusely from head wounds, Dambia eventually attracted the attention of two constables. Sergeant Barigi roused Griffin from the office and he sent armed police out to search for O'Brien. After attending Dambia's wounds, Griffin sent a letter to Monckton at Tamata and messages to the villagers telling them that if they saw O'Brien they were to report to Kokoda and if 'he shot at them they were to spear him'. Two police took notices to Yodda to display at the stores telling the miners that they were 'perfectly justified' in shooting O'Brien if he failed to stand or go in front of them to the police.

Rochfort wrote to Griffin telling him that O'Brien had visited his camp and left without saying where he was going. Co-operation between some miners and government officials ended at that point. On 2 June Griffin went to the camp of B. Reynolds and G. Arnold on Klondike Creek. Arnold's cook, Lavai-i, told one of the police that Arnold had fed O'Brien for several days and Griffin heard that Arnold had threatened to kill another employee who helped the police. Reynolds declared that if O'Brien asked for anything he would get it, not because he sympathised with O'Brien but because O'Brien would shoot anyone who refused his demands. Monckton, after investigating Lavai-i's statement, decided that although Arnold's behaviour was suspicious there was no clear evidence that he had assisted O'Brien. Government officers kept a watch on tracks leaving the goldfields and on shipping on the north coast. Among the miners there were rumours that O'Brien had been seen in Yule Island, German New Guinea, Brisbane and Singapore; but nothing definite was ever known and he may have died in the Northern Division.

Some miners, already resenting the fact that O'Brien had been guarded by the police, objected when Griffin sent armed police to hunt him. Les Joubert, Clunas and Clark's storekeeper at the Yodda, wrote a note to Rochfort:

> The A.R.M. is not satisfied in putting us under black martial law. One of the black police actually sent word by my cook last night saying that no white man should go out at night for fear they would be shot in mistake for O'Brien. I sent up and advised them in most emphatic way that on no account were they to bang bullets about in this vicinity, unless they knew what they were shooting at. It seems that the police are getting as good as their boss at issuing proclamations.

At the store the next day twenty-three miners decided to tell the Minister for External Affairs of their grievances. They objected to a white man being forced to work with 'a black policeman as boss standing over him with a rifle'. The policeman, 'the naked savage of yesterday', was as good or perhaps better than could be expected but 'it acts like a dagger in the heart of a white man when he knows that the poor ignorant savage is placed in authority over him by his own fellow white men'. The miners said that the evidence of 'every black witness' was totally unreliable and any white man charged with a crime should only be brought before a magistrate 'in the hearing of white men'. The letter was signed 'on behalf of the miners' by Martin Gallagher and Dan Horan.

Barton forwarded the miners' grievances to the Governor-General with the assurance that the 'better class miners' held a 'very different opinion'. O'Brien, said Barton, was a troublesome criminal who would not have been permitted to stay on a goldfield by miners in a more civilised country, but the Yodda miners were afraid of him. Not infrequently, Barton observed, one met miners like O'Brien, men who expressed 'the utmost repugnance for the natives' yet had 'the least reluctance in making use of native women to satisfy [their] sexual appetite'. Barton enclosed details of O'Brien's appearances in court at Tamata and Kokoda, the record of an inquiry into his escape and point by point a refutation of the miners' case written by Monckton in the prose he was later to use for dramatic effect in his books.

Monckton said that O'Brien had eaten the same food as Bellamy and Griffin and had been well-treated by the police. He agreed that

> 'black savages' be not placed in authority over white prisoners. I hardly think, however, by this the miners can mean the members of the Armed Native Constabulary, to whom the present security of life and property on the gold-field is entirely owing.

While he regretted that 'natives' did not always tell the truth in court, he thought they were not alone in this respect. Trained and experienced officers, he said, were better fitted to judge the value of evidence than miners 'by whom the evidence has not even been heard'. Having reminded His Excellency that Clunas and Clark's place of business was licensed to sell intoxicating liquor, Monckton supported his superior's belief that the miners' letter did not express opinions generally held on the field.

Barton forwarded two other statements from the Yodda miners to the Governor-General. The first was a letter from Rochfort to the miners advocating that they petition the government condemning the 'superabundant energy' officers displayed in hunting O'Brien, who was known to be guilty of only minor crimes, while the murderers of Jassiach and Blackenbury, killed three years before on the Gira, remained free. The second was a statement by 'residents of the Yodda' who did not condone the offences of, nor sympathise with, the escaped prisoner, O'Brien. It was signed by seventeen men, the first being William Little, member of the Legislative Council, and included were some of the best known miners on the field: Mark Royal, William Parkes, Matt Crowe and Hugh Clunas.

In August Senator Staniforth Smith of Western Australia gave a long and eloquent exposition of the O'Brien case in the Federal Parliament during which he was occasionally interrupted by the chairman attempting to maintain a quorum. Smith was unique among members of the Federal Parliament. He had travelled widely in British New Guinea and written to advertise its attractions and wealth. He had walked from Bogi to Kokoda, he was at Kokoda station while O'Brien was evading debt-collectors, and, prostrated by fever, he had been carried part of the way back to Buna by conscripted villagers. Smith objected to Griffin's notice at the Yodda stores which 'outlawed' O'Brien and pointed to the wider implications of O'Brien's treatment. 'The whole government', said Smith, '... rests on one word, "prestige".' Forcing O'Brien to cut scrub, 'an occupation which no one but a native ever undertakes' would 'revolutionize the views of natives in regard to the power and prestige of the white man [and make him] an object of contempt rather than of respect ... as he should be'.

In the brief debate which followed Thomas Playford of South Australia said that O'Brien was 'unmistakably a bad lot' and he had not been under the control of black gaolers. Smith replied that O'Brien may not have been the fiend described by Playford and as he escaped by attacking a black guard presumably there had been one. One Senator, John Gray of New South Wales, cut through the arguments by saying that he could not see 'why a man, even though he is not white, should not have the necessary power given to him to uphold the law'. No Senator rose to support him. The papers relating to the administration of justice in the case of Mr O'Brien, British New Guinea, were tabled, printed and forgotten.

Speaking of the man who was soon to be his rival for the Lieutenant-Governorship of Papua, Smith told the Senate that the miners respected Murray, and would accept a report from him. When asked for an opinion on Griffin's declaration that the miners had the right to shoot O'Brien if he resisted arrest, Murray decided that it had no basis in law.

By the time the two most vigorous Royal Commissioners reached the Yodda they found that the miners were not greatly interested in the events of eighteen months earlier. In their Report the Commissioners quoted a statement made by Horan in the presence of the miners: 'We [the miners] never said a word about the O'Brien affair until the man was outlawed by a notice posted by Mr Griffin, A.R.M., at the store ... Now it has been established there is no such law on the statute book they are satisfied.' The Commissioners decided that the government deserved to be censured on only one point; it had made too little effort to recapture O'Brien.

Two of the Royal Commission's general recommendations arose from the O'Brien case; that white police should be employed and juries should be introduced. The Commissioners noted the 'strong feeling of resentment with which the majority of the white population regarded possible arrest by one of the Armed Native Constabulary'. They accepted that government officers used the police against white men in exceptional cases only; but even these were to be avoided. The Commissioners re-stated Smith's argument that the superior prestige of white men was essential for good government and all white men could be reduced by the actions of one: 'no matter how little a particular white man may deserve the respect of the native, it is still necessary in the interests of all white men that the natives should not be put in a position where respect for the ruling race will be jeopardized'. While most miners accepted that a 'boy' sometimes responded best to 'physical argument', O'Brien's brutality had alienated nearly all the diggers on the northern fields. Remembering the incident forty years later, one of the Gira miners said that only Patrick Finnegan sided at all times with O'Brien. Finnegan's concern was tribal: 'Joe is no good, but his name is O'Brien — and we must respect the name.' O'Brien gained the support of about half the miners after his arrest: they thought that no white man, for his sake and for the sake of all white men, should be shackled, forced to work under black guards, and on his escape be a target for village spears and police guns. For all men at Kokoda O'Brien's treatment was indeed revolutionary. Yet many miners still chose not to defend O'Brien. Frank Pryke expressed their attitude:

> O'Brien left himself open for it and in fact had been looking for it some time previous to the affair and I don't think that I could manage to bring myself to sympathise very much with a man of the O'Brien stamp under any circumstances.

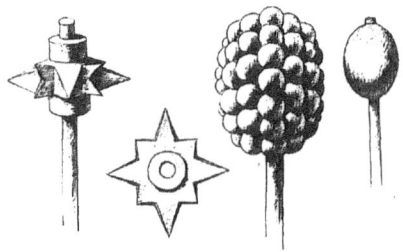

On Christmas day 1909 the police arrived at Kokoda with the overland mail from Port Moresby; in letters taken on to the Yodda the miners learnt that gold had been found on the Lakekamu River to the west of Port Moresby. The next day two miners and twenty-two labourers paused at Kokoda to transfer labour agreements before beginning the walk over the 'Gap' to Port Moresby. On the last day of the year a second group of eight miners, about 100 labourers and police guides left Kokoda for 'Port'. By the middle of January only two miners remained on the Yodda: the storekeepers had closed up and shifted their goods to Buna for shipment south. On the Gira miners and labourers gathered at Ioma, some waiting for a boat to Samarai and others preparing to walk to Kokoda and on to Port Moresby, a distance of about 170 miles. The police mailmen could reach head-quarters in twelve days, but it was a longer trip for the miners, forced to camp early to allow the slowest man to reach shelter before dark.

Watching the exodus the villagers thought that the *gavamani* too would leave. Sergeant Bakeke on a patrol to the Kumusi from Kokoda to reassure the people that the government officers would stay and government laws would still be enforced, found that many people had abandoned their villages for temporary bush shelters. The weaker clans feared that they would be raided by stronger groups. On the Mambare Village Constable Barigi was asked to counter similar rumours that the foreigners were leaving and the old warfare would begin again. There were no raids, the government stayed, and eventually a few miners returned. In most years after 1910 there were two or three miners on the Gira and Yodda. Dave Davies, Jack Murphy and Bob Elliott were camped on the same creek on the Gira in 1919. Davies was one of three miners who escaped by raft from Mambare beach in 1897, Elliott had been a pioneer prospector in the area, and Murphy had been known as One-Eyed Jack ever since his eye had been damaged by gunshot fired by an Orokaiva on the Yodda in 1904. Matt Crowe and Bill Parkes were the only miners on the Yodda in 1925. Crowe had taken out the first claim on the Yodda. Parkes, sometimes known as Red Bill or The Leaning Tree to distinguish him from William (Sharkeye) Park and Andy Park, had first taken gold from the Yodda over twenty years before.

The Yodda revived briefly in the 1930s when Yodda Goldfields Limited operated a dredge on the upper Mambare. The company solved some of its transport problems by building an airstrip just to the west of Kokoda Station. An aircraft coming from Port Moresby had to be at a height of at least 7000 feet to pass through the Owen Stanleys, then go into a spiral, dropping against a folded wall of mountains, to land on the vivid green of the Kokoda 'strip'. In its best year, 1939-40, the company produced £12,053 of gold, but it was already in decline by the time World War II came to the Pacific.

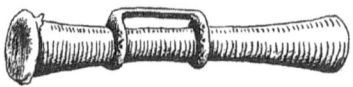

The most violent encounters between villagers and foreigners in British New Guinea took place in the Northern Division. Every year between 1895 and 1910 some community in the Northern Division or along the Waria fought the foreigners. Given the fighting tradition of the Orokaiva, a peaceful meeting between them and aggressive, confident strangers was unlikely. After the death of Clark no miner entered the Mambare unprepared for violence; when Green, Fry and Haylor were killed the miners became more inclined to meet any show of opposition with rifle fire. Government officers, sometimes following and sometimes in advance of the miners, used the language of war. It was a language which came easily to some because they went to New Guinea after assisting the Empire to defeat the Boers in South Africa. For ten years the vocabulary of war was apt in particular clashes, but in general there was no war. No government officer or policeman on patrol was killed. The language disguised the imbalance in power and gave an excuse, even a nobility, for what were otherwise brutal acts of punishment. When Elliott's police surrounded a village, there was little fighting, and a lot of shooting.

Table 10
Papuan goldfields
Total Gold Yield from first working to 30 June 1926*

Name of goldfield	Gold yield ozs	Value of gold £	Division	Date field proclaimed
Louisiade	138,049½	253,557	South-Eastern	20 May 1889
Murua	200,348	693,105	South-Eastern	6 November 1895
Gira	67,242	253,414	Northern	5 November 1898
Milne Bay	14,230	49,987	Eastern	6 December 1899
Yodda	76,822	287,090	Northern	31 July 1900
Keveri	4770	17,737	Eastern	6 August 1904
Lakekamu	37,170	138,822	Central	13 December 1909
Astrolabe	3299½	13,500	Central	21 December 1906
Total	541,931	1,707,212		

*With the flourishing of the mines at Umuna on Misima after 1926, the Louisiade was the premier field by 1941. (From *Handbook* 1927, p. 150.)

Moreover the fragmentation of the 'sides' made the conflict something other than a 'war'. The Beda formed different relationships with the Orokaiva, Koiari and Chirima peoples. They traded with the miners, who valued and tried to protect the relationship they had formed with them; labourers traded with them, stole from their gardens and fought them; and government officers protected them and gave the police orders to fix bayonets and charge them. Orokaiva villagers fought one another with great ferocity. The foreigners sometimes acted together and sometimes independently. Groups of runaway labourers struggling to survive robbed both miners and villagers; the police worked for government officers, themselves, their own people, or groups with whom they had formed an alliance. There was not a war but a series of encounters between divergent groups and shifting alliances. Violence was frequent; but if the police, miners or officials had rifles then they had the power to wound or kill while the bravery of their opponents became futility.

Events on the Yodda and Gira Goldfields rarely disturbed Australians. Little news reached them, and when it did it was usually about incidents which had taken place at least a month earlier. To Australians who remembered the death rate on north Queensland and Western Australian goldfields, the events in the Northern Division were not unusual — and the miners found less gold. Later writers have tended to look at the history of Papua New Guinea from the centre and through the eyes of the white and important. They have looked at the correspondence between Port Moresby and Australia to comment on the plans and reports of the central government and the behaviour of Sir William MacGregor and Sir Hubert Murray. But the outstation records are the main written source for a history of the Northern Division and much of the violence took place after MacGregor left in 1898 and before Murray assumed office in 1907. Until recently historians have tended to look at the peoples of Papua New Guinea only as the recipients of a policy expounded in overseas capitals and Port Moresby and Rabaul. Had historians been concerned about the most dramatic and significant experiences in the lives of the peoples of Papua New Guinea, they would have looked at the Northern Division between 1895 and 1910.

Joe Saruva of Papaki village and a student at the U.P.N.G. supplied information about his people's memories of contact with miners and government patrols.
Reports by Armit, Elliott and Walker of their violent patrols are in the *Annual Reports* of 1899/1900 and 1900/01. Walker's comments on the Kumusi peoples and their reaction to miners leaving for the strike at Cloudy Bay are from his report of 23 November 1901, Tamata station papers. The quotes from Robinson are from his diary, Papua New Guinea Archives. Other reports of patrols over the Northern Division plains are among the Bogi and Kokoda station papers.
Ross 1923 wrote of his research into the cause and prevention of malaria.

The building of the Buna-Kokoda road is recorded in the Kokoda station papers. Meek 1913, Monckton 1934 and Kokoda station papers refer to the killing of Meek's labourers. Elliott's patrol to Beda, March 1904, is with the Bogi station papers.

The account of relations between the Seragi and the miners and government officers is based on the Tamata (Ioma) station papers. Cawley, *Annual Report 1922/23*, and Chinnery and Beaver 1917 comment on the Seragi. Both use 'Ali-Tahira' or 'Aili-Tahari' as alternative names for the Seragi.

Chinnery 1931 located villages and language groups on the Waria. Pilhofer 1915 wrote of his trip up the Waria and down the Bulolo in 1913. Pilhofer has an excellent map. Apart from the Ioma station papers, *Annual Reports*, the Prykes's letters and diaries, the file 'Delimitation of the Boundary between Papua and German New Guinea, 1903-1910', C.R.S. Al, item 14/4329, Australian Archives, has references to the miners by government officials, newspaper cuttings and A. Darling's report on his and Crowe's pioneering prospecting expedition. Van der Veur 1966a has a detailed account of the marking of the boundary.

Elliott's statement that the miners would have been better off at home is with the Bogi station papers. Frank Pryke 1937 recalled the Beda track in the poem, 'To a Mate and Brother'.

The figures for the numbers of labourers who died or deserted are from *Annual Reports*. Green wrote the story of Lihoiya and Yarumeku in his letters and in the *Annual Report 1895/96*. Cases of desertion were noted in the Cape Nelson and Samarai station papers as well as in the papers of stations in the Northern Division. The examples of deceptive recruiting are from the Samarai station papers. Copies of Bellamy's articles for the *Grey River Argus* are in the library of the U.P.N.G. Randolf Bedford wrote an article on the Northern Division goldfields, *Sydney Morning Herald*, 16 December 1905: he thought men deserted for trivial reasons. The Reverend Copland King's article is in the *Sydney Morning Herald*, 17 October 1908.

Stone-Wigg wrote about the eagerness of men to join the police and the morality of those who had already joined in his diaries now held in the library of the U.P.N.G. Stone-Wigg's published comment on the miners was quoted by Tomlin 1951 and his private note on a meeting is in his diary, 30 June 1904.

Cases which went before the Central Court are listed in Register of Criminal Cases, Central Court of British New Guinea, Papua New Guinea Archives. Murray's record of the cases which he heard is in his Notebooks and he referred to many cases in his diary.

Several Yodda miners appeared before the Royal Commissioners and their evidence was printed with the report in *Commonwealth Parliamentary Papers*, 1907.

Copland King's record of his meeting with the miners to discuss morality is in Box 1 of the Anglican mission records, library of the U.P.N.G.

Monckton 1921, 1922 and 1934 revealed a great deal about himself; and so did Griffin 1925. Gibbney, *Australian Dictionary of Biography*, Vol. 3, outlined Armit's life. Black 1957 wrote a biography of Bellamy.

The basic source for the section on Joe O'Brien was 'Papers Relating to the Administration of Justice in the case of Mr. O'Brien, British New Guinea, *Commonwealth Parliamentary Papers*, 1906, Vol. 2. Other information was taken from Kokoda and Tamata station papers, Report of the Royal Commission of 1906, *Commonwealth Parliamentary Debates*, Monckton 1922, Griffin 1925, Lett 1943, D.H. Osborne, *Pacific Islands Monthly*, May 1946, Roe 1962.

The exodus from the Yodda and Gira is recorded in the Kokoda and Ioma station papers and in *Annual Reports*.

SIDESHOWS

Paddle handle, Massim, after Haddon 1894

10

Milne Bay

nothing very exceptional

Nearly five years after the Cairns and Cooktown prospecting expeditions had looked for gold on the northern arm of Milne Bay the miners in Samarai learnt that James Lindon had found payable gold to the south-west of the head of the bay. In May 1899 Lindon arrived in Samarai by canoe to obtain treatment for a wounded jaw. In a sworn statement he told Matthew Moreton, the Resident Magistrate, that he and his prospecting partner, Jack Gray, had had their shotgun stolen, their camp stoned, and he had been speared while using his revolver in desperate defence. Later the Kumudi villagers told the investigating officers and police a different story. A Kumudi man in Lindon and Gray's hut had called out that he was being beaten by the white men. The Kumudi then began to stone the camp. Gray fired hitting a man in the shoulder, and when Lindon stepped outside, the wounded man's brother speared him. The villagers had not stolen a gun. Gray, interviewed near Kumudi, told a story which gave more support to the villagers than to Lindon.

Without waiting for his jaw to heal Lindon left Samarai. Until then the incident at Kumudi had just been a topic to enliven the conversation of the miners staying at Billy the Cook's or camping in the sago roof huts near the waterfront while they waited for a boat to Woodlark or the Gira. Now they guessed that Lindon, known to some of them from his days on the Western Australian goldfields and already an experienced New Guinea prospector, had made a strike. It is also likely that some of the government party on their return from Kumudi told outsiders that Gray and another miner were working gold. The miners had been unable or unwilling to tell the government officers whether the find was payable. The men who followed the rumours found that Lindon was 'on gold'. By the end of June seventy-eight miners were on the field. Twenty-eight of them arrived on the *Ivanhoe*; they had left Cooktown for the Mambare but, having heard of the new strike in Samarai, persuaded the captain to turn into the bay. The normal rumours exaggerating the richness of the find spread

Map 11 Milne Bay Goldfield

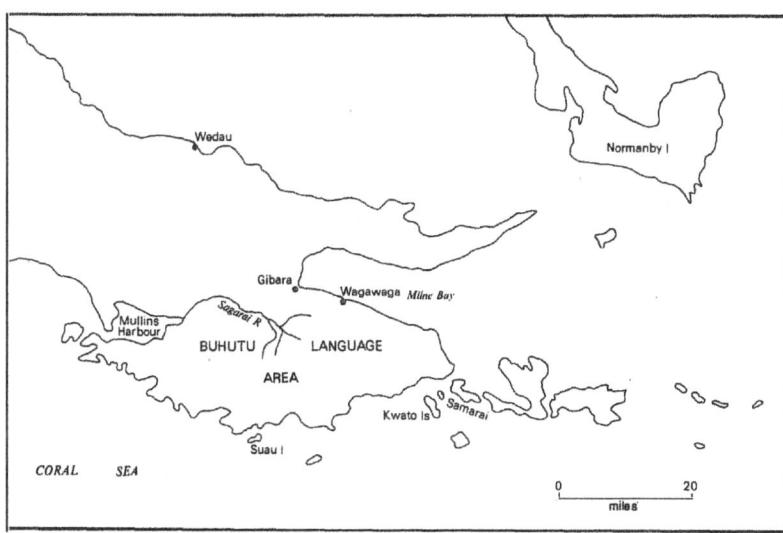

through the mining community. A sample of quartz sent to Australia was said to have yielded 60 ounces to the ton. Here was a field to support 10,000 men: Samarai was 'jubilant', and the missionaries 'considerably anxious'. But probably never more than 100 white miners worked on Milne Bay at any one time. Even before the end of 1899 Milne Bay was 'cooling down'. For the next five years twenty or thirty alluvial diggers attempted to make a living there; but with no chance of getting rich. The great advantage of the Milne Bay field was that it was close to a seat behind the iron lace balcony of a Samarai hotel. With stores, a spell and local food readily available, men were prepared to work at Milne Bay for about half the average return of the miners on the Gira and the Yodda. By the end of 1906 Milne Bay had produced only 5 per cent of Papua's gold.

The diggers left their boats at Gibara Landing, a few hundred yards from the mouth of Gibara Creek. The main workings were on the upper Sagarai River, which flowed west into Mullins Harbour. Broad and sluggish where it passed through swamps into the harbour, the Sagarai headwaters reached back into the steep ridges close to the southern shore of Milne Bay. The track from Gibara to the field was wet, steep and short: most of the mining camps were within five miles of the stores at the landing. Sometimes the south-east winds drove big seas up the bay and held rain clouds against the mountains, and at other times, especially during the north-west, the winds failed, leaving the bay as flat as the water in the sago swamps; but in fair weather the canoes, whaleboats, and schooners of the Samarai 'mosquito fleet' could reach Gibara in a day.

The people of the Milne Bay coast had known a variety of missionaries, government officers, recruiters and traders for twenty years before the miners began moving up Gibara Creek. Within the loosely linked communities of Massim peoples spread through the south-east, the villagers from the head of the bay were 'great fishers and traders'. Captain Moresby admired the craftsmanship of the men who made and sailed the 50-foot canoes:

> They use a great oval-shaped mat sail, and handle it so skilfully, that when we met them at sea, and the 'Basilisk' was going five knots, they easily sailed round us, and luffing under our lee were with difficulty prevented from boarding us while under weigh.

By 1899 the villagers from the head of the bay were taking their canoes to Samarai to sell food and buy goods in the stores. A few men had gone to the Mambare but no more would go, others worked on boats and around the town of Samarai, and many had come under the influence of the London Missionary Society teachers from Kwato. From his arrival in the area in 1891 Charles Abel had gathered young people at Kwato to train them as Christians, craftsmen and cricketers. During a holiday declared in 1899 to honour his first visit to the south-east, Lieutenant-Governor Le Hunte watched the Kwato cricketers defeat a team selected from Samarai's white community. But Abel saw enemies as well as opponents in the bare-foot pyjama clad drunks of Samarai looking for an easy life and in the harder men in high collars or working men's flannels prepared to push aside or enslave the local people in their determination to make money.

Away from the coast the Buhutu of the Sagarai Valley were also Massim people. They traded with the Suau of the south coast and the Wagawaga on Milne Bay, but few foreigners had disturbed their home villages. In 1899 the Buhutu were neither converted by missionaries nor controlled by government officers.

During the first two years of mining the Milne Bay peoples behaved in much the same way as the islanders in their encounters with the diggers. They visited the claims, sold food and did some carrying and mining. Relations between villagers and diggers were changed by the Buhutu stealing from the camps and the miners retaliating with brutal, summary punishment.

Alexander Symons, the Assistant Resident Magistrate at Samarai, heard complaints on a patrol to the goldfields at the end of 1900 that people from the Sagarai River were stealing everything they could carry away, including tents and working gear. The police and carriers he sent to arrest the people thought guilty returned with seven prisoners taken after a fight in which one of the government party and one villager had been wounded. The thieving continued and in April

1901 Symons took tougher measures. He burnt three villages where he found stolen goods and the police, attempting to take prisoners, fired several shots. In his report Symons did not say whether they hit anyone, and he mentioned only briefly the actions of four miners who were also trying to recover stolen property.

Table 11
Milne Bay Goldfield
production and population

Year	White Miners	Indentured labourers	Production (ounces)
1899/1900	73	292	4000
1900/01	20	80	1503
1901/02	20	80	1503
1902/03	34	136	1748
1903/04	28	112	1680
1904/05	28	112	1440
1905/06	18	72	1000
1906/07	7	28	357
1907/08	4	27	211
1908/09	1	0	170

After Charles Abel had made seven trips to the head of Milne Bay and spoken to witnesses, he wrote to Moreton at Samarai and Le Hunte, then in Australia, adding to Symons's brief report of his second patrol. Symons and four miners, Jack Gray, Harry Morley, Bob Lindsay and Steve Woolf, Abel said, had burnt thirty-eight houses, destroyed gardens and graves, shot two men and a woman as they fled, and another man, Sipiliei, who had accompanied the patrol, was roped, led away by the four miners, and later found shot through the head. Sipiliei, a leading man of the area, had previously acted as a guide for Abel and for government parties. Abel also revealed a bond of deception linking Symons and two of the miners. A few weeks after the shooting and burning in the Sagarai Valley, Jack McLean, the Gibara storekeeper, was murdered. McLean lived with Bi, 'the Queen Bee', a beautiful and notorious woman remembered by a miner as being part-Samoan and by a government officer as coming from Dutch New Guinea. At the inquest into McLean's death three Papuan witnesses told Dr Cecil Vaughan that Morley and Lindsay had killed McLean and beaten Bi, but at the trial of the two miners they said that Bi had killed McLean. Other Papuans told Abel that the evidence given at the inquest had been true; and when he asked the three men why they had changed their story for the court they said they had been intimidated and then rewarded by Symons. Abel assumed that Symons had protected the miners because they could so easily have incriminated him by revealing the truth about the patrol through the Buhutu villages. Even if Symons had not personally acted illegally Abel believed he had neglected his duty by failing to control those under his command and taking no action against men guilty of brutal crimes.

By threatening to make the whole affair public Abel forced Moreton to hold an inquiry. After a long delay the four miners were brought to court to answer for their actions against the villagers. To the applause of the white citizens of Samarai crowded into the court, Judge Winter decided that three of the miners deserved no punishment. In their defence they had said that they were acting under the instructions of a government officer. Winter ruled that while they were mistaken in their belief, it was still an adequate excuse. The crowd demonstrated their disapproval when Winter sentenced Woolf to six months' imprisonment for shooting a woman in the back at a distance of 5 yards. Woolf, the only non-Australian among the miners before the court, asked the judge if he called this British justice. Winter accepted that Symons first learnt of the miners' actions from Abel's letters although there were several witnesses ready to say that Symons had talked of the shooting soon after his return to Samarai. At the end of his inquiries Winter concluded: 'These rumours, when sifted and reduced to distinct charges, and to the evidence that supported them, showed that considering the circumstances, nothing very exceptional had taken place.' It was a statement remarkable for its qualifications and lack of substance. After all, the 'rumours' were supported by corpses and the burnt wreckage of villages. But Abel was most disturbed by a comment made by Winter in court: 'Racial feeling is so general and so strong in this country, that I cannot regard the defendant [Woolf] morally culpable in taking the life of a native.' In another case Michael Bowler, a Milne Bay miner, came into Samarai boasting that he 'had done for another of the d...d niggers' who had been stealing. Bowler was fined £5 and imprisoned for three months. The man shot was not a Buhutu villager and he was probably not a thief; two points of fact which may not have disturbed Bowler or altered the decision of the court.

These events happening before an Australian judge, in courts crowded with Australians, on the eve of the transformation of the Possession from a British Colony to an Australian Territory, made others ask with Steve Woolf whether they were watching the end of British justice. On leave in Australia to recover from fever, Abel went to Melbourne to see Alfred Deakin, the Attorney-General in the first Australian Government. Deakin assured him that in speaking on the bill to transfer British New Guinea to Australian control he would state in 'unmistakable terms' that in dealing with New Guinea

Australia would 'think first for the well-being of the aboriginals'. Since he believed that 'the labour party dominate[d] Australian politics' Abel also saw John Watson, the leader of the twenty-four Labor members in the Australian parliament. Having heard some of the speeches in defence of the White Australia Policy, Abel asked Watson whether 'the labour party saw any place in the world for a man with a black skin'. Watson replied that while he believed in an Australia for white men he would advocate the advance of natives in their own lands. Abel left 'jubilant'. He was comforted by the declarations of goodwill, and for other reasons he decided that the Papuans would be saved by the Labor Party. There would, he thought, never be enough white men in New Guinea for their interests to override the interests of the white citizens of Australia. The Labor Party would be a powerful force retarding economic development in New Guinea: it would never allow industries using white capital and cheap black labour to grow and compete with Australian factories and farms.

Abel was also reassured by Le Hunte's mild actions. He shifted Moreton and Symons to posts of lesser importance and promoted Campbell, 'a much better stamp of man'. Campbell toured the Milne Bay villages telling the people that, the old order had gone. The Buhutu were in need of reassurance for some had abandoned their homes to shift away from the diggings. Miners who had abused Abel and threatened him with revolvers became less hostile and he began to preach again in Samarai of a Sunday evening. On Gibara some of the miners, finding that they had helped protect McLean's murderers and convict the Queen Bee, forced Lindsay and Morley from the field. No government officials were brought to court, the Papuan witnesses were not charged with perjury although it was obvious that either at the inquest or at the trial they had lied, and none of the four miners was again called to answer for his actions in the Sagarai Valley or in the Gibara store. Woolf was gaoled for another lapse on the Gira. Winter resigned in 1902. In his defence it ought to be remembered that he was criticised by MacGregor for being too lenient with Papuan offenders. He was a man of kindly prejudice.

What distinguished the Milne Bay field was not the events in the Sagarai Valley, but the fact that they became matters to be disputed in public and decided by the courts. The men shot on the northern rivers were anonymous to all outsiders; Sipiliei was not. In a sense Winter was right, nothing very unusual had happened at Milne Bay. Later O'Brien was exceptional, but until then no miner in the Northern Division had been taken to court for mistreating villagers; they were called to answer only for their treatment of labourers, Papuans with their names in government records. The Milne Bay miners appreciated the comforts of Samarai, but the closeness of the port and the existence of the 'fine church' in Wagawaga meant that they were more likely to have to account for their actions.

In 1909 there was only one white miner on Milne Bay. He had no indentured labourers but employed local men who worked for two or three months and then returned to their villages. The Gibara storekeeper equipped other local men with boxes and tools to rework old ground. It was a practice followed again by Henry Dexter in 1929 when copra prices fell. Of the twenty men he supplied with tools seventeen returned with gold. In a week of fine weather they brought in up to £80 worth of gold. Sometimes an inland man would appear at the store counter and take from his hair a parcel of gold wrapped in rag. After it was weighed he might produce a second and a third, the earnings of other miners. When he came to make his purchases he took a longer time than coastal people and he was more reluctant to use the few words of English that he knew. Europeans kept returning to Milne Bay to work reefs and peg dredging leases; but it is doubtful if any made more than they spent. By contrast Papuan miners who invested time and labour won £1200 in 1937-38. Like the islanders the Milne Bay people were able to turn to alluvial mining when their other sources of cash declined.

Henry Dexter, Reminiscences of a 'Gin-soaked' Trader, Pacific Manuscripts Bureau microfilm, recalled his years as storekeeper at Gibara in the 1920s and 1930s. He also collected and recorded information about the early days on the field. Charles Abel and Frederick Walker wrote of developments on the goldfield and missionary-miner relations in Papuan Letters, London Missionary Society Archives, microfilm, National Library of Australia. Charles Abel 1902 described the peoples of Milne Bay, and Russell Abel 1934 referred to his father's response to events on the goldfield. Dutton 1973 has a map showing the distribution of languages in the Milne Bay area. Moresby 1876, Belshaw 1955 and Seligman 1910 have material about the Milne Bay peoples during their early years of contact with foreigners. Most reports by government officers are in *Annual Reports* and the Samarai station papers. Monckton 1921 wrote his account of Symons's punitive patrol. Cecil Abel, Charles Abel's son, and John Smeaton, his son-in-law, have spoken to me about Charles Abel.

11

Keveri

a magnificent valley
and an intense interest in killing

Three days' walk from where Frenchy and Jimmy the Larrikin's copper armour had failed to protect them from axe-blows to the head a track passed Mount Clarence on the west and crossed the watershed of the Owen Stanley Range. To the east of the track Mount Suckling rose to over 12,000 feet, the highest point in south-eastern New Guinea. North of the divide the creeks drained into the Adau River, which escaped through a gorge at the northern end of a broad valley to meet the Moni and flow on as the Musa. Wet with sweat from a hard climb, travellers looked across a 'panorama ... of great beauty', the Keveri Valley. The floor of the valley was lightly timbered grasslands cut by sharp spurs and watercourses. In dry weather the creeks were clear streams winding and falling along rock-strewn beds, but when rain washed the hills they changed to roaring clay-stained torrents blocking all work and movement in the valley.

Foreigners called the people of the valley Keveri. They lived in small ridge-top villages, three or four tree-houses and fighting platforms rising above a cluster of another four or five houses within a stockade. The Keveri hunted, collected breadfruit, okari nuts and pandanus fruit, and cleared and cultivated the slopes above the grasslands. Within the Bauwaki language community, which spread south of the watershed, the Keveri traded and fought with people speaking related languages near Cloudy Bay. But by their decorations and the way they wore their hair twisted into a mass of trailing plaits, the Keveri men showed that they also had some association with peoples to the north. Travellers thought the valley secluded, fertile, healthy and peaceful: but F.E. Williams, the one anthropologist to see the Keveri in their homeland, concluded that the 'salient feature of [their way of life] was nothing other than an intense interest in killing'.

The 'killing' brought government officers to the valley. In 1899 Albert English, the Assistant Resident Magistrate at Rigo, twice visited villages recently raided by people said to have come from Keveri. At the end of the year English and Joseph Blayney, the Resident

Magistrate for the Central Division, spent four days in the 'magnificent valley', but they saw little of its inhabitants and made no arrests. In 1900 about sixty warriors from Keveri and Dorevaide, south-east of the valley, attacked Merani, killing nineteen people. English, Blayney and Francis Barton, Commandant of the Armed Native Constabulary, made another patrol north from Cloudy Bay. After arresting three Dorevaide men the patrol moved on across the watershed. Although one or two men from the valley had already been away to work as indentured labourers, most of the Keveri kept well away from the patrol, and again none was arrested.

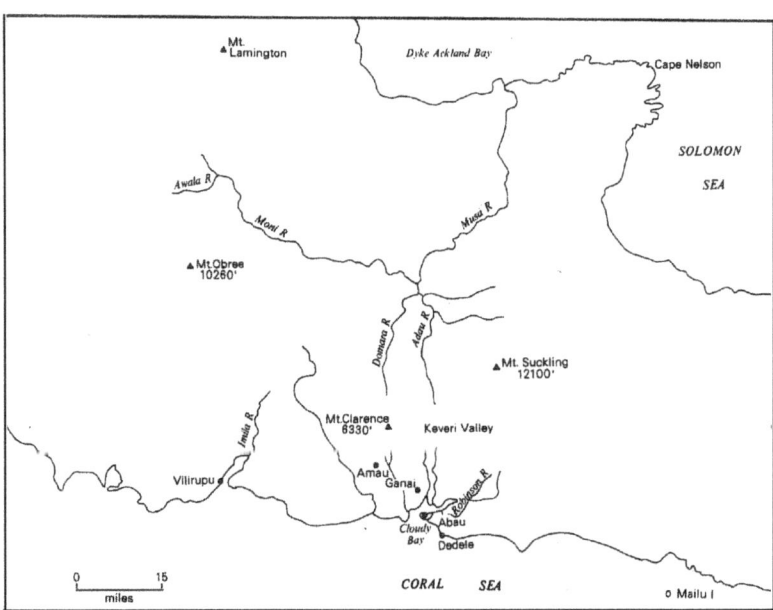

Map 12 Keveri Goldfield

Fighting and trading parties from Keveri may also have learnt about the prospectors who had approached their lands from the Musa. William Simpson, after his second trip up the Mambare, tested the Moni and Adau Valleys at the end of 1896. Penetrating well beyond the highest point reached by MacGregor and an estimated 25 miles up the Adau, Simpson must have been close to the northern edge of the Keveri Valley before he turned back to try the Moni. On his return to New Guinea from Australia in 1897 Simpson went back to the Mambare; but the fine colours he had seen on the Adau encouraged others to go up the Musa. Frank Pryke was one of twelve prospectors who recruited carriers in the D'Entrecasteaux Islands and then ascended the Musa in a whaleboat and two canoes. The carriers having deserted in the whaleboat, the prospectors persuaded local villagers to help shift their gear about 20 miles up the Moni. Ignoring the warnings of the villagers, the prospectors attempted to raft downstream and they and their stores were tossed overboard. Frank Pryke spent three days without food making his way back to camp and 'nearly pegged out'. When the prospectors reached the beach they found a note from Moreton telling them that he could not pick them up in the government schooner as he had just received news that Green had been killed at Tamata.

Having tried the north Queensland goldfields Pryke teamed with George Klotz for a trip to the Gira and across to Finnegans Creek on the Yodda where they 'did fairly well'. After recovering their health in a camp at Gabagabuna at the head of Milne Bay, Pryke, Klotz and nine carriers went north across the ranges along the northern arm of Milne Bay to prospect the country behind Bartle Bay. Like the north Queensland miners who had preceded them, they suffered from fever, constant rain and inadequate food, and found no gold. In Samarai they heard that a group of miners had picked up signs of gold inland from Cloudy Bay. Although the reports were not encouraging Pryke and Klotz gathered a team and three months' stores, and took a cutter along the south coast. When they arrived at Cloudy Bay they found that the miners there had decided that they were unlikely to find anything worthwhile in the area, but Pryke and Klotz thought they would prospect the country near the Domara River. In May 1901 they put in a box on a small ravine: when they left for Samarai three months later they had 370 ounces.

Men set out immediately from Samarai, and as the news spread to the camps at Milne Bay and the more distant fields the demand for a place on the boats increased. Whittens established a store at Iaba, just inland from Cape Rodney and about 40 miles from the field. At first selling straight from crates, the storekeeper had hardly completed a grass roof and packing-case counter before the miners were drifting back saying that the field was a duffer. More than a hundred men had landed at the beach; not all had bothered to go inland and none had made a worthwhile strike. The ravine had held an isolated patch; for years it had been acting as a natural sluice box.

Pryke and Klotz on their second trip to the area were among those who failed, but Frank Pryke and his brother Dan returned to the area in 1902 and after prospecting about the head of the Musa found gold in the Keveri Valley. Frank and Dan worked at Keveri in 1903 and 1904, and Frank and Jim, a younger brother, went back there again in 1905. At the end of ten months' work in the valley in 1903 Dan banked 256 ounces in Cooktown while Frank went to Samarai, 'Imbibing and playing cricket at Kwato' before taking Whittens' schooner for another recruiting trip through the D'Entrecasteaux Islands. But other miners, and the Prykes in other years, took less profit from the valley: they made 'a few quid over exes', or, looking at the amount of stone to be shifted, decided there was a 'crust in her with a big team'. The greatest number of miners at Keveri was about twenty in 1904; the next year there were only five. By the end of 1907 the miners had completely deserted the field. But when little gold was being found in other areas a few men went back to Keveri, and even in 1940 there was still one miner in the valley. The miners built stores at Ganai and used local villagers and their own labourers to carry rations and equipment over the range. The Keveri miners rarely saw a government officer or any other outsider.

The isolation had disadvantages. In 1904 Luke Soich's leg was shattered by a stone thrown out by dynamite being used to cut a race. Frank and Dan Pryke and eleven labourers carried him to the coast in three days. 'Coming down the main range today', Dan wrote, 'it looked impossible to carry a man on a stretcher but the boys were really splendid.' The labourer sent ahead found a boat in the bay, and Soich was in the Samarai hospital six days after the accident, but he died within a few hours of an operation to amputate the damaged leg. Nearly all the white residents of Samarai witnessed the burial of old Luke at the Logea Island cemetery, and Campbell recorded a tribute in the station journal: Soich was 'one of those good sterling miners who are respected by both white men and the brown alike for their honesty and straightforwardness in all their dealings with either race'.

In spite of their reputation for fighting and killing, the Keveri lived in economic harmony with the miners. On his last trip into the valley Frank Pryke was greeted by a boisterous crowd, and several men asked anxiously about Dan, who had gone to Australia. After he had worked on nearly all the alluvial fields of Papua, Frank Pryke decided that he never met a 'more friendly lot of coloured men' than the Keveri. They carried and mined, but like the Beda they were most important to the miners as suppliers of garden crops. The miners also added to their stores by 'shooting fish' (throwing explosives in pools) and hunting pigs, wallabies and cassowaries across Keveri lands. With no mosquitoes, cool nights and plenty of food Keveri was a favoured spot; except that there was little gold.

Peace between miners and villagers did not mean that the Keveri and neighbouring communities lost their interest in killing. Raiding parties attacked villages within the area or up to six days' walk away, killing ten or more of the inhabitants, taking anything of value and throwing the dead back into their burning houses; but more frequently a small group of warriors went out and killed some man, woman or child found alone. The lone victim might be killed by *mimi*: he would be held down, suffer assault and sorcery, be released and die within a few days in his home village. Europeans tried to explain the violence in the area by suggesting that before a Keveri man could marry or have any standing in his own community he had to earn the right to wear the badge of the killer, the beak or tail feather of the hornbill. The explanation was inadequate for often the man who emerged after a period of purification wearing the hornbill feather was not the killer; his display was part of the ritual to deceive the spirit of the dead man and prevent it from taking vengeance on the killer.

The Abau detachment of the Armed Native Constabulary
PHOTOGRAPH: H. DEXTER

While the miners were witnesses to lament and triumph in the villages their labourers were sometimes more directly involved. Local men working for the miners were at risk and indentured labourers who deserted were likely to die. In 1903-4 fifteen runaway labourers were killed. The labourers briefly became raiders when Kruger and the Prykes' labourers shot three Barua men from south of the watershed. In court the labourers admitted that their first story about only firing in self-defence was untrue.

The miners did not appreciate English's attempts to arrest the men who raided coastal villages. He shot pigs, fed his police and carriers on the gardens and left the valley almost deserted. By disrupting trade between miners and villagers the government patrols cost the miners time and money, and Frank Pryke went on to make a general condemnation of the government's 'way of civilizing the niggers':

> I think that they should either leave them alone to settle their own feuds or else when they start out to give them a lesson they should give them a proper one, and not make a farce of it by taking a few harmless ones who are either too old or too young to be in the mischief and giving them a few months or a few years while the real culprits almost invariably escape.

Pryke was right about the ineffectiveness of the government's actions for the Keveri continued their raids into the 1930s.

George Arnold, the only miner on the field in 1912, fought in defence of the village of Ona-Audi. A raiding party armed with spears, shields and clubs ran past his camp and on towards Ona-Audi. A man from the village called for his help, and he responded when he saw a fleeing woman caught and clubbed to death. According to the *Papuan Times* Arnold grabbed his 'Winchester and managed to put in some good work — the only work that is effective in such circumstances'. But Charles Wuth, the Acting Resident Magistrate at Abau, reported that Arnold had killed one man only while seven men, six women and one child from Ona-Audi had died. Occurring ten years after gold was first worked in the area, the shooting at Ona-Audi was the only reported time that a miner on the Keveri Goldfield shot a villager. The Keveri were no less violent than the Orokaiva, but they lived in small communities engaged in feuds and swift raids; it was a type of fighting which did not need to include the foreigners.

Nor did the Keveri fight the government, in spite of the irregular government patrols which passed through the valley pursuing men wanted for murder. Guilty men evaded the patrols, but when surrounded or surprised they were normally handcuffed without a shot being fired. Leo Flint, after serving several years as Assistant Resident Magistrate at Abau, wrote:

> The Dorowaida and Keveri people give continual trouble. I am certain very few of the adult male population have not been in gaol for murder. Some have returned home after serving a sentence for murder, and then within a few months, committed the same crime again. Yet in other ways they are amenable to the law. They clean roads, erect rest houses, and assist with carrying etc.

Murray said he knew of no other area where men who had served a term for murder came before the Central Court charged with another murder, and asked Flint if he had exaggerated. But Flint was able to name men and cases to support his statement. Fortunately for the peace of government officers the Keveri were too few and too far away to force them to question the value of their basic practices.

The Keveri Valley is now almost deserted. By reducing the numbers of fish, birds and game animals in the valley, and bringing diseases, the miners may have contributed to the decline in population. George Arnold in 1923 said that when he first mined in the area twenty years earlier, there were twice as many people in the valley and there was much more food for sale. Yet between 1923 and 1940 there was little, if any, further fall in the population. The main factor in the final depopulation of the valley was emigration. In 1940 274 people lived in the valley: the movement out was just beginning. Influenced by missionaries or government officials or for reasons unknown to outsiders the Keveri had stopped fighting. The missionaries were from the Kwato Extension Association, two of them were sons of Charles Abel and the others were Papuans who had passed through the schools he had founded. Many Keveri, after listening to the missionaries, condemned their own past. The men understood the 'message' for they had learnt some Motu in Port Moresby gaol. The Keveri gave up their dances, decorations, songs, plaits, bark belts, sorcery and killing, for calico *ramis*, short hair, family prayer, washing, shaking hands, the word of *Darava* (God), peace, a hope of everlasting life, and a chance to participate in the new life being experienced by communities on the coast. They shifted first from their ridge-top villages to Eoro and later to Amau south of the watershed. Keveri, the miners' 'happy valley', held no permanent community.

Humphries, *Annual Report 1922/23*, and Williams 1944 described the valley and its people. L. Armit, *Annual Report 1913/14*, has some ethnographic notes. Dutton 1971 mapped the distribution of languages in the area.
   The early patrols by English and Blayney were noted in the *Annual Reports* for 1899/1900 and 1900/01. Simpson's letter to MacGregor reporting his prospecting trip up the Musa was printed in the *Annual Report 1896/97*. Frank Pryke wrote an account of his 'First trip to New Guinea' in a coverless notebook, Pryke papers. The expedition was reported in the *Sydney Mail*, 1 May 1897. Frank and/or Dan Pryke's diaries for 1901, 1903, 1904 and 1905 are with the Pryke papers. Entries are brief and the diaries do not cover all months of those years. Fred Kruger in an obituary for his mate and fellow miner described the discovery at Cloudy Bay, *Papuan Courier*, 22 October 1937. Frank Pryke's reference to the friendliness of the people is from his poem, 'Keveri'. Collinson 1941 recalled his time as the manager of Whittens' store at Cloudy Bay. Robinson, diary, recorded his walk into the valley. Some wardens' reports were printed in *Annual Reports*. The Abau station papers recorded events after 1911. The *Papuan Times*, 22 January 1913, reported the attack on Arnold. The extension of mission influence in the valley has been written about by Russell Abel. They broke their Spears, typescript, Abel papers, library of U.P.N.G.; Vaughan 1974; Wetherell 1973; and Williams 1944.

THE LAKEKAMU

Shield, Gulf of Papua, after Haddon 1894

12

Two Ounces a Day and Dysentery

it grieves a man to lose one of them especially if he is a good boy

By 1909 gold was difficult to find on the alluvial fields of Papua. Twenty-four miners and over 300 labourers were still on the Yodda, about the same number were on the Gira and a few were prospecting on the Waria. In the south-east one miner was working on Milne Bay, men returned intermittently to Keveri, and on Sudest, New Guinea's first payable field, the villagers were still panning a few pennyweights. On Misima foreigners and villagers shared the little gold being taken from the east of the island. Three or four of the white miners in the Louisiades were investing in plantations and trade or trying to raise capital to work lodes of unknown extent; and the others no longer cared for the values of the communities they had left. On Woodlark alone each boat brought one or two new men. The 'reefers' were optimistic. Served by crushing batteries at Kulumadau, Busai and Karavakum, they had won 3537 ounces in the previous twelve months. The alluvial miners had taken about half as much, but they were not hopeful of increasing production.

The Papuan Government introduced three schemes to assist the alluvial miners. Some Yodda miners argued that below the conglomerate on the valley floor they would again find rich alluvial. To test their belief the government granted them £150 and passed the Gold Mining Encouragement Ordinance which allowed the granting of a reward claim to anyone boring through the false bottom of a field to locate more gold bearing ground. A second ordinance introduced into the Legislative Council by Little permitted the Lieutenant-Governor to pay up to £1000 to the discoverer of a goldfield if it supported not less than 200 miners of European descent for eighteen months. The payment was in addition to the normal reward claim equal to forty times the area of one man's claim. The government also agreed to pay £800 to finance a prospecting trip into new country. No one collected the cash reward for opening a new field and it was another twenty years before worthwhile gold was again found on the Yodda; but the government prospecting party opened up country where a few men pegged claims yielding over 2 ounces a day, more miners were disappointed, and many labourers died.

The miners on the upper Yodda met at Clunas and Clark's store where ten voted for Matthew Crowe to lead the prospecting party and five favoured Frank Pryke. Further down the Yodda all eleven miners on Finnegans Creek who registered a vote decided to support Crowe. Billy Ivory (with the covering assurance, 'you can Rely on this been correct') sent in the Gira and Waria vote:

> we had a meeting hear. But all that was not to tired to vote sent in a voting Papper to the store which had to be oppened on a giving Date. The Result I am sending to you. You will see By it that Mr F Pryke is Easialy the man hear. Now the quearely is will he except it has he is out on the waria yet no one knows wheather he will or not ...

The Woodlark miners decided that they would accept the decision of the men on the mainland.

Believing that the Waria was finished and not looking forward to shifting a lot of earth in the Keveri Valley for little gold, Frank Pryke was indeed eager to open new country, but when the votes from the Yodda and Gira were added, Crowe had just defeated him. In Samarai Crowe offered to toss Pryke for the leadership, Pryke declined, and Staniforth Smith, the Director of Mines, formally appointed Crowe leader of an expedition to 'search for valuable minerals'. Crowe then had the right to choose his mates and, rejecting Staniforth Smith's offer to secure a professional geologist, he named Frank and Jim Pryke.

Matt Crowe was about forty-seven in 1909. His six-foot-four frame was becoming more bent and his tongue sharper. There were few men on the goldfields who had not been lashed by old Matt's tongue. After Frank Pryke retired to Sydney and a hectoring wife he wrote:

> Few people understood Old Matt
> Or his keen sarcastic wit
> But I had early found out that
> He was straight and full of grit
> Few as good as him are found
> And I should surely know
> For many years I knocked around
> As mates with long Matt Crowe

Crowe left the Victorian police to mine in Western Australia and the Yukon before going to British New Guinea. He took out the first registered claim on the Yodda, opened the Waria with Arthur Darling and prospected widely across the headwaters of the northern rivers.

Map 13 The Gulf of Papua

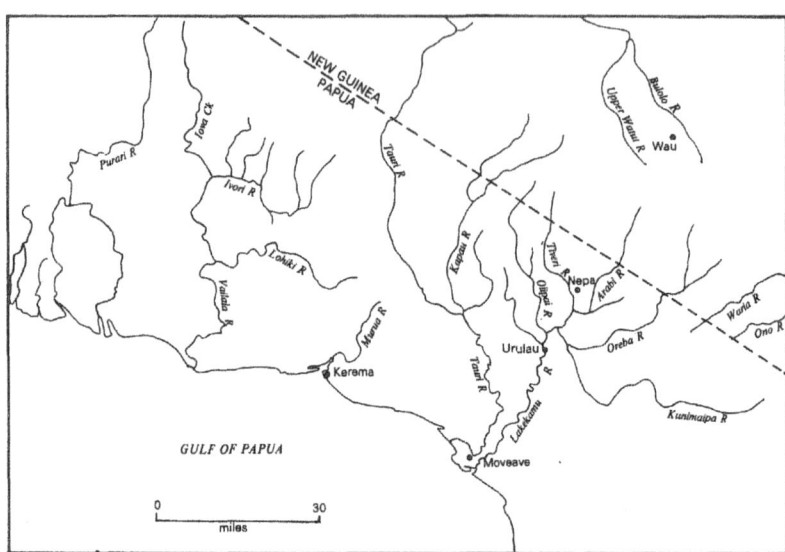

Frank Pryke was thirty-seven, just above medium height and strongly built. Although no longer the young man who had run at athletic meetings, competed in fire brigade demonstrations, and played rugby in Goulburn, New South Wales, he was tough, tireless in the bush, and genial and generous towards all men. The son of an English migrant who spent nearly all his working life on the goldfields of Victoria and New South Wales, Frank Pryke looked for work in New Zealand and then went prospecting in Western Australia with an older brother, Dan. Frank and Matt Crowe first met in Western Australia. After the failure of the expedition up the Musa in 1897 Frank had carried his swag through the north Queensland fields before returning to New Guinea to join the early miners on Finnegans Creek. Dan had taken reefing claims on Woodlark, but he joined Frank at Keveri after Frank and Klotz made the first rich strike in that area. When Dan left New Guinea in 1904 to marry and become the proprietor of the Royal Exchange Hotel in Armidale, New South Wales, Jim joined Frank for one year before they shifted to the Gira and then the Waria. Jim was thirty-five. Like Frank he was strong, reliable and quick-witted; but he was less restrained, he had a touch of the larrikin. He played as a forward for a country side in Goulburn against the English rugby touring team of 1899. The miners of the Northern Division had voted with good sense when they selected Matt Crowe and Frank Pryke.

In June 1909 Matt Crowe, Frank and Jim Pryke and thirty-five indentured labourers with stores for six months left Port Moresby on the *Merrie England* for the headwaters of the Tauri River in the Gulf Division. The labourers, recruited from Milne Bay and the islands, signed-on for £1 or 10s. a month. Each was granted a Special Arms Permit giving them the right to carry a gun or rifle. In heading for the eastern Gulf, the prospectors were following the wish of most of the miners who attended meetings to select the expedition's leader. The miners' choice of country was based on a little knowledge and a lot of speculation. Much of the mainland east of Yule Island had been crossed by prospectors, and those pockets still unknown were accessible. As Billy Ivory put it: 'a Private Party Can Prospect anything on the N E But west of Port Moresby Say 100 miles is beyond a Private Party to expensive'. The rivers draining into the eastern Gulf rose close to the source of the Aikora and the Waria, and it was possible that the formations releasing gold into the northern streams were duplicated on the southern slopes. Two recent expeditions had increased interest in the southern rivers. James Swanson and one of his sons, who had previously tried the rivers behind Yule Island, went up the Vailala and the Tauri, and Charles Higginson, Resident Magistrate of the Gulf Division, obtained a 'few substantial colours' from a dish taken from the upper Tauri in 1908. There was, Higginson reported, a vast tract of country which an experienced miner could prospect from a base on the upper Tauri.

From a camp he located as 'fifty mile beach' Crowe sent a note down river: 'if we get into touch up here with Natives we may be able to go on for months'. In September Murray attempted to find the prospectors and replenish their stores. From Ernest MacGowan who was trying to establish a copra, fibre and cotton plantation on swampy ground just above Moveave village, Murray learnt that Crowe and the Prykes intended to form a base camp at a point where three tributaries joined the Tauri. Murray found the long-abandoned camp and travelled by canoe and foot up the tributaries without seeing any recent evidence of the expedition. He returned to MacGowan's where the prospectors were waiting for him. They had explored the headwaters of the Tauri then gone east, made sago and canoes and come down the Lakekamu. The prospectors told him that they found good indications of gold in the east, and they were anxious to return. They also told Murray that they had frequently seen sturdy, light-skinned people armed with small bows and wearing grass skirts and bark cloths. The prospectors knew that they had been watched constantly by these people who had shot a lot of arrows at them. They 'did no harm'; but neither would they trade. Murray loaned the prospectors a launch from the *Merrie England* to take them back up the Lakekamu; he was not to see them again until December. By foot, raft and canoe, the party prospected the headwaters of the Lakekamu from the Tiveri to Mount Lawson, frequently finding 'colours'. Again the mountain people were hostile. On the western branch of the Tiveri they 'sniped' at the intruders driving their 'boys back' and on the middle Tiveri the 'boss boy', Wagawaga Dick, was killed by an arrow. On their way down the Tiveri, Crowe and the Prykes pegged a reward claim on a small tributary: a box put in just below the claim averaged 2 ounces a day.

In Port Moresby, after six months in the field, the prospectors reported that they had found enough payable ground for the miners then in Papua, but they warned that they had found nothing to justify a rush from Australia. The field did appear to have one advantage: it was only one day's walk from navigable water and the miners hoped it would be a place where they, rather than the storekeepers, would profit. By confidential dispatch and coded telegraph from Thursday Island, Murray advised the Minister for External Affairs to caution impetuous Australians against joining the rush: the extent of the field was uncertain, it was difficult to reach and the 'natives extremely hostile'.

The expedition had carried the hopes of the Northern Division miners, and they were eager to learn of its success. In Samarai they took little notice of Frank Pryke's warning that the amount of gold was limited; they were confident that as soon as the 'crowd' got there, more payable ground would be uncovered. With fresh stores and their labourers' contracts endorsed for mining and carrying in the new area, the miners waited for a passage. Boats which had serviced the Northern Division fields shifted to the south coast and the Lakekamu. The movement to the west hastened Port Moresby's eclipse of Samarai as the chief port of the Territory, but 'Port' did not gain the miners' affection. Frank Pryke wrote: 'Port Moresby is one of the most miserable places to spend a few days that even I have yet struck. I would sooner spend a week at Kulumadau or Tamata. There is still only one pub there ... .' The patronage of the Lakekamu miners helped support another: the Papua Hotel opened in 1911.

Hopeful amateurs from Australia where the 'rush [had] been boomed a bit too much' crossed to Papua. The first fifty to arrive in Port Moresby had the money to get to the field but forty-nine soon returned to Port: without stores, equipment and a 'team' they were unable to work. By the end of February Murray reported that about 120 miners had come from Australia and already about eighty had left the Territory. Some had been misled by reports of claims yielding 1 and 2 ounces a day. They did not realise that where fifteen labourers could recover $1\frac{1}{2}$ ounces a day, a white miner working alone could only put enough earth through a box to obtain 1 or 2 pennyweights. Murray telegraphed the Minister again asking him to warn diggers through the press that they should not come to Papua unless they had three months' supplies and £100 in cash. On the Lakekamu and in Port were destitute men. After they signed a statutory declaration the government gave them an order for meals at Tom McCrann's Port Moresby Hotel and provided them with a free passage to Cairns.

To reach the field, miners and labourers travelled by launch and schooner from Samarai and Port Moresby. The best known of the launches was Whittens' *Bulldog* which, impeded by floods, low water, breakdowns and snags, made irregular runs up the Lakekamu. The mouth of the Lakekamu was deep, but sometimes the south-east beat up high seas and some boats preferred to enter the Tauri and then take the 'cut' through the low-lying sago country into the broad stream of the Lakekamu. The miners had two or three days sitting on the deck, the sun sometimes on their back and sometimes in their eyes, as the boat followed the broad curves of the river. By the second day they saw the first high ground, the Kurai Hills, rising above the tall vine-covered timber on the river banks. At night they could camp on the broad sand beaches which stretched for over a quarter of a mile on the inside of bends when the river was low. Piles of broken, bleached tree trunks showed the power of the stream in flood. Having passed the junction with the Kunimaipa, they went up the narrower, faster-flowing Tiveri to the landing close to the point where the Tiveri is joined by the Arabi or Aiv Avi. The boats went no further and in the dry season men and stores sometimes had to be unloaded several miles below the landing. If no launch were available at the Lakekamu mouth, a miner anxious to find gold and escape the mosquitoes might persuade Moveave canoe men to make the eight-day trip against the current to the Tiveri landing. Tiveri was one day's walk from the field. The road, crossing many creeks and cut through thick undergrowth, was flat except for a steep spur before the descent into the valley of Ironstone Creek. Crowe and the Prykes had pegged their reward claim a mile above the junction of Ironstone and Rocky Creeks. New arrivals took up most of the ground along the two creeks. Lyons, the Warden and Resident Magistrate, told Murray that by February there were about 100 white miners and 500 labourers on the field. In April, when there were 120 miners and over 800 labourers on the Lakekamu, the field had reached its height. From the ridge overlooking Ironstone Creek it was difficult to see the workings, the tents and the bush houses because 'the men lived, like fish, groping about at the bottom of the deep green sea of forest, a hundred feet removed from the light of day'. Races and flumes carried water to the teams ground-sluicing and shovelling the alluvial into the boxes. Lines of carriers constantly brought rations from the Tiveri landing where two storekeepers had set up businesses. Two other stores were built on

Ironstone Creek and the government declared over 12,000 acres Crown land and reserved it as a site for a town. Both stores on the field and one of those at the landing were licensed. For celebrations, beer was 3s. and whisky 10s. a bottle; for sustenance, flour was 15s. a 25-pound tin, and golden syrup 1s. 6d. a tin. The prices were as high as those on the Northern Division Fields. The warden's office, officers' quarters, gaol, police barracks, carriers' shelter and dispensary of the Nepa Government Station were built on the slopes overlooking the field.

On 25 January the reward claim was measured and registered; after nine labourers had worked for five hours the box was cleaned up and $2\frac{1}{2}$ ounces were taken out. The reward claim continued to average over 2 ounces a day for several months, but the richest strikes were made in an area known as the 'Jeweller's Shop' at the head of Rocky Creek. One miner with a team of eight men won 195 ounces in less than four months: the highest daily earnings were about 7 ounces. In April Lyons thought that perhaps seventy miners were on payable gold, although their costs were high and they were employing larger teams than on other fields. After six months over 200 white miners and 1100 labourers had come to the field; 61 miners and 677 labourers were still on the Lakekamu. They had produced about 3000 ounces of gold.

Stone adze, Kukukuku, after Blackwood 1950

The Papuan Government supervised the Lakekamu more closely than any other field; and the Lakekamu became a tough test of its willingness to look after the welfare of its subjects. On 11 December Murray reported the arrival of Crowe and the Prykes in Port Moresby, on 13 December an area of 768 square miles at the head of the Lakekamu stretching to the German New Guinea border was pro-claimed a goldfield, and on 28 December Murray left Port Moresby for his first visit of inspection. Having looked at the claims on Ironstone Creek, Murray and Matt Crowe led a party to make contact with the people who had attacked the prospectors on the upper Tiveri. They found houses, gardens and tracks but no people. Murray said he had expected no more, but he hoped that his 'reconnaisance in force' might make the villagers hesitate before attacking any stray miners. When he left Tiveri on 10 January, two assistant resident magistrates,

Norman Bowden and George Nicholls, a troop of police, and about twenty miners remained at Ironstone Creek. At the end of January, Lyons, who had served as Resident Magistrate and warden on the goldfields in the Northern Division, and Dr C.C. Simson, who had just resigned after one year as Chief Medical Officer, joined Bowden and Nicholls at Nepa. When the Reverend Henry Newton of the Anglican Mission offered the services of two nurses for the Lakekamu, Murray accepted and he also authorised the immediate construction of a tent field-hospital until timber was supplied for a permanent building. 'It is to be hoped', Murray wrote to the Minister, that '[the Lakekamu's] death roll will be less than those of other fields'. On 22 April Murray telegraphed the Minister from Cooktown: 'dysentery raging'. During February fifteen labourers and one miner had died. To shelter the sick, the miners provided sixteen labourers and a supervisor to construct a 'native hospital', while the police built a 'European hospital ... a commodious structure of native material'. But as the epidemic intensified, buildings, supplies and staff were inadequate. In March fifty-four labourers died. On one day six labourers and one policeman, Corporal O'ori, died. On 28 March Bowden wrote in the station journal: 'After working till 5p.m. we suddenly remembered that it was Easter Monday. The holidays were not celebrated here'. On 1 April, eight labourers died; on 2 April, seven died; for the month, seventy-two died. Police and government carriers helped bury the dead. When a white miner died the Resident Magistrate read the service while the sympathisers stood about the grave. Soon the rise above Nepa Station was being called Graveyard Hill. In May fifty-three died; but by the end of June when another twenty-three had died, the epidemic was almost over. In the six months to the end of June, 1101 labourers had worked on the field: 255 had died, and of these, 231 had contracted dysentery; 415 had been admitted to the hospital established in February, and 160 had died there. Dysentery epidemics had occurred frequently before on plantations and goldfields in Papua, but none had persisted for so long or killed so many. Only four Europeans died from dysentery. One who survived its symptoms, Chas Lumley, Whittens' storekeeper, said that he had had the 'shittttts' so badly that he had feared there would soon be nothing left of him to push his pen.

Murray visited the field at the height of the epidemic. Sixteen men had died on the three days before his arrival, but the police guard of honour which received the Lieutenant-Governor at Nepa Station may have assured him that his government could maintain order in the face of disaster. After inspecting the field with Little, Murray announced that the field would be closed to further recruiting. Whitten and Nelsson, the storekeepers, asked Murray if less severe measures could be taken, but Murray thought all alternatives inadequate. Without enough members of the Executive Council to form a quorum, Murray had to wait until his return to Moresby to put the restrictions into law. By proclamation on 21 April 1910 Murray banned all recruiting for the Lakekamu Goldfield, and under the Native Labour Ordinance he published a regulation to prevent a spread of the epidemic. All labourers leaving the field were to be taken direct to Port Moresby where they were to be examined by the Chief Medical Officer, and no labourers were to land or return to their homes without his permission. In May the law was strengthened by further proclamations stating that 'No native may be removed to the Lakekamu'. The previous proclamation had allowed an employer who had recruited labourers before 21 April to take them to the Lakekamu.

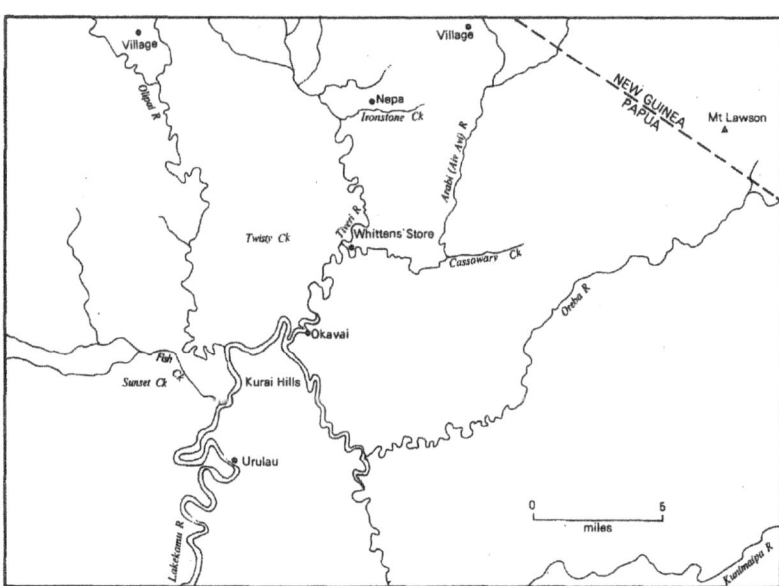

Map 14 Lakekamu Goldfield

Under the labour regulations which came into force in August 1909 government officers had the power to see that labourers were supplied with minimum rations, a latrine in 'a proper sanitary condition', and a 'suitable dwelling' with the surrounding area 'free from weeds and refuse of every description'. A government officer could direct any labourer to enter hospital and employers of more than ten labourers had to keep a stock of basic medicines on hand.

The legislation was more adequate than the stores, staff and knowledge available to combat the epidemic. In 1910, Dr Julius Streeter, the Acting Chief Medical Officer, issued instructions on the treatment of dysentery. He recommended the isolation of the patient, and the disinfecting of buildings and the careful disposal of waste. If a case were detected early and there were signs of constipation Streeter recommended 'a good dose' of a half to a packet of salts and a further four to eight doses in the next twenty-four hours. But if a case was not seen early, then it was better to give the patient a 'dose of castor oil' laudanum and ipecacuana (an emetic and purgative). Where practicable, the 'most rational plan of treatment', said Streeter, was 'to irrigate the bowel' by giving an enema.

The instructions to isolate the patient and ensure general cleanliness would have helped prevent the spread of the epidemic. The laudanum may have given the patient some relief from pain, but the other medicines recommended by Streeter would have hastened his decline. The false prescription arose from a misunderstanding of the effects of the disease: it was believed that the dysentery 'germ' ulcerated and inflamed the bowel so that it became blocked. It was the doctor's task 'to clear the passage' and when this had been accomplished the patient was 'out of immediate danger'. In fact, the use of purgatives and emetics was likely to increase the loss of fluid which doctors later believed it was their task to prevent. The shortage of drugs deplored by the medical officers during the early weeks of the epidemic may have been to the advantage of the suffering labourers.

Simson, who had accepted a temporary position on the Lakekamu, stayed only long enough to see that the government's apparent foresight gave him few advantages. The epidemic had arrived before the promised tent hospital, the timber for a prefabricated building, and the nurses. When Streeter took over in March he found two European patients in tents near the government station and, about 200 yards away, eighteen labourers in 'three dirty, dilapidated native buildings', a poor supply of medicines, and a staff of one untrained European attendant and three untrained Papuans.

Before going on leave in June Murray could assure the Minister that he had agreed to give Streeter everything he asked for except a pair of handcuffs. This was true; Streeter's problem was getting stores and staff to the goldfield. There was often over a month between incoming mails and it was difficult to get bulky supplies to Nepa. Because the government did not have sufficient stores to feed the hospital patients, rations had to be bought from the traders on the field, and they were not always well stocked. At the end of May and July rice, the basic food of the labourers, was unavailable. When the *Merrie England* was damaged in a storm and had to go to Australia for repairs it was even more difficult for the government to get staff and equipment to the field. A portable house to be used as a hospital arrived on 6 April, a marquee to house European patients on 28 April, the nurses Combley and Nowland, whom the Anglican Mission had offered in January, arrived in June, and Nurse MacDonald, who was delayed in Australia during April while it was decided what colour dress she should wear, finally reached the Lakekamu in October. At the height of the epidemic Streeter had to use the resources already on the field. The miners and the labourers built more bush shelters and he employed more unskilled attendants. Satisfied with the work of the builders, Streeter had little sympathy for his Papuan orderlies: 'Natives', he reported, 'are of little use as wardsmen, as they are unreliable and careless, and are in a state of constant fear when in attendance on infectious cases.' He might have conceded that their fears were well-based.

Government officers were probably most effective when they enforced the regulations made under the Native Labour Ordinance to prevent the spread of disease. Murray claimed that he had asked Streeter to go round the field periodically to discover and remove the causes of the epidemic. This was, said Murray, the only time he had attempted to direct the work of the medical officer. Streeter refused: he argued that the inspections would be useless and in any case he was physically incapable of doing the travelling. But at least after April the three field officers made frequent tours of the camps, instructing the miners to make adequate shelters, clear the surrounding undergrowth, get rid of old tins and rubbish, dig proper latrines and send all sick labourers to hospital. In June Bertram Brammell, the Commissioner for Native Affairs and Control, inspected the goldfield and reminded the government officers to see that each labourer was supplied with a blanket, that his quarters were dry and warm and the floor was above ground level, and that where there were more than ten labourers, the quarters were to be inspected daily by the employer or another European. The steep gullies, dense undergrowth and over 200 inches of rain which fell in the first year the field was worked made it difficult for employers to provide reasonable conditions for their labourers, and government officers took few employers to court. On 20 April Nicholls visited the camps on Ironstone Creek and ordered four labourers to hospital; on 25 April, two of the employers, Fred Kruger and Peter Dowell, were each charged with failing to send a labourer to hospital. In July Dowell was summonsed for not having a pit latrine, and W.S. Pinney was charged with removing a labourer from hospital without permission. While prepared to pay labourers above the minimum of 10s. a month and provide more than the prescribed minimum rations (if they could get them), the miners had to be constantly reminded about keeping their camps in order and they were reluctant to send labourers to hospital. It was not merely the inconvenience or loss of the labourers' services. Murray wrote that sometimes when a miner told a 'boy' he had to go to hospital, the labourer wept, divided his possessions among his friends, arranged for some things to be sent home, and went to the hospital to die. It was 'pretty tough', Frank Pryke said, when a labourer asked his employer to hit him on the head with a tomahawk rather than send him to hospital. In these circumstances some miners allowed sick labourers to stay in camp, particularly when employer and employee had been together for a long time. And there was some evidence to support the beliefs of the labourers and the actions of those miners who did not send labourers to hospital. Over one-third of the men admitted to hospital died there, whereas when Robert Elliott and Dave Davies left in November because their labourers' contracts had expired, they took with them the twenty-six men that they had brought to the field: none had died, none had deserted, and the eighteen who had contracted dysentery had recovered in camp before government officers enforced the regulation making hospitalisation of the sick compulsory. The case of Elliott and Davies was exceptional but it helped miners believe that they were justified in letting the sick stay in camp. In a letter to Dan, Frank Pryke wrote with his normal honesty of the forces affecting those miners keen to work valuable ground, concerned about the welfare of their labourers, and distressed at the 'little hell' around them:

> Living here is very expensive as you have to buy a lot of medicines and luxuries for the nigs to keep them alive at all and then you cannot get much work out of them as it is not safe to drive them and I tell you it grieves a man to lose one of them especially if he is a good boy. I have several of our old boys here. Bete and Gelua amongst them. I had a big contest to save Gelua as he was laid up for a month ... .

Gelua and Bete both survived to be paid £9 for twelve months' work.

The government struggled to prevent the spread of the epidemic from the goldfield. The first regulations controlling the movement of labourers from the Lakekamu were replaced in July by a more general regulation under which an area could be proclaimed a 'centre of epidemic disease'. Labourers leaving a centre had to be taken either to Port Moresby or Samarai where the master of the vessel had to report to the Chief Medical Officer. If the C.M.O. detained the labourers, then the employer continued to be responsible for paying for their rations. In centres of epidemic it was compulsory to send sick labourers to hospital; previously the employer could wait until he received a specific direction from the government officer. Captains of ships on which cases of dysentery occurred had to sail direct to Port Moresby or Samarai. Penalties under the new regulations were greater than they were for most other breaches of the Native Labour Regulations. Employers could lose the right to employ labourers, be fined £50, or be imprisoned with hard labour for six months. When Captain A.C. Reid was fined £50 for failing to go directly to port after a labourer on board the *Kia Ora* contracted dysentery Whittens protested strongly, claiming that

Reid (and most other captains) was ignorant of the regulations. Whittens argued that Reid had attempted to contact the doctor in Port Moresby, and that the regulation was unreasonable in requiring a captain to make an accurate diagnosis of the ills of his passengers. The Executive Council compromised and reduced the fine to £25. In April 1911 a new regulation was introduced under which any camp or plantation on which a case of dysentery occurred immediately became an 'infected centre' and comprehensive provisions for the isolation of the area came into force. But by this time the government was more concerned with outbreaks of dysentery in the Port Moresby area where eighty-one labourers had died between July 1910 and March 1911.

Although the epidemic on the Lakekamu appeared to have reached its height in April 1910 this was not clear to Staniforth Smith, acting as Administrator in Murray's absence. Smith told the Minister that while the number of deaths had fallen in June so had the total number of labourers. Smith concluded: 'after a period of five months, the epidemic is raging as seriously as ever'. The next day he left Port Moresby for the Lakekamu taking with him the Chief Medical Officer and the Government Secretary so that he could hold a meeting of the Executive Council on the field if necessary. 'If no improvement is manifest', he wrote, 'I shall, in spite of the large interests involved, have no hesitation in closing the field ... '. In the event of the field being closed, the sick were to be moved to high ground where they were to be cared for. All other labourers were to be taken to camps at the mouth of the Lakekamu where they were to be picked up by the *Merrie England* and shifted to a quarantine area on Fishermans Island near Port Moresby. But when Smith reached the Lakekamu on 10 July he found only six men suffering from dysentery and nine convalescing. He decided not to close the field. At a public meeting the miners agreed that 'The government had done all they could to meet the difficulties and overcome the epidemic of dysentery'. Either Smith suppressed some evidence or the miners decided there was no point in being critical after the event. Frank Pryke in his letters had been strongly opposed to the government's plans to take labourers from the area and thought Streeter might have 'put up a record here as he must have lost at least 200 cases in a little over six months'.

By September there were no cases of dysentery, but there was an increase in the incidence of beriberi. Labourers complained of pains or heaviness in the limbs, then partly lost the use of their legs. Sometimes there were few outward signs of illness and an employer might think the labourer was malingering; but after suffering acute distress the patient died. Beriberi had occurred in other parts of New Guinea but its cause was unknown. The most common explanations were that it was a result of either dampness or some diet deficiency. When it appeared on the

Table 12
Lakekamu Goldfield Death Rate

Total number of labourers on 30 June		Deaths in preceding 12 months
1910	643	258 (6 months only)
1911	428	57
1912	507	26
1913	141	8
1914	386	1
1915	224	11 (including 5 in a mining accident)

Lakekamu Streeter was not in fact certain that the labourers were suffering from beriberi and he preferred the diagnosis, 'Epidemic ideopathic polyneuritis'. In admitting that his conclusions were based on inadequate evidence, Streeter pointed out that he had carried out no post-mortems, had frequently been ill with malaria, and although he had asked for a microscope none had been supplied. He thought it appropriate to remind his superiors of the 'famous epigram of Dr. Ross, a mighty enemy of tropical diseases. ... "The success of Imperialism depends upon success with the microscope"'. Streeter noted that on the Lakekamu most who had the disease had also suffered from dysentery; they were what the miners called 'hard-workers', so perhaps over-exertion was a factor; the sick tended to come from particular areas; and the inclusion of rice in their diet seemed to have no ill-effects. While calling for further investigation, Streeter thought that the disease may have resulted from 'mineral poisons' in the water. But Dr. R. Fleming Jones, the Resident Medical Officer at Samarai, informed his colleagues that studies in the Malay States had established the cause of beriberi: the rice theory was right after all. Each year, said Jones, a new explanation was put forward but the rice theory, like Banquo's Ghost, could not be put to rest. Now it had been shown that beriberi resulted from the absence of some substance in milled rice. Smith instructed the Resident Magistrate at Nepa to increase the size of the government gardens, and he arranged for peas and lime juice to be brought from Cairns. But the disease persisted. From September 1910 until July 1911, fourteen labourers died, and the following year two died of beriberi and another two deaths were classified as 'mouth disease', presumably also caused by diet deficiency. The Medical Officer on the Lakekamu wrote that beriberi 'still seems to lurk in our midst, and no doubt will continue to do so until the use of polished rice as a stock food is abandoned'. Yet the regulations concerning rations remained unchanged: it was still possible for an employer to feed his labourers on a diet of $10\frac{1}{2}$ lbs rice, one lb biscuits, and one lb meat per week. The government's continued approval of a diet largely dependent on rice without specifying its type was all the more surprising in the light of events in other parts of the territory: in 1910-11, sixty-seven labourers died of beriberi and on Woodlark Island 131 cases were reported among 574 labourers. But at least on the Lakekamu, the advice, 'isolation of the sick', concern with the comfort and cleanliness of camps, and time, led to an improvement in health.

It seemed that the effectiveness of the government's policies was to be tested again in 1912. On 27 December 1911 Dr William Giblin found that one of Whittens' labourers from Tiveri had dysentery. Giblin reported that the rest of the labourers were healthy, and the employers kept adequate supplies of disinfectant. On 3 January three labourers had dysentery and the man admitted to hospital on 27 December had died. The police cleaned out their barracks and disinfected their blankets. Lionel Armit inspected camps, reminded miners of the regulations about the control of dysentery, and sent Lance-Corporal Lagoni to Port Moresby to inform the Government Secretary of the outbreak. When one of Henry Fletcher's men contracted dysentery Fletcher burned his old camp and moved to a new site. The attempts to contain the epidemic had little effect. By the end of the month thirty-two labourers had been admitted to hospital with dysentery. By gathering at Tiveri for Christmas miners and labourers had spread the infection widely. But the disease was apparently of a mild form; few died and the epidemic had passed by the end of February. Now malaria, which had been absent from the field when the miners first arrived, became 'abundant in every camp'. Stagnant pools were drained or covered with kerosene to no effect: the malaria depressed and debilitated miners, government officials and labourers, but it rarely killed.

Bark belt, Toaripi, after Haddon 1894

In mid-1911 Giblin reported that the work of a doctor on the field was 'extremely light, in fact, monotonously so'. A year later, Frederick Rorke, Giblin's successor, left the field and the hospital was closed. To that date the Lakekamu had produced about 17,500 ounces of gold worth £65,600 at a cost of 331 labourers' lives: or 53 ounces of gold worth £200 for each dead labourer. When the dysentery had just passed its peak, some of the old miners told Staniforth Smith that it had been worse on the Gira: they were probably right.

News of Papuans dying on the Lakekamu caused members of the Commonwealth Parliament to examine briefly Australia's stewardship. The outlawing of Joe O'Brien was the only other event to stir the members' interest in the behaviour of men on the Papuan goldfields. In August 1910 William Higgs, the Labor member for the Queensland electorate of Capricornia, asked whether it was true that 300 out of 1000 labourers had died, what caused their deaths, who employed them, what were they paid, and what rations they were given. Egerton Batchelor, the Minister for External Affairs in the Labor Government, replied: 269 had then died, he listed the causes of death, he did not know the names of the employers, the labourers were paid from 10s. to £1 a month, he quoted the minimum rations prescribed under the native labour regulations, and in the government's defence he pointed out that hospitals, medicines and trained staff had been provided. Later in the debate on the estimates, Higgs made clear what lay behind his questions. A journalist, Higgs had previously sat in the Senate from 1901 to 1906 where he had been one of few members to oppose the transfer of British New Guinea to Australian control. 'Now', he said, 'we have no right to use our superior strength to make wealth out of the natives of New Guinea.' He went on to support his conclusion with a rambling but consistent argument. Imperialism was in the economic interest of one class in the metropolitan power and to the detriment of the people in the colony. Capitalists, said Higgs, not content with low interest rates in the old country, wanted to invest abroad where they could obtain labour for 2s. 6d. a week. The labourers were 'treated as slaves', poorly housed and fed on a monotonous diet which was responsible for the epidemics of dysentery. Now the companies wanted to recruit not just men, but whole families to live on plantations, a move which would lead to the break-up of the villages and a decline in population. If Australia's administration was 'primarily in the interests of the natives', then the government, said Higgs, should see its money spent to assist Papuans and not the companies who 'exploited these guileless children'. The Papuans were not lazy, he claimed, they were skilled craftsmen and anyone who doubted this could have a look at their art in the Melbourne Museum: they should be taught trades so that they 'might produce wealth for themselves in co-operatives'.

Batchelor, who had visited Papua, answered Higgs much as Murray would have. The Papuans could not be left in 'a savage condition'. It was in the interests of the Papuan himself that he was being encouraged to work: the government aimed to develop his agricultural instincts and the only way to do so was to permit his employment. Government officials ensured that the labourer freely and knowingly entered a contract and that he was treated humanely by his employer. 'The Government', said Batchelor, 'would be no party to the surrender of the natives to capitalists.' But in two interjections, Higgs showed that he wanted not protection of the Papuans, but a fundamental redirection of policy. Did the government, he asked, spend any money teaching Papuan men a trade or educating their children? Batchelor was not prepared to say offhand that the government spent nothing directly on education, but he could have told members that the government provided no schools for Papuans and that he had just received a letter from Staniforth Smith stating that 'practically the whole of the white residents, including officials' objected to 'natives and half-castes' attending a proposed government school in Port Moresby. Later Higgs pointed out that acceptance of the government's arguments depended on whether you thought 'the native should work for himself or for the capitalists'.

The implications of Higgs's interjections may not have been clear to the House. When Batchelor claimed that government was in the interests of Papuans, he was saying that government officials were instructed to see that employers met the minimum conditions set by the government: but Higgs thought that if a government acted in the interests of Papuans it would give them opportunities for advancement so that they did not stay low-paid, unskilled, indentured labourers. Higgs had a vision of the development of Papua by Papuans: Batchelor spoke of development by Australians made benevolent by government supervision. At the time Higgs spoke most indentured labourers were employed on plantations, but the group who had initially aroused his interest were the teams employed by alluvial miners. The diggers staked by trading companies, hopeful of finding good dirt and getting full shammies before the 'boys'' contracts run out, the battlers of Kalgoorlie and the Palmer, had become capitalists and exploiters in Papua. At the close of the debate, Higgs suggested that the Minister consider 'giving the white residents an elected representative ... in this House, and, further, of giving the natives a native representative when one is found able to speak our language intelligently'. For the next thirty years few other members saw that the government's proclaimed intention to act 'primarily in the interests of the natives' meant only that it would prevent the worst abuses of the people in its territories.

The few other members who joined the debate wanted the government to be more vigilant in its protection of the Papuan labourers. Two members pointed out that the death rate among labourers was higher than it had been among the kanakas in Queensland. All were concerned that 'no stigma shall rest on the reputation of the Commonwealth in connection with the administration of Papua'. Several members assumed that the Papuan was a 'child of nature' who lived in a 'happy village' and therefore it was not 'civilisation' to take him to labour on plantations or mines where the death rate was high. Advanced in the government's defence, this assumption undermined Batchelor's position for it was basic to his argument that 'the natives are very much better off, even as they are treated on the worst labour plantations, than they were originally'. In further defence of Australia's actions in Papua, Batchelor pointed out that the death rate in 1909-10 was exceptional and that more people had died in the villages. He may well have been right. For several years there had been outbreaks of dysentery in the villages and, in 1909, both dysentery and whooping-cough had killed many people east of Yule Island. In the next year, when Papuan men had to decide whether they preferred the village or the plantation, many chose the plantation: in 1909-10, 2407 were engaged for plantation labour, and in 1910-11, 4514 signed-on.

While the ban on recruiting labour for the Lakekamu was in force, the goldfield's population steadily declined. Labourers who died, deserted, or were paid-off when their contracts expired could not be replaced. In July 1910 there were about 600 labourers on the field; by the end of the year there were only 343. On Ironstone Creek where there was 'good gold' the miners 'deplored the absence of labour to enable them to carry on'. Following advice from the Resident Medical Officer the government revoked the proclamation against recruiting in November, and a few miners left immediately to recruit new teams. Knowing that news of the death rate on the field had been spread widely in recruiting areas by government officers distributing trade goods to the next of kin, miners thought it might be difficult to persuade men to sign-on. But in January Davies returned with twelve labourers, the first new recruits. He had been away two months; Lyons thought his trip 'singularly successful'. Frank and Jim Pryke found recruiting 'pretty tough' but they signed-on thirty-six men and they paid them the same wages as the previous year, 15s. a month. By March the *Bulldog* was landing over forty labourers at Tiveri on each trip, and the sight put 'fresh heart into the miners'. There were over 400 labourers on the Lakekamu in June 1911, and 500 the next year.

Table 13
Lakekamu Goldfield production and population

Year	Miners (30 June)	Labourers (30 June)	Production (ounces)
1909/10	61	643	3000
1910/11	38	428	8000
1911/12	33	495	6500
1912/13	14	141	5000
1913/14	29	386	4000
1914/15	22	224	3000
1915/16	27	340	2715
1916/17	14	280*	3400
1917/18	5	55	194
1918/19	5	46	350
1919/20	4	32	500

* The total number of labourers on the field during the year. The wardens could not obtain accurate figures for production.

In spite of the miners' fears most recruits continued to come from the D'Entrecasteaux Islands. In 1912 the warden reported that some were signing-on for their second term on the Lakekamu and 'nearly all had experience on the older fields of the Territory'. For almost twenty years now the eastern islanders had worked for alluvial miners, and a few had probably worked gold for themselves. They knew their trade, responding to the *taubada's* (master's) enthusiasm when rich alluvial was going through the *bokis*. Their aptitude convinced the miners that the 'east end boys' were in some way more intelligent and better fitted for mining than western men. When a number of Milne Bay labourers deserted in 1913, the Assistant Resident Magistrate thought it was because not enough gold was being found. 'These boys', he wrote, 'like to work where there is gold, and if they find that after a week or a month's work there is nothing to show for it, they get very dissatisfied.' Kiwai, Bamu, Purari, Goaribari, Orokaiva and Paiwa men also worked on the Lakekamu. Sometimes they were the first to recruit from their area: they had to learn to use the shovel and cook strange foods as well as master the skills of bush camp and mine. Under experienced miners, recruits normally settled quickly to the work, although problems arose with those groups who had no language in common with the *taubadas* and there was no interpreter on the field: they could not learn what was expected of them. At the end of 1911 W. Newland was getting an ounce a day and he attempted to employ 'raw, Western Division men', but gave up in despair after a day and a half. Recruits from Paiwa on Goodenough Bay were on the field for a year before a Paiwa-speaking policeman arrived to tell them that they were not supposed to run away and live by robbing the camps. Because men from the Western Division, the Gulf, and the Northern Division had little experience of working gold, and some miners were prejudiced against them, they were often employed only as carriers, taking the stores from Tiveri out to the camps or, when the river was low, meeting the *Bulldog* and packing the rations to the Tiveri landing.

There was a lot of 'dead work' on the Lakekamu before the final washing of the gold, and the miners employed bigger teams than on other Papuan goldfields. Heavily timbered areas had to be cleared, and a head of water taken from upstream to give sufficient pressure, particularly if the claim was on the terraces above the valley floor. Races up to a mile in length had to be dug following the valley contour, crossing gullies on flumes made of hollowed logs or bark, and tunnelling through spurs. Within six months the ground between Ironstone and Rocky Creeks had 'been torn out of all recognition' by men equipped only with pick, shovel and axe.

The labourers, at first living in tents or rough shelters, built substantial bush houses with wooden sleeping platforms 'close fitted', according to regulation, 'so as to prevent draughts'. After the first year the only store on the field was Whittens at Tiveri, and labourers had to carry all the stores from Tiveri to the claims; a one day walk to Ironstone and several days to the camps further out. To add variety and quantity to the stores the 'shooting boys' brought in wallaby, cassowary and pigeon to put with the rice. When the *Bulldog* failed to arrive and stores were short some labourers camped at the swamps, making sago for those working the claim. During the first year the heavy rains frequently washed out bridges and roadworks and flooded the claims; yet there were still times when there was not enough water to run all the races and the teams just 'dodged along'. Sunday was, by regulation, a day of rest for all; a point impressed on miners and labourers in 1911, when the Resident Magistrate prosecuted two employers for working their teams on the Sabbath. At Christmas police and labourers competed in sports and danced at Nepa and Tiveri.

When they made periodic inspections Resident Magistrates and officers of the newly formed Department of Native Affairs and Control asked labourers if they had any complaints, the labourers normally had little to say. But their habit of deserting was evidence of dissatisfaction. In the first eighteen months after the opening of the field 180 labourers were charged with desertion at Nepa, 179 were convicted and one case was adjourned for want of an interpreter. The rate of desertion was high, partly because of the fear of dying of dysentery, but it was still less than it had been on the Gira. Normally three or four policemen were out looking for deserters, who were not easy to find. Labourers from the

Lakekamu showed the same skills and determination as escapees from the Northern Division and Woodlark. They were picked up at Moveave, nearly 70 miles from the mining camps, and along the coast to the west at Kerema and to the east at Kairuku and Manumanu. In 1919 two men were arrested for desertion at Efogi. They had travelled about 250 miles to reach their home village on the Port Moresby-Kokoda track. When William Murray reported that two of his labourers had made rafts and gone down the Lakekamu, Humphries assumed there was little chance of catching them: they would not go to their home village on Kerema Bay but to Port Moresby or some other station and sign-on under another name. Some labourers must have been surprised at the tenacity of the system they entered by signing-on. In April 1918 two policemen brought back from Kerema four labourers reported missing by Andrew Gillespie in September 1915. In the following year Gillespie continued to use the law to his advantage, charging several of his labourers with absenting themselves from work or failing to 'show ordinary diligence'.

While Resident Magistrates were charged with seeing that all labourers knew what they were doing when they signed-on, some labourers only learned of the intransigent nature of a contract when they decided they had had enough of working on the goldfields. On 10 November 1916 Humphries, after gaoling two of James Preston's labourers for desertion, and discharging one, wrote in the station journal: 'It is quite a common thing for Preston's boys to turn up at the Station (carrying their swag and gear) and to say that they have come because they don't want to work for "money"'. In reply to Humphries's questions the labourers said that Preston did not hit them, and they were well fed. Humphries then attempted to explain to them that they had to work for Preston or be sentenced to hard labour on the station and then return to Preston for more work. Two days later the labourer discharged on the 10th came back to Nepa: he still did not want the 'money'. Then another turned up carrying his pannikin, plate, spoon, mosquito net and blanket. He claimed he was sick but he appeared normal to Humphries who watched him prepare to settle at the station. By 20 November Preston had only two labourers left: his Bamu recruits had learned that under the indenture system a re-negotiation of the contract was impossible. The alternatives were to suffer or run. Escape was simple for the Western Division men: they entered the vast sago swamps to the south of the station where food was plentiful and capture difficult. The country was similar to that near their home villages. One man, Humphries reported, had camped for days in a hollow tree trunk from which he had watched the passing police. Other deserters lived in abandoned camps and in 1910 Lyons reported that five deserters had been living for three months in the bush near Tiveri, fed by the Kiwai carriers based there. Not all deserters were so fortunate. In 1910,

A 'team' of labourers from Orokolo on the Lakekamu Goldfield, 1914
PHOTOGRAPH: FRANK PRYKE

Two labourers bringing in a cassowary to feed the team, 1914
PHOTOGRAPH: FRANK PRYKE

Pegolo, a Milne Bay labourer, was wounded and his two companions were killed by men who attacked them on their journey to the coast, and in 1911 Moveave villagers were said to have killed some 'runaways'. But generally the life of the deserter on the Lakekamu was far less hazardous than on the Northern Division fields where the way of escape lay through more numerous and warlike peoples.

There was also less violence between employers and labourers, partly because the Lakekamu was more closely supervised by government officers than any of the other alluvial fields. In 1913 Lagalaga was committed for trial for assault occasioning grevious bodily harm to James Mulholland. But when Mulholland died in the Port Moresby Hospital the charge was changed to manslaughter, and Lagalaga was sentenced to five years' hard labour. The case was unique: any other cases of labourers assaulting the *taubadas* were minor ones dealt with in the Resident Magistrate's court. Instances of miners savagely assaulting labourers were also infrequent, and there is no evidence to show that such cases were not taken before the Central Court. In fact the senior government officers, Murray, Campbell, Herbert and Bramell, were determined to see that the law was observed and that the courts would receive and consider evidence from Papuans. At least twice field officers took miners before the central court for gross mistreatment of labourers. In August 1911 Lyons committed Henry Jeffrey for trial for assaulting Ewarupa, an indentured labourer, and 'doing him bodily harm'. While collecting evidence, holding a preliminary hearing and sending one European and nine Papuan witnesses to Port Moresby was a burden for the government, the costs for Jeffrey were greater. He was ground-sluicing a claim at Robertson's Gully just to the north of Nepa when he was instructed to go to Port Moresby for trial. 'As the notice was so short he simply had to let all his work go. The last day's boxing he did before commencing ground-sluicing yielding him 4 ozs. gold.' In Port Moresby Judge Herbert found Jeffrey guilty of assault causing bodily harm, and fined him £20. Jeffrey returned to the field where two months later he was again fined for assault, although the records do not say whom he assaulted. His antagonisms were not exclusively inter-racial. Later in 1912 the warden had to resolve a dispute between Jeffrey and another miner and he was involved in a fight with Lumley at the store. On this occasion Lumley conceded that he was at fault: 'I started it through being sober & touchy & thin skinned & just recovering from a near approach tothorrers [to the horrors]. Anyhow Jeff finished the bastard fite. My beak aint set straight yet & I never put a mark on Jeff.' Shortly after Lumley 'picked a row with Joe Sloane [in Port Moresby] and got a devil of a hammering'. If the miners sometimes engaged in 'physical argument' among themselves they could not be expected to refrain from throwing punches at their labourers. The question is not whether miners struck their labourers, but whether they treated them brutally.

Miners at Sunset Camp, Lakekamu Goldfield, 1914
PHOTOGRAPH: FRANK PRYKE

Having shown in its treatment of Jeffrey that it was prepared to take men to court in spite of inconvenience and some miners' prejudices, the government faced a more complex and serious case in 1915. In May the Resident Magistrate, Eric Oldham, while questioning some men employed by J. Regan, mentioned the name of another labourer, Jiaro. One of the men present remarked, 'He is dead — Jerry killed him'. 'Jerry' was William Kelly, a miner who had had some of Regan's labourers transferred to him earlier in the year. At the time, Jiaro's death had been reported and as Regan said he had been ill, Oldham had made no further investigations. Now a different story emerged. Oldham took statements from two labourers who claimed to be witnesses. One, Miara, said:

> Before Jerry hit one boy named Jiaro plenty time with a big stick. The stick was about as big as my arm. He hit him so much that plenty excrement come out along behind. When Jerry finish hit him Jiaro

fell down and Ai-i-ia and me picked him up and put him along fly house [tent].

Miara believed that Jiaro had been beaten because he left a bag of rice on the road while carrying goods up from the store, and when Kelly told him to get it he had refused. Jiaro died the day after he was beaten. Ai-i-ia's evidence agreed almost exactly with that of Miara. Other labourers had knowledge of the case, but one group was from Paiwa and Oldham had no interpreter, and the others from Goaribari had completed their contracts and left the field. When Murray learned of the case he instructed Oldham to take evidence from the Paiwa labourers and 'take such steps against W. Kelly as the evidence warrants'. The Resident Magistrate in the Delta was asked to contact the Goaribari labourers, but he reported that there was no point in doing so because the Goaribari labourers had been paid off before Jiaro had died. (On this point, the Resident Magistrate is contradicted by the dates in the Nepa station journal.) In September Oldham committed Kelly for trial charged with murder. In the Central Court the charge was reduced to causing grievous bodily harm and Herbert ruled that Kelly was not guilty. Herbert or Murray may have reduced the charge to avoid having to call a jury, mandatory when Europeans were tried for crimes punishable by death. After his acquittal Kelly decided that he had been persecuted by Oldham and threatened to sue the government, but Murray assured the Minister that Kelly had no case against the government. In fact Murray thought that Oldham had acted correctly in charging Kelly with murder and although the case against Kelly was not a strong one, he was sure that had Herbert found Kelly guilty of murder no court would have quashed the conviction.

Minor cases of assault were also infrequent. Only two miners were convicted in the Resident Magistrate's court in the four years 1910-13. As in their policing of the regulations controlling the spread of dysentery, the government officers were more inclined to warn than prosecute. For example, Oldham recorded in the station journal for April 1914: 'Mr. C. Castleton, cautioned as to his treatment of natives.' Later in the same year, Thomas Murray, Regan and Sloane were also warned. None was taken to court, but James Wallace and Murray had their permits to hire labourers from Whittens' cancelled, a more severe punishment than the normal £5 fine imposed by Resident Magistrates.

Apparently indulgence could be as much a crime as severity. In 1917, Humphries inspected Jerry Ford's camp at Sunset Creek and reported:

> As this man's age increases he becomes less fit in my opinion to control natives. They sleep in his hut, help themselves to his stores,

and in fact just do as they like. He was put under a Prohibition Order in Samarai, but that has not prevented him from getting liquor ... he is irresponsible and incorrigible.

Ford was fined for a breach of the Native Labour Ordinance, but the labourers' contracts were not cancelled.

The labourer who went to the Lakekamu was most likely to appear in court because he ran away, and a few went before the magistrate for failing to work hard enough or for stealing or fighting. A government officer would also send him to Port Moresby if he had venereal disease, or have him sent home if he were too ill to work. But he would also have learnt that the court sometimes had power over the *taubada*. Each year one or two miners were fined for working their labourers on Sunday or not having a pit latrine, or for ill-treating labourers. In the courts the labourer could speak against his employer, and weight was given to his evidence. This was not the case in the other major Australian territory. In the Northern Territory European juries heard all criminal cases and there, Mr Justice Bevan said, when a European was charged

> with an offence against a coloured man, no matter what the evidence may be, the matter is decided on the question of colour ... The injustice is made more glaring by the fact that it is not that 'coloured' evidence is disbelieved *qua* 'coloured' evidence, but it is disbelieved because it is offered against a European; the same 'coloured' evidence against a 'coloured' man will be accepted, and, rightly, acted upon.

Murray later recalled that employers had reacted in horror when they found that the government intended to enforce the Native Labour Ordinance and accept Papuan evidence in the courts. But both Murray and Batchelor were incorrect to conclude that the government therefore acted primarily in the interests of Papuans. What it had done was make employers behave as enlightened self-interest should have directed them. The government attempted to ensure that the 'boys' did not run away, were fit to work, and unlikely to deter others from trying the same experience. But in forcing employers to act in their own interests, Murray's officers gave Papuans rights which were not granted to non-whites in the Northern Territory and which were resented by some employers in Papua.

Mining on wet cliff faces was sometimes a dangerous occupation. In November 1910 there were two accidents: in the first a man was killed and another injured, in the second a man's leg was crushed and he died after it was amputated by Giblin at Nepa. The death of only one other labourer killed in a mining accident was recorded in the station journals until 1915 when J. Reilly ... had a terrible accident in his claim whereby five boys were killed'. Taking all the police and prisoners Chisholm went to Reilly's claim but the men were buried under many tons of earth and their bodies were not recovered until the next day. Under the Queensland Mining Act of 1898, an inquiry had to be held before four experienced miners, and another miner assisted the warden in the conduct of the inquiry. The report on the accident at Reilly's claim was referred to the Director of Public Works, who found no evidence of negligence. But the application of the Queensland Act to Papua made possible an abuse not considered by those who passed the legislation: in Papua the men who heard and assisted the inquiry were more fellow-employers of Papuan indentured labour than they were fellow-miners. The difference may explain the failure of the warden and the four miners to agree in a second inquiry in 1915. Tomowi-u-ia, employed by Sloane, had his leg broken, and when the miners came to apportion blame they decided that the accident was due to 'the negligence of the native labourers'. The warden, showing more concern for the labourers' welfare and less respect for their ability, thought that Tomowi-u-ia's leg was broken by a fall of rock (not a point of dispute) and that Sloane should not have left the removal of the rock 'to the intelligence of his native labourers'. While the miners had given themselves absolution, Sloane had demonstrated his personal concern for Tomowi-u-ia by accompanying him over 100 miles to Yule Island where he had him put on a boat for the Port Moresby Native Hospital.

Only one European was killed in a mining accident. (Not that they led uneventful lives: one was shot by a Tiveri storekeeper, one went mad, and another was drowned.) Early on Saturday morning 7 September 1912 D. James came into the station 'in an excited state' and reported that one of Edward Jones's labourers had just arrived and said his boss had blown one of his hands off with dynamite. Oldham, Reilly, three police and twenty labourers left for Jones's claim. Later other miners set out to help. Believing that Jones's camp was within a day's walk, Oldham and Reilly took no stores and when they failed to reach Jones by Saturday night they were forced to camp without food or tents. The next day when Reilly became exhausted, Oldham, the police and labourers went on and had Jones ready to shift when three other miners and their labourers arrived. They carried Jones to Tiveri and then took him by boat to the mouth of the Lakekamu where they caught the *Bulldog*, but the *Bulldog*, short of benzine, was forced to return to Motumotu. Oldham then took the Kerema whaleboat and after stopping overnight at Keveri village they reached Yule Island, where the mission sisters attended Jones before Oldham commandeered the *Maimera* in the absence of its skipper. Half way to Port Moresby they sighted the *Kia Ora* which towed them into Port. Jones was taken to hospital where he was operated on but died of tetanus. Oldham and the police had covered about 250 miles in eight days. Commending Oldham, Lyons wrote: 'I think this performance is unprecedented so far as Papua is concerned ... .'

Wooden club, Kukukuku, after Blackwood 1950

At first concentrated at Ironstone and Rocky Creeks, the miners soon began prospecting in neighbouring valleys, cutting the tracks originally marked by Crowe and the Prykes. Early in 1910 Fred Kruger began mining at Cassowary Creek, a tributary of the Arabi River about 14 miles east of Tiveri. After nearly a year's work Kruger was said to have taken out 400 ounces of gold. By June 1911 twelve other miners were working claims at Cassowary and they spread out in what became known as Big Cassowary, Little Cassowary and Preston's. To the north of Ironstone gold was found on the Tiveri, Fish Creek and Robertson's Gulley. In the west there were five miners working at Tailend Creek or Thirty-five Mile on the upper Olipai early in 1911, but they abandoned the workings as the cost of packing stores was higher than the returns. More substantial finds were made west of Tiveri at Twisty Creek and south-west of the Olipai at Mosquito and Sunset Creeks. Miners concentrated in particular areas following rumours of big 'wash-ups'. In January 1914 most miners were on Twisty Creek: the next year they were out between Mosquito Creek and the Tauri. But from 1910 to 1918 there were always some miners on Rocky and Ironstone Creeks. On the reward claim the Prykes and Matt Crowe had won over 1400 ounces of gold in their first twelve months and after they abandoned the claim in December 1911 it was re-worked by several miners.

From early in 1911 when the Lakekamu was reopened to labour and it became clear that the field offered an income for only thirty or forty diggers, the miners were keen to see new country opened up. In September 1911 the miners on the Lakekamu elected Frank Pryke to lead a prospecting expedition financed equally by the miners and the government. Pryke selected Bob Elliott and Charlie Priddle to go with him. In December the prospectors and forty-one labourers established a base 120 miles up the Vailala. Finding that the Iova, a turbulent north-western tributary flowing through narrow gorges, was impossible to prospect Pryke, Priddle and thirty carriers turned east and crossed a series of gullies where they found fine 'colours'. Although they saw gardens they met no people until 20 December when, at a point Pryke thought was close to the German New Guinea border, they came to a village of twenty or thirty houses. The people accepted tobacco and appeared friendly, but as Pryke led the party on a track away from the village he was suddenly confronted by five or six men standing on a rock. He walked forward making signs to them to put their bows down; but one man released an arrow. Pryke shot and killed his attacker and the prospecting party 'shook tribe up generally'. The arrow had entered Pryke's chest 'just below left nipple and travelled down towards left kidney'. Frank pulled the arrow out then became 'pretty ill'. Fearing he might die he dictated a note to Priddle describing what had happened and largely absolving the villagers from blame. After camping for two days the party started back for the base camp carrying Pryke on a stretcher. The return journey took nine days. While Pryke recovered, Priddle and Elliott tested the Ivori, another branch of the Vailala. By their return Pryke had recovered sufficiently to accompany Priddle up the Lohiki then south across country to the government station at Kerema. Frank Pryke's recovery was only slightly less dramatic than that described by Murray in his *Annual Report*: 'Mr. Pryke is a man of iron nerve. An arrow went nearly through his body, and would probably have killed any one else: Mr. Pryke, however, simply pulled it out and went on with his prospecting.' While waiting for a boat back to the Lakekamu the prospectors tried the Murua River, but found only 'colours'. In March the Lakekamu miners learnt that the expedition had opened no new field. In spite of a second clash with villagers on the Lohiki and much tough travelling all the carriers completed the return trip on the *Bulldog*.

In April 1912 the miners raised another £150 to support a prospecting expedition and the government again provided an equal amount. Avard Newcombe, a Canadian-born engineer who had spent several years on Papuan goldfields, was chosen as leader and he selected Gordon Robertson and A.G. Hicks as companions. For three months Newcombe, Robertson, Hicks, and 'thirty-eight boys' tested the upper Tiveri, Olipai and Fish Creek. Having crossed to the Tauri they made a double-canoe and went downstream to Moveave. Before ascending the Tauri and returning overland to the Lakekamu they prospected the Maiporo and Kororo Creeks to the west. The villagers of the upper Tiveri and Olipai again showed they 'desired no intercourse with the outside world'; they accepted no trade goods left for them and their arrows wounded three carriers. In his report Newcombe, B.Sc., regretted that he had found no new goldfield for it would have helped bring 'the native population into touch with civilization and [done] something toward making them useful members of the community'. 'All hands' on the Lakekamu were 'sadly disappointed' at the expedition's failure. In October John Butler, a miner at Cassowary Creek, reported that he had made two trips west to the St Joseph River but had found neither gold nor people.

By the end of 1912 there was no talk of big 'wash-ups' and many were marking time, employing their teams on well-worked ground. Frank and Jim Pryke had left the field to invest in businesses in New South Wales and Lumley sent them news from the Lakekamu:

> Things are dam crook up here. No gold, no water & in fact you can feel a state of buggerization in the atmosphere. The blokes had a meeting on the Field last Sunday & sent an ultimatum demanding a reduction in prices all round of about 40%.

But some men still had cash for when Mulholland's leg was 'cronk' and would not heal, 'the crowd sent a list around to get him out of it for a few months. £117/-/- was hit up for him'. Later in 1912 Lumley wrote that the Lakekamu was 'dam near done for bar tucker shows'. He did not know 'where the crowd [would] go next'; he himself had already left for Woodlark. Some miners wanted to try the Fly River but the cost of an expedition would have been high. Their hopes of a new field depended on Crowe, James Preston, Edward Auerbach and William Park, who had gone up the Markham valley in German New Guinea. In Rabaul before crossing to the mainland they learnt that Piastre had won the Melbourne Cup: Crowe and Auerbach had put their money on Duke Foote. Two days after leaving the mission station on the Huon Gulf they clashed with New Guineans and they fought with them nearly every day for the rest of the trip. The Australians prospected on south New Britain before returning to Australian territory.

By 1914 Crowe was back on the Lakekamu building canoes to go up the Kunimaipa, and not finding any gold there, he joined another expedition to the head of the Tauri and Olipai.

In 1914 the miners were given a chance to prospect the Fly. On his return from an exploratory survey of the Fly and Strickland Rivers Murray brought alluvial samples which revealed a few grains of gold, and then in May Sir Rupert Clarke, a Victorian property-owner, businessman and Papuan planter, financed a well-equipped expedition. Frank Pryke had learnt, after a year as a tobacconist, seller of fancy goods and proprietor of a billiard saloon in Moree, New South Wales, that 'selling penny and halfpenny articles [was] a mighty slow way of accumulating a fortune', and his partner was 'as mad as a dingo'. Jim too was 'talking Papua again' so the Prykes needed little persuasion before they joined Clarke as prospectors. With Clarke and Archie MacAlpine, the manager of Clarke's Kanosia plantation, the Prykes explored the Black and Alice, upper tributaries of the Fly and then after Clarke and MacAlpine returned to Daru the Prykes dragged canoes and scrambled up the steep valleys of the Tully and the Alice. They had gone further up the Fly than any previous expedition, and although they

Labourers from Milne Bay with canoe that they have made to allow miners to prospect the tributaries of the upper Fly River, 1914. Frank Pryke is on the left.
PHOTOGRAPH: FRANK PRYKE

Prospector trading with people on the upper Fly, Pryke expedition, 1914
PHOTOGRAPH: FRANK PRYKE

washed a few 'colours' they found nothing worth working. On their way down the river the Prykes stopped to trade just below D'Albertis's Attack Point. At first the people appeared friendly but when they became anxious to get the prospectors ashore, offering women as an inducement, the Prykes became suspicious and attempted to move the launch into deeper water. Immediately a burning stick was thrown onto the awning of the launch and a shower of arrows followed. The Prykes returned the fire. Among the prospecting party five labourers received arrow wounds. Jim Pryke was 'scratched' on the stomach and an arrow went through Frank's forearm. On shore the prospectors found the body of one Fly River man shot through the heart. They destroyed canoes and houses and cut down coconut trees before going down to Daru. Apparently still preferring Papua to town life in New South Wales the Prykes returned to the Lakekamu in November and went up the Olipai.

Early in 1916 Henry Fletcher reported finding gold inland from Vilirupu to the west of the old Keveri field in the Central Division. He was granted a reward claim one day's walk from the landing on the Imila River. It was the first reward claim granted in Papua since the opening of the Lakekamu six years earlier. About twelve other miners joined Fletcher but none found sufficient gold to keep them at Vilirupu. Most of the Lakekamu miners had waited hoping that extensive new ground would be opened near Fletcher's claim but it was not; and in 1917 Sloane, Arnold and Robertson reported that they had prospected much of the Keveri country without success. All three were restless prospectors. Besides accompanying Newcombe's expedition in 1912, Robertson had gone with Arnold to prospect between the Tauri and Kerema in 1914. In the same year Sloane and Preston had investigated country on the west of the Lakekamu.

Often prospectors were beyond the government's protection, but in 1913 when Murray learnt that M. Walsh was to prospect in the upper Tiveri he instructed Patrol Officer Frederick Chisholm to establish a police camp in the area. During the two months Chisholm, the police and their prison labourers maintained the camp two diggers, Smith and Arnold, prospected the area. Later Murphy and Davies accompanied Chisholm beyond the headwaters of the Kunimaipa to the north-west of Mount Lawson into German New Guinea, and Jeffrey and Harry Ariotti ('dago but good worker') went with Oldham to the north-west of the Olipai. Several times the villagers fired arrows at Oldham's patrol and when a carrier was wounded the police fired twenty shots in reply. Ten days later Jeffrey and Oldham each fired a warning shot. Murray wrote on the cover of the patrol report for the Government Secretary to transmit to Oldham: 'I do not consider it advisable that G. O.'s should take gentlemen who are not members of the service upon a patrol when there is likelihood of a collision with natives ...' It was good advice although it ignored the fact that for four years miners had roamed the Arabi, Tiveri and Olipai without benefit of government presence.

Between 1909 and 1916 the Lakekamu miners had penetrated the valleys between the headwaters of the Vailala in the west and the St Joseph in the east. Major expeditions had also gone up the Fly and the Markham and some miners had returned to old alluvial areas of the Northern Division and the south-east. They were often in the country either unknown or vaguely known to government officers. On government patrols the Native Armed Constabulary were intermediaries between carriers and officers and their guns protected the patrol. In the bush miners and carriers were mutually dependent and with fewer resources they were more reliant on their own skills. Where possible they traded with strange peoples for food, made sago and collected bush foods, cut canoes and rafts on river banks and tested river beaches for gold with boxes made of hollow logs or bark. Although several expeditions lasted over three months the death of only one labourer was recorded on a prospecting trip. By contrast eleven carriers died on the Staniforth Smith government expedition in 1911: this was an exception but it was not unusual for carriers to die on long government patrols.

Frank Pryke's revolver. It 'barked beautifully' on the Fly.
National Library of Australia

On 20 August 1914 Oldham wrote in the station journal: 'News of European War to hand.' It was sixteen days after Germany invaded Belgium and thirteen days after Murray had signed a declaration notifying citizens of a 'state of war between the United Kingdom and the German Empire'. The war had no immediate effect on the actions of men on the Lakekamu. Government officers were more concerned about stores not arriving and signs of laxity in the Armed Native Constabulary. Most of the rice ordered in February did not arrive until September and without rice government officers could not go on patrol. When Constable Aviri allowed a prisoner to escape he was given one month's imprisonment with hard labour for his carelessness, and when Constable Fonu permitted another to escape the penalty was increased to two months' hard labour. The miners were worried about the dry spell which interrupted work and made it difficult for the *Bulldog* and the *Mayflower* to reach the landing. The Lakekamu miners also shared in the general interest in dredging prospects in Papua. Experts and investors came up from Australia to inspect likely areas and several of the local miners pegged dredging leases on the Tiveri and at Cassowary Creek. But when Chisholm resumed patrolling in October he discovered evidence which made miners and officials think that Nepa might be the scene of the first international battle fought on Australian territory.

On his second day out from Nepa Chisholm came across a camp on the Arabi recently abandoned by a large German expedition. He sent two constables back to warn Oldham, who relayed the message to the miners and sent a report to Port Moresby. By the time Chisholm returned Oldham had police, prisoners and miners' labourers digging trenches around the government station, mounting guards and conducting short reconnaissance patrols. On 10 October, three days after he had sighted the German camp, Chisholm, the miners Sloane, Preston and Swanson, three armed constables and five carriers set off to look for the Germans. Guides from a village on the Arabi led them to a more recent German camp and on their second day out from Nepa they caught sight of the Germans on the divide between the Arabi and the Tiveri. Estimating the strength of the German party at two white men, twenty-five police and seventy carriers, Chisholm sent a message back to Oldham asking for more men. Before the reinforcements arrived the Germans moved on, leaving three police at the camp to look after some sick carriers. 'These I captured', Chisholm reported, a task which should not have been difficult because apart from their recent fight with local villagers they did not know they were at war. Chisholm sent one of the captured carriers with a letter to the leader of the German expedition informing him that war had been declared and all German possessions in the Pacific had been captured, and inviting him to come to Nepa. The carrier did not return. Two days later when Oldham arrived with seven miners and seven police, the Australians set out to find the Germans. After several days following their tracks, Oldham decided that his party was three days behind the Germans who were 'making their way in a Northerly direction as fast as they possibly can'. On the Tiveri the Germans had burnt two village houses, a punishment, Oldham thought, for people who had followed their custom of firing arrows at all who entered their lands. The village gardens too had 'suffered severely. The Germans apparently fed their police and carriers out of them'. After complaining of the 'disgusting state' in which the German expedition had left village houses Oldham felt obliged to leave trade goods to compensate for the behaviour of these other imperialists. The Australians returned to Nepa and Oldham sent the five prisoners to Port Moresby where Murray, uncertain of their status, did not know what to do with them.

The German expedition had left Morobe in March to map the country along the Papua-German New Guinea border. Its leader, Hermann Detzner, gave a different account of his meeting with the Australians. Detzner said that when his head of police, Konradt, who spoke no English, went to meet carriers bringing up supplies from Morobe he found a note held in the hand of a dead policeman. On one side the note said: '10 tins Bully beef, 8 tins vegetables, 5 tins hard bread, 4 tins butter, 12 tins tobacco etc'. On the other side it said:

> To the Officer in charge of the German forces. I have to inform you that war has been declared between Great Britain and Germany on August 4. 14. In order to avoid unnecessary loss of lives I advise you to come in as soon as possible to the Nepal Camp [sic] at the Lakekamu-Goldfield which you will reach after five days march and to surrender there with all your men. You will be treated as an officer and a gentleman. Two native policemen and carriers I took along as prisoners of war.
>
>   Chisholm
>   Officer in charge of the British force

Detzner said he then learnt that the Fatherland was at war with its hated rival, England, while he wandered in the far corners of the earth. Writing in 1920 Detzner recalled the questions immediately raised by the note. France he knew must be involved in the war but what of Austria, Italy, Russia and the United States? What had been the immediate cause of the fighting? Perhaps the note was just a cruel joke by an Australian miner or recruiter. How could he explain the behaviour of the Australian official who arrested two inoffensive men, made no attempt to contact the main party, and left only a ridiculous note? Detzner decided that although Chisholm had a five-day start he would take a selected group of men, find him and obtain a full explanation. Having reached the edge of the ranges above Nepa Detzner turned back: he said that one of his patrols was fired on and he did not want to walk into an ambush. He rejected any plan to attack Nepa because he did not want the Papuans to see white men fighting each other, a spectacle which in any colony could be followed by grave consequences. Detzner thought that the wisdom of his decision to withdraw was confirmed when his two carriers, having escaped from Nepa, arrived. They reported that the two captured policemen had been sent by boat down the Lakekamu and that the force which had left the note for Detzner consisted of two white men and many Papuan police. Detzner's presence, they said, had caused great excitement at Nepa and Chisholm had asked that two extra companies of police be sent to the Lakekamu.

Detzner's account disagrees with that of Chisholm and Oldham on many points, and the reports made by the Australians are generally easier to believe. For example Chisholm reports the capture of three carriers and three policemen and the sending of one carrier with a note to Detzner, Oldham notes the despatch of five prisoners from Nepa and Murray acknowledges their arrival in Port Moresby. It seems that there were five prisoners, not four, and that none escaped. Detzner's story of his encounter with the forces from Nepa appears to be a fitting preliminary to the later fantasy of his journeys to the vicinity of Mount Hagen. Alone of the German officials, Detzner refused to surrender to the Australian occupying force. After the war he wrote that he had carried the German flag deep into the previously unexplored central highlands; a claim he eventually admitted was false. Presumed dead by the Australian troops, he had waited out the war in the known, if not comfortable, lands around the Huon Gulf.

In December 1914 Chisholm found numerous old German camp sites on the upper Oreba and the village men who had been 'unarmed and friendly' in 1913 seized their weapons and the women urged them to fight. Eventually they recognised Chisholm's patrol and indicated that they had thought the Germans were returning. 'By the signs they made', wrote Chisholm, 'the Germans took all they wanted from the gardens, and gave them nothing in return.' In March and April 1915 Chisholm again patrolled the upper Arabi, Tiveri and Oreba Rivers to make 'friends with the natives and see if any Germans were moving in the area'. While Chisholm saw no more Germans, Armed Constable Kaptin on sentry duty at Nepa thought he did. When three policemen arrived unannounced from Kerema, Kaptin opened fire, hitting Constable Naboko. After holding an inquiry Oldham decided that Kaptin had fired without first challenging. Constable Naboko was the only casualty in the defence of Nepa.

The war had little other effect on the Lakekamu. Of the thirty-nine miners who were on the field either in February 1914 or May 1915 only three certainly enlisted, Jim Pryke, Ryan and Castleton. At least two other miners who spent long periods on the Lakekamu, Newcombe and Joubert, also joined the army, and Lumley, Whittens' storekeeper, served in the Light Horse. Government officers were more inclined or more of an age to fight for king and country. Chisholm and Oldham both enlisted and of the earlier Lakekamu officers, Keelan and Giblin served overseas. Castleton, Chisholm, Pryke and Ryan were killed. Newcombe died on the troopship taking him overseas. A teacher in a county school before leaving England, Claud Castleton travelled widely in eastern Australia before working on the Lakekamu. The citation for his Victoria Cross said: 'Sergeant Castleton went out twice in face of … intense fire and each time brought in a wounded comrade on his back.' He went out a third time and was killed. To both Pryke and Lumley the

Lakekamu improved with distance. From the mud and snow of the training camp on Salisbury Plain Pryke thought the English would have been well advised to let Kaiser Bill have the place. In a 'rotten bloody hole' near Jericho Lumley, regretting his eagerness to leave his wife and his struggling trading and pearling interests, wrote wistfully of the 'land of the coconut' and 'Tommercrann's' (Tommy McCrann's hotel in Port Moresby). At least he had the satisfaction of using his specialised knowledge: his experience with dynamiting fish made him a valuable member of demolition parties and he believed he could cure the troops' malaria with Dalwhinnie whisky.

The Lakekamu miners were given a chance to vote in the conscription referenda of 1916 and 1917: the only time Europeans in Papua voted in an Australian poll. By the second referendum in December the wet season had set in and Humphries had to wade and swim to reach the eleven miners still at scattered camps at Sunset, Mosquito Creek, Belfield's Gully, Fly Gully, Palm Island and Tiveri. The votes were carried in a bag tied to Constable Ekau-hu's hair. Constable Wali then took the votes to Port Moresby, leaving by raft on 15 December and not returning to Nepa until the middle of January. Unlike most Australians the white residents of Papua voted in favour of compelling young men to fight Turks and Germans.

The Lakekamu declined between 1914 and 1918. The war may have made stores more difficult to obtain and more expensive, and a few miners had left 'for the front', but generally the war had not caused the decline in the number of men on the Lakekamu. They left when the amount of gold they could recover no longer gave them an adequate return above the cost of their stores and the wages they paid to their labourers. The war may have delayed attempts to operate dredges on the Lakekamu. In the wake of the development of Bulolo the Tiveri Gold Dredging Company was able to raise the capital to place a small dredge on the Lakekamu and keep it working from 1934 until 1939. But Guinea Gold No Liability, which carried out an extensive testing program from 1934 to 1936, decided the field was not rich enough to support dredges.

When the *Bulldog* arrived at Tiveri on Sunday 8 December 1918, after an absence of two months, Assistant Resident Magistrate C.R. Muscutt, the only government officer still at Nepa, learnt of the armistice declared a month earlier in Europe. There were only eight other Europeans on the field, six miners, one storekeeper and Mrs Priddle who had arrived on the *Bulldog* to join her husband at Cassowary Creek. Isolated, rarely assisting or restraining either miners or villagers, Nepa station became an embarrassment to Murray who was reluctant to withdraw his most inland station. Their actions partly excused by a prolonged shipping strike in Australia, government officers in Port Moresby forgot about Nepa. In May 1920 when the storekeeper brought the mail up from Tiveri Muscutt received his first news of the outside world for five and a half months and the police were given eleven months' back pay. Muscutt wrote that he found it much easier to live on the police and prisoners' rations than to go without news. In May the storekeeper learnt that Whittens had written in February instructing him to dismantle the store and pack the galvanised iron ready for shipment to Port Moresby.

After the store closed the Lakekamu might be a place for prospectors, 'hatters' and speculative companies, but it was no longer a goldfield. Humphries, who saw the field in its declining years, recreated the importance of the store:

> Everything radiated from the Store. Every expedition, every prospecting party, started from the Store. It was the hub of the Field. There you could drink anything, buy anything, and gamble anything. There you could 'hire a team' (carriers), there you could have a spell. If it was the latter you wanted, it would cost you nothing. Meals were always free at the Store, and there was the ground to sleep on. What needed man more? If you died there, as men did — well, there always was someone to bury you, and there was not always someone to do it on the claim.

The storekeeper listened to everyone 'rousing' and received orders written on 'dysentery wipe', but when there was no rice, no mail and no news at the store he might give the waiting carrier a bottle of Dalwhinnie for the *Taubada*. Scrawled across the bottom of Whittens' accounts Lumley wrote invitations and scraps of information: 'Got a crowd in. Newcombe, Arnold, Belfield, Reilly, Franklin, Hendry & Clare. Wrestling and boxing is now being discussed. Why dont some of yer come down & ave amenegs.' Or '3 Rice 37/6, Fruit 5/3, 1½ Biscts 1/-£2-3-9 ... . The Doc is also on the shicker & tried to do it to Kruger's gin.' If the miner himself made the trip he could listen to the 'charming strains of Mme Melba at her best on the Phony graph' while he drank and yarned. If he was still dissatisfied he could fight the storekeeper; and in Lumley's time he had a fair chance of winning. The 'deadhouse', an incomplete shack, was available for those needing to sleep off the effects of various forms of excess.

For Christmas 1920 only one miner, William Ward, walked in to Nepa. Ward, the first white man Muscutt had seen for about five months, was the only miner still on the field and his team was then making a canoe on the Olipai to take them to the coast. The next year the three agents of government, the white officer, the police and the prisoners, left Nepa. Periodically alluvial miners returned to the area, their interest stimulated when Jack Hides reported finding gold on the Tauri in 1930 and by the later activities of the dredging companies. In 1940 A. Bethune and H. Garbutt were still making 'a comfortable living' on the Lakekamu, and Bethune, who had worked on the field from soon after its opening in 1909, returned after the second world war.

The results of the ballot to choose the leader of the prospecting expedition are preserved in the Australian Archives, C.R.S. G 70, Papua Lands and Mines Dept etc. Mines Papers, Annual Single Number Series 1907-27, item 1907/87. The file includes correspondence between Staniforth Smith and various people about the selection of the prospectors, a report written by Crowe while on the expedition, and some early comments on the field by government officers. Frank Pryke in a letter to Dan, 4 April 1909, mentioned the vote, and he and Jim wrote about the expedition and working the field in later letters. In the Lett papers, folder three, there is a letter by Chas C. Deland(?), 10 August 1944, outlining Crowe's early life. The verse is from 'Matt Crowe' by Frank Pryke, *Poems*, p. 46. Higginson's report of his patrol up the Tauri is with the Kerema station papers. Murray in his dispatches to the Minister reported on the progress of the prospectors, the discovery of the field and the development of mining. He also informed the Minister about important legal cases and enclosed statements from witnesses. The Commissioner for Native Affairs reported direct to the Minister and the reports of his officers on labour conditions are in dispatches. The Nepa station papers are detailed and almost complete. The journals run from January 1910 to June 1915 and from July 1916 to December 1920. The patrol reports are from 1913 to 1920. As government officers came to the field with the first miners, Nepa was close to the workings and the officers' main duty was supervising the field, the government records for the Lakekamu are fuller than for any other Papuan field. Murray and the Government Secretary sometimes commented on journals and patrol reports. The annual reports of the warden, resident magistrate and medical officer were published in *Annual Reports*.

Dr Anthony Radford, then at the U.P.N.G., provided information about past and present methods of treating dysentery.

Lumley's brief and colourful notes were scrawled across the bottom of Whittens' accounts sent with stores to the reward claim on Ironstone Creek. Frank Pryke preserved them and they are with his papers.

The various amendments to the Labour Ordinance and the regulations to control dysentery were published in the Papuan Government *Gazette*.

In the six years 1910 to 1915 five cases of assault were referred from Nepa to the Central Court. Lagalaga, Jeffrey and Kelly account for three. The others may have been cases of labourers fighting amongst themselves, or fighting villagers, or European miners fighting each other.

Mr Justice Bevan's comments on the practices of juries in the Northern Territory were printed in *Commonwealth Parliamentary Papers*, 1914-17, Vol. 2, pp. 1783-4.

Frank Pryke kept a rough diary on the prospecting trip to the Vailala. The diary, which includes the note Pryke dictated to Priddle, is in the National Library. There is a more carefully written account in the Mitchell Library, and he also wrote the report published in the *Papuan Times*, 20 March 1912. The *Papuan Times*, founded in 1911, occasionally carried other items of news from the Lakekamu. Newcombe's report of his prospecting expedition is in the *Annual Report 1912/13*. Edward Auerbach recalled the trip up the Markham in *Pacific Islands Monthly*, March 1940. Frank Pryke's diary written on the Fly is in the National Library and his typed report is in the Mitchell Library. Murray to Minister, 7 October 1914, enclosed a sworn statement by Pryke about the conflict with Fly villagers. Government officers' reports of patrols to the Vilirupu field are with the Abau station papers, eg. Bastard, April and June 1916.

Detzner 1920 wrote of his exploits and Biskup 1968 tested his accuracy. Dr William Gammage, U.P.N.G., checked the Australian War Memorial records for references to men from the Lakekamu.

Humphries 1923 wrote of the Lakekamu in decline. The quote is from p. 34.

13

No Meeting

a salute of skewers

Miners entering the mouth of the Lakekamu passed a village on the right which they sometimes called Toaripi; Motu sailors on *hiri* trading expeditions to the Gulf called the same place Motumotu, and the people who lived there thought of themselves as belonging to two villages, Mirihea (on the beach) and Uritai (inland). About thirty years before the first miners went up the river a group from Mirihea had moved east to establish a new village on the beach at Kukipi. The Mirihea, Uritai and Kukipi were Toaripi people. About 4 miles from the coast on a creek coming into the Tauri from the east was another large settlement of around 2000 people. Miners and government officers used the Motu name of Moveave for the big village, but again the people thought of themselves as belonging to two villages, Heavala and Heatoare. Although speaking the same language and conscious of a common heritage, the Moveave and Toaripi feuded until peace was imposed by missionaries and government officers in the 1880s. The Moveave, from the oldest settlement and jealous of their land rights, limited the movement of the Toaripi up the Lakekamu and Tauri; but before they were employed on European-owned boats the people from the beach villages knew about 300 miles of the Papuan coast. They sailed south-east to the villages beyond Port Moresby where they had close links with other sea going and trading communities, and they went west to the Kikori delta.

The Moveave-Toaripi had a rich ceremonial life centred on the cycle of *semese* festivals and the construction of *eravo*. Up to 60 feet high at the front and sloping downwards to be about 20 feet high at the back, the *eravo* 'temples' could be constructed only by a people with an economy and social organisation which gave them the time and unity to carry out large projects not concerned with immediate survival. Tall and muscular, the Moveave-Toaripi appeared to outsiders to be fierce and independent peoples. After he visited them in 1893 MacGregor said the Moveave were 'a strong tribe, and used to feel very confident in their own strength'. The Reverend James Chalmers, who lived at Toaripi from 1888 until 1892, wrote that his hosts had been 'the terror of all the other tribes' from the Gulf to Kerepuna. Where they had not been well-received they had

> killed every pig they found and robbed all the plantations, and wound up by turning the houses into w.c.s ... . In one afternoon they killed thirty-six men. women, and children at Kabadi, and at Partanu, inland of Hall Sound, a few years ago they made a nearly clean sweep of the village.

The Moveave- Toaripi cultivated sago in the vast swamps as their basic food. The beach people were short of good land for other crops, but the Moveave gardened on the banks of the Tauri and Lakekamu where they grew yams, sweet potatoes, taro and bananas and cut canoe logs. In 1893 MacGregor found gardens over 20 miles up the Tauri and he saw temporary houses ('hunting seats') to about 36 miles. The Moveave travelled the upper Lakekamu, and miners and government officers took over the Moveave names for the river and its tributaries: the Olipai, Tiveri and Arabi.

From the 1880s missionaries, government officers and traders visited the Moveave-Toaripi. The missionaries alone sustained their interest. From 1884 South Sea Island teachers of the London Missionary Society lived in the villages, and by 1910 there were 191 pupils on the rolls of the two schools in Moveave. Attendance was irregular, perhaps because the students and their parents saw little use in prolonged studies which gave no advantages except enabling a scholar to say a few words in English and read four gospels of the New Testament translated into a language which foreigners thought was Toaripi. The crafts, songs and beliefs of the people had been strongly influenced by the missionaries, but in 1910 the old cycle of festivals was still important.

Traders came for sago, copra and sandalwood, and at various times maintained stations in the area. On the Tauri MacGowan cleared a block which gave little sign that it would nourish the coconuts, cotton and other crops which he had planted. Generally the Moveave-Toaripi had decided not to work as indentured labourers. In 1909-10 only 186 men from the entire Gulf Division signed on while over 5000 Papuans from other areas decided that they would accept the recruiters' terms. Both missionaries and planters had offered glimpses of another economy to those who stayed in the villages: MacGowan employed a horse team to plough his land before planting, and the Reverend Edwin Pryce-Jones on pastoral visits bicycled along the sand-bars.

Government officers visited the area occasionally after Robert Hunter made a bloody entry into Moveave in 1887. He led a punitive expedition against the Moveave after they killed a South Sea Island teacher, his child and five Toaripi who had accompanied him up the Lakekamu. The visits became more frequent from 1906 when the Gulf became a separate division with its headquarters at the new station at Kerema. Soon the government officers thought they knew the people well enough to give them a corporate personality. Charles Higginson said in 1908 that he found the coastal people

> a very well behaved lot, compared with other parts of the Territory. ... The chief fault I find with them as a whole is their refusal to sign on for labour. This is particularly noticeable with Toaripi and Karama and Waima villages. The sooner some means of making the tightly-laced and ornamented young bucks go out and do some work, instead of as now, loafing around the villages, the better.

Other officers also said they were lazy. Wilfred Beaver thought the people of the Eastern Gulf had a 'rooted dislike to work' and James O'Malley wrote of the 'lordly male who spends his life in "magnificent idleness"'. To explain apparent indolence, some officers observed that now warfare had ceased the men could no longer perform their traditional task as warriors, and as food was always available from the sago palms they did not have to work hard to survive. Beaver thought it fair to add that the Gulf men were no lazier than 'all the other natives from Samarai to Cape Possession', and O'Malley conceded that the Gulf man made a 'good armed constable'. By 1910 several Moveave-Toaripi had served in the Armed Native Constabulary.

Apart from fruitless attempts to persuade the men to go away to 'work', the government officers who took the whaleboat east from Kerema on visits of inspection were most concerned with keeping the villages clean. In August 1909 Henry Ryan at Moveave 'shook the VCs [village constables] up about the state of the village and ordered it to be cleaned'. He was pleased to find Motumotu in better order, but the government rest house there was dilapidated and the surrounding area had been used as a rubbish dump. He selected a new site and the villagers promised to build a 'decent home'. In February 1910 Ryan instructed the people of Motumotu village to 'turn to and clean it up', and again in April a corporal and four constables were detailed to supervise the Motumotu while they cleaned their village. Moveave was 'none to [sic] clean' either, and Ryan ordered the village constable to see that it was improved immediately. The government's concern with hygiene and Ryan's administration of betel-nut tea did not save the Moveave-Toaripi from the whooping-cough and dysentery epidemics which occurred in 1909.

Government officers supported the mission teachers trying to make their students attend school regularly. In Motumotu Ryan called the village constables together and impressed on them their duty to see that the children went to school, and he reported that at Karama the Assistant Resident Magistrate 'punished some of the children for not attending the school regular'. The combined influence of the church and state was limited: in 1910 at Heatoare where there were ninety-six registered students the average daily attendance was forty-three, and at Heavara from ninety-five registered students the daily average was only seventeen.

In their thirty years of close association with foreigners before the discovery of gold on the Lakekamu only one major change had been forced on the Moveave-Toaripi: the ending of warfare. The advantages of other changes had been pressed on them and they had seen many new ideas and objects displayed before them; but generally they had been able to take what they wanted and resist what seemed of little value. The villages were only intermittently connected to Kerema and Port Moresby. Attempting to govern by making one or two inspections a year, and unable to speak the villagers' language, government officers became concerned with what they could see, explain easily and knew to be 'good'. They looked first to see whether the village was clean. It was a form of government which might interrupt village life, but not transform it. In 1910 the Moveave-Toaripi were in danger of being overwhelmed.

Motumotu became a staging point for the goldfields. Early in 1910 tons of stores and over 1000 miners and labourers were landed on the beach. Government officers, who had the power to compel Papuans to carry, immediately recruited twenty men from Motumotu and Moveave to help establish the new government station on Ironstone Creek. But in April when the government wanted more men to work at Nepa they could not be obtained. Sergeant Kasari told Ryan that the Moveave had fled rather than recruit for the goldfield. Later at Karama, Ryan found the men equally unwilling to go to Nepa because they knew of 'the boys dying there'. To confirm the reports of those Moveave-Toaripi who had been on the field there were many deserters living in the villages, some of whom were sick with dysentery. It now became one of the duties of village constables to report runaway labourers to government officers, or if they were men with the authority of Village Constable Lai of Motumotu they themselves sent the deserters back to Nepa.

To maintain at least 300 labourers, miners, police, prisoners and government officers on the field until 1917 traffic on the river was heavy, but the Moveave-Toaripi continued to select from the parade of foreign goods and ways. Although few worked as indentured labourers many went to Nepa. Some worked as crewmen on launches, but most paddled their long, double-hulled canoes against the current to Tiveri landing. Hired to transport men and stores, Moveave canoemen also took garden food and betel nut to sell to the labourers or exchange for tobacco. When the number of men on the field declined and the launches made infrequent trips, the Moveave did more of the carrying on the river. By July 1920 all goods going up-river were taken by the Moveave, and Whittens' Tiveri storeman made his last trip downstream on one of their canoes. Organised by the government agent at Kukipi, the Moveave made one trip a fortnight. At times they moved much cargo. On 3 March Muscutt reported the arrival of twenty-seven men bringing stores and the next day another fourteen men landed eleven more 180-pound bags of rice and four bags of meal. The Moveave were each paid one pound of tobacco for the trip.

Apparently only lightly touched by the goldfield, the Moveave-Toaripi may have been deeply influenced by what they had seen and experienced. The Reverend Herbert Brown came to live among the Toaripi speaking people of Moru in 1938, learnt their language, and stayed in the Gulf for over thirty years. He has recalled conversations with Moveave men who travelled with him across old mining areas:

> As I walked the overgrown tracks in the company of the Heavara men who had worked there years before, listening to their reminiscences, it was apparent that their experiences had had a profound effect on them. The crowds of men, piles of stores, the sudden burst of activity directed towards what was then to the Elema a remote, unknown region, gave them a glimpse of power with which magic could not compare. In the villages the traditional way of life — seclusion, the Bull-Roarer, the Oioi, and the Semese — continued as before, but beneath the surface minds were being prepared for a decisive break with the past.

The acceptance by the Moveave-Toaripi of a cult movement which spread from Vailala in 1919 was a sign of the depth of disturbance in the coastal communities and an agent of further change. Entranced leaders described by the Pidgin term, 'the-head-he-go-round-men', prophesied the arrival of their ancestors in a *sisima* (steamer) bringing food, tobacco, knives, axes and calico. While imitating the rituals of white culture and desiring its products and power, members of the movement expressed resentment against foreign intrusion. Old ceremonies were abandoned and objects made by craftsmen of the old ways were treated with contempt: 'Throw 'em away, bloody New Guinea somethings', a Moveave man told F.E. Williams. The movement was at least partly a rejection of a way of life which had apparently left the people inferior and an attempt to grasp the source of the foreigners' power. The growth and decline of the goldfield on the Lakekamu had not caused the Vailala 'madness', but its form and intensity in Moveave-Toaripi had been influenced by the events which followed the discovery of gold at Ironstone Creek.

In the 1930s the activities of the dredging companies and reports of gold on the Tauri quickened traffic on the river. Men from the coastal villages transported stores and sections of dredges, and they saw the airfield built at Bulldog by the mining companies. Influenced by missionaries, government officers, miners, traders, other Papuans and their own perception of the world, the Moveave-Toaripi changed. In the 1920s they abandoned the practice of secluding the young from the rest of the community, in the 1930s they held their last great *semese* festival, and more men began to go away to work. During World War II nearly all able-bodied men were taken from their villages to support Australian and American troops and fight the Japanese; and the Lakekamu, the feeder for the Bulldog track to Wau, carried a greater tonnage than ever before. The Moveave won more honours in the Papuan Infantry Battalion than any other village.

The Moveave-Toaripi returned from the war determined to build larger and stronger political institutions; and by starting a sawmill and engaging in coastal shipping they hoped to secure greater rewards in the new economy. Their positive response to the war was partly because they had been prepared by their experiences of the previous fifty years. They had suffered other shocks and they were already searching for a way to improve their position in the new order. After long delays the sawmill was built at Moveave and the Toaripi Association purchased the *SS Kukipi*, but the limited success of these ventures left men wondering whether or not the new way had been found.

The rapidly changing response of the Moveave-Toaripi as new opportunities opened to them can be seen in one Kukipi family: Hasu Morauta worked on the *Bulldog*, his son, Morauta Hasu, was a leader in the Toaripi Association and a Local Government Councillor, and *his* son, Mekere Morauta, graduated in economics from the University of Papua New Guinea and joined the public service.

Drum, Gulf of Papua, National Museum of Victoria, 1887

In 1910 small communities of Kovio people lived inland to the south and east of the Lakekamu. In earlier times they had been harried by their stronger neighbours, the Moveave in the south and the peoples from the mountains to the north. Government officers at Nepa first became aware of their existence in 1913 when Moveave sago-makers reported the presence of strangers on the river. Oldham went downstream from Tiveri and met the Kovio one day's walk east from a point just below the Kunimaipa junction. Related by language and custom to the Mekeo on the St Joseph River, some Kovio had visited the government station at Kairuku, and they were prepared for their encounter with Oldham's patrol and his Moveave companions. The development of the goldfield and the station at Nepa gave the Kovio another point of contact with the outside world and protected them from raids from the north. At least during 1918 and 1919 groups of Kovio led by Village Constable Peuma made several trips to Nepa bringing sago to exchange for tobacco. The presence of the goldfield and their freedom from attack probably encouraged the Kovio to move westward on to the less isolated and more productive lands along the main rivers.

Numbering only seventy children and seventy-five adults in 1954, the Kovio have been too few and too far away to attract more than brief visits by white men. New ideas have come more from Papuan missionaries, Kovio who have been away, and contact with other Papuans on the rivers. In 1972 the Kovio occupied two villages, Okavai near the junction of the Oreba and Kunimaipa, and Urulau on the Lakekamu. The Moveave tolerated their presence at Urulau although they claimed the land on which the village was built. At Urulau a Kovio man, trained by the United Church in New Britain and married to a girl from New Ireland, conducted a school. It was unrecognised by, and in fact unknown to, the national education system.

The men who had contested the right of Crowe and the Prykes' prospecting expedition of 1909 to move through their lands were Kapau speakers, the most numerous and south-eastern of the Anga language group. To miners and government officials they were the Kukukuku; the Moveave called them the Iariva. Never having had to think of themselves as a group, they had no word to distinguish themselves from other men. The origin of the term Kukukuku is obscure, but some of the early officials in the Gulf believed it was a Motu term of abuse applied by the *hiri* traders 'in the same lordly way they renamed the villages, rivers, and other parts of the coast'. Through their habit of raiding coastal villages to kill and carry off the dead, the Kukukuku became known to government officials in the 1890s. In 1900 Blayney and

Amedio Giulianetti, the government agent from Mekeo, investigated the killing of five people at Kerema and learnt from the survivors that the Kukukuku were a wandering people who occupied the interior from the Vailala to the Lakekamu. But it was not until the government station was established at Kerema that Griffin made contact with the Kukukuku. Using Hawaiu people from the Vailala as intermediaries, Griffin had brief and peaceful meetings with two Kukukuku communities in the Lohiki Creek area. In 1907 the Hawaiu conducted Charles Higginson into the presence of the Kukukuku and he told them that they were not to make any more raids on the coastal peoples. Two months later the Kukukuku attacked Lovera village just to the east of Kerema station and killed two men and a girl. Although he was uncertain if the raiders were the same people that he had spoken to earlier, Higginson took the raid as defiance of his orders and sent the police to pursue the Kukukuku. Getting between them and their homes, the police shot and wounded three or four of the raiders. Higginson was now confident that raids on the coast had ended and that the Kukukuku could await further instruction on proper behaviour while the government consolidated its position along the coast.

At the time the prospecting expedition went up the Tauri in 1909, government officials did not know how many Kukukuku there were nor how far their lands extended. They believed they were at least semi-nomadic and while they were aware of their skills as bowmen and bushmen, they found it difficult to explain the terror these 'fierce little people' caused among the coastal villages. 'All sorts of weird queer stories are woven round them and their habits', said Higginson. 'It is well the Kuku-kuku do not know their powers, as I am sure if one Kuku-kuku came out on to the beach he could chase the local population for as long as he liked to run after them.' For some coastal peoples the Kukukuku were not men but evil predators who watched constantly for their victim to relax his guard and then struck with unlimited and random savagery. Already they were known to Europeans as the 'famous Kukukukus', but government officials had no doubt that they would soon control them. 'This type of native', Murray wrote, 'may be brought to a state of comparative civilization without any great difficulty by the exercise of tact and patience.' He was too confident.

Matt Crowe and the Prykes had spent ten years travelling and working in country among hostile communities; but on the Tiveri they encountered a more tenacious people than they had found elsewhere. The Kukukuku attacked frequently and without warning. After the killing of Wagawaga Dick on the Tiveri the prospectors could not make an immediate counter-attack because those who had released the arrows were protected by the dense scrub and broken country. They buried Wagawaga Dick sitting up 'New Guinea fashion', but, as Jim said, no doubt the Kukukuku dug him up and ate him, also New Guinea fashion. The Kukukuku 'set us again the next morning', Frank said, 'but we had the best of that argument as we had got up before daylight and had gone to meet them'. The Kukukuku were not intimidated. 'The nigs around here', Frank acknowledged, 'are the pluckiest lot I have run across in this country, they don't mind one or two getting topped over'. They refused all attempts to trade or talk and persisted in saluting the camp with an occasional flight of 'skewers'.

When Murray visited the goldfield in January 1910 he attempted to 'enter into communication' with the Kukukuku. Accompanied by Vavasua, a Moveave man who could speak their language, and Matt Crowe, Murray went beyond the point where the prospectors had been attacked. But while the members of the patrol thought they were being watched as they inspected gardens and houses the Kukukuku gave only further evidence of the 'gift of invisibility'. Occupied with the administration of the goldfields, government officials then suspended their attempts to civilise the Kukukuku.

Although there was a village within one day's walk of Nepa station, miners and officials had almost no contact with the Kukukuku for the first three years of the field's existence. Government officials on visits of inspection and police out looking for deserters sometimes saw them briefly, and two deserters and one of Frank Pryke's labourers were killed by 'bush natives'. The Kukukuku may also have been responsible for looting some of the camps left deserted during the day. By their attacks on Pryke's prospecting expedition to the Vailala in 1911 and on Newcombe's party in 1912 the Kukukuku maintained their reputation for fearlessness and treachery. The one serious attempt by a government officer to communicate with them was a failure. In September 1911 Lyons went up the Olipai to Thomas and Swanson's camp. The miners reported that on their way from Tiveri they had seen Kukukuku making sago but they had run away. The miners left tobacco at a shelter used by the sago-makers but it was not accepted; their lack of interest in the gift was not surprising as unknown to the miners the southern Kukukuku did not smoke. Thomas went with Lyons's patrol to an area occupied by Kukukuku gardens and houses where the police surprised and seized a young girl and a woman 'of fine physique ... and of bronze hue'. Unable to pacify them, Lyons held the woman's hand and she 'put it to her mouth not in gallantry but to bite it'. When released she fled, leaving the trade goods which had been offered to her. That night the Kukukuku showered the camp with arrows and Lyons ordered the police to clear the area with rifle fire. Members of the patrol picked up twenty-five arrows which had been fired into the camp: Lyons doubted that any of the attackers had been hit by rifle fire. The next morning Lyons examined the houses and gardens and, having taken some tools

Kukukuku warrior, Lakekamu Goldfield, 1914
PHOTOGRAPH: FRANK PRYKE

and plants, left trade goods in exchange. The events of Lyons's patrol were repeated many times: Kukukuku were seized and held for brief periods; but the most frequent exchange was between rifles and bows.

By the end of 1913 government officials had a little knowledge of the peoples whose lands they had occupied. A community of about fifty lived on the Arabi one day's walk from Nepa. Up the Tiveri and the Olipai were several hundred people, while across on the headwaters of the Tauri they were more numerous. They were not, it now seemed, nomadic, but lived in scattered hamlets of two or three houses on the ridges next to their gardens. The groups of cone-roofed houses were often within hailing distance of each other. The hamlets, government officers noticed, were kept clean and the surrounding area was planted with decorative crotons and betel palms. Extensive gardens were planted with sweet potato, taro, bananas and sugar cane; sago was cultivated in lower and wetter areas. Fighting men carried a bow, stone club and wooden shield. For clearing new gardens and preparing the ground for planting they used a stone adze or axe. Not being potters, the women used bamboo for holding water and cooking, and string bags were used for carrying between house and garden. Hunters set animal traps in the bush and built fish weirs on many of the streams, and while government officials saw no Kukukuku south of the junction of the Olipai and the Lakekamu, they saw the shelters which they had used while on hunting trips through the lowlands before the goldfields had placed a barrier between them and the south.

Much of the information about the Kukukuku had been collected by Chisholm, who arrived at Nepa station when supervising the few men left on the goldfield took little of his time. Murray too now decided that establishing friendly relations with the Kukukuku should be 'one of [his officers'] principal objects'; and after he had examined the Nepa station journals for September and October 1913 he reprimanded Oldham and Chisholm for the number of days when both men were on the station: 'One or other should always be out'.

More energetic than many other officers, Chisholm was certainly trying to bring the Kukukuku under the influence of the government. From a camp on the upper Tiveri he made a series of patrols to nearby villages. On 2 July he captured a man, a 'well proportioned, and very clean looking woman' and a boy of seven. All were released immediately but the experience apparently did nothing to reduce the hostility which the Kukukuku felt towards the foreigners. On the 4th the police surrounded a group of houses and managed to seize a man and some women and children. Chisholm and the police again tried ineffectively to calm them. (Chisholm collected a sample of hair which he attached to his patrol report.) Later the police saw a party of eight out hunting and managed to capture two whom they took back to Nepa, but they escaped that night. Oldham had thought the two men were unafraid of the police and interested in the objects they had seen at the station. Nepa had then been on their lands for three and a half years and they were the first Kukukuku to enter the station grounds.

When Chisholm was first posted to Nepa Murray had warned him that he was to do only 'what [was] strictly lawful'. Now he wrote on Chisholm's patrol report for July: 'Seen: I suppose when PO Chisholm speaks of "capturing" natives ... he does not mean that they were taken against their will'. When the Government Secretary conveyed the Lieutenant-Governor's message to the field officer, he may have had difficulty translating irony into advice.

On patrol Chisholm continued to evade arrows, seize the unwary, and attempt to calm them with gestures and proffered presents. The field officers had no other way to begin communications. In the Nepa area there were no people who could act as interpreters for the government and the Kukukuku were apparently neither curious about the intruders nor anxious to possess their goods.

But the small community on the Arabi was vulnerable. In November 1913 they accepted presents from Chisholm and then in January 1914 the police met a man and woman who chewed betel nut with them and afterwards led them into the village. Another ten men came into the village; they appeared to be weak and many were suffering from sores. Oldham presented the man who first met the patrol with a knife and he gave the others boxes of matches. When Chisholm returned to the Arabi in May he again distributed presents, and on the second day the Arabi permitted him to wash and dress their sores. After first refusing they assisted Chisholm to compile a brief list of Kukukuku words. But they would not go back to Nepa with him. A month later a group of Arabi men appeared at W. Brandon's camp and stayed overnight. Soon afterwards a group visited the Tiveri store where they accepted presents and in July they made their first free visit to Nepa station. From that time Arabi men visited Nepa irregularly; sometimes up to six months passed between visits.

During the six years before the abandonment of Nepa, relations between the Arabi and government officers did not advance beyond nervous visits to Nepa and requests for steel tools. In 1915 Chisholm thought the Arabi might be prepared for another step and invited two men to join the Armed Native Constabulary, but they declined. Government officers also attempted to accustom them to working for rewards. The Arabi delighted in the efficiency of the tools which they used to clear the scrub around the station, but were not enthusiastic about the new system; they were eager to take their pay and go. In turn the Arabi made a request of the government. Seven men and a boy came in from Arabi. They indicated that the Olipai were hostile to

Group of Kukukuku ('More of the nice boys'), Lakekamu Goldfield, 1914
PHOTOGRAPH: FRANK PRYKE

Patrol Officer Fred Chisholm trying to compile a Kukukuku vocabulary, 1914
PHOTOGRAPH: FRANK PRYKE

them and then pointed to the police rifles and the Olipai mountains: Armit assumed they were trying to tell him to direct the power of the police against their enemies. The deputation stayed overnight at the station and Armit put them to work at seven in the morning. When they knocked-off at noon he paid them in beads, knives, empty bottles and mirrors. The Arabi were unarmed while on the station, but when they passed a miner's claim a few hundred yards from the station all were armed. Armit assumed they must have hidden their bows close to Nepa. After one more session of working for reward the Arabi stayed away for six months. When they reappeared Humphries found their constant requests for steel 'almost annoying'. The Kukukuku were not satisfied with their role as mendicants either: they burst the wall of the tool store and took all the government's knives and axes. Humphries was indignant: 'After my treating them kindly and using every effort to gain their confidence, they steal all I have and creep away in the night.'

The theft gave the government officers a new way of exerting their power over the Kukukuku. When four men visited the store in April 1917 Humphries arrested two for stealing from the government store and sentenced them to six months' imprisonment to be served in the Port Moresby gaol. How Humphries determined their guilt, or explained the proceedings of the court to his prisoners, is unclear. The two men left on the *Bulldog* for Port Moresby where, it was hoped, they would learn some Motu and return able to explain to their own people the aims of the government and the benefits to be gained by intercourse with the outside world.

While the government officers had made slight progress with the Arabi the Tiveri and Olipai peoples remained hostile. In 1915 the Olipai killed one of Robertson's labourers and wounded a man working for Kelly. Chisholm, investigating the cause of the attacks, found that the labourers had been making sago from palms probably cultivated by the Kukukuku, who had acted in defence of their food supply. The Olipai fired on Chisholm's patrol when it entered a deserted village, hitting a constable in the shoulder. Chisholm called out to his attackers and showed them knives and tomahawks, but they replied with another flight of arrows. After the police fired a volley to disperse the bowmen, the patrol withdrew.

Several equally ineffective patrols followed. But Murray insisted that the 'efforts to effect these arrests must not be relaxed'. By November 1916 when Patrol Officer Cyril Cameron left on a patrol to Olipai, Humphries was able to predict that he was 'sure to be attacked'. In fact the patrol, caught on a narrow ledge while stones were rolled on them, was lucky to survive. When they finally entered the village it was again deserted and as they left they were fired on by unseen bowmen. Constable Lai-i-woi was hit in the arm. Humphries decided that the next attempt would be a night raid.

In May 1917 the police led a rush down a spur and over log barriers into a village. Protected by the gloom of a wet evening, Humphries hoped his police could trap the Olipai in their houses. By the time Humphries reached the village the police had fired 'one or two volleys', two men and three women had been captured, and the police were struggling desperately with one wounded man. Humphries bandaged the wounded man and released a woman who begged to be allowed to attend to him: 'The woman helped him to his feet and they walked slowly away, and I never saw them again ... he was shot through the lungs, through the liver and through the stomach, through the head and through the right arm.' Humphries took three prisoners to Nepa: a young woman, 'Mary' Bundowi, he held at the station and two men he sent to Port Moresby charged with murder. As there was no one in Port Moresby who could interpret for the court, the prisoners were never taken before a judge. Disturbed by Humphries's patrol report, Herbert Champion, the Acting Administrator, asked what instructions had been given to the police, why they fired and why had such 'shocking injuries' been inflicted on one man when at most one or two shots fired in self-defence should have been sufficient. Humphries, alone at Nepa, duly carried out an inquiry into his actions and those of his police.

Of the four Kukukuku sent to Port Moresby, one man died, and another was released after three months, possibly because he was older and government officials thought there was no chance of his acquiring a knowledge of Motu. At Nepa Humphries found Mary Bundowi's habits disgusting: she was, he decided, so akin to an animal that he thought 'the task of humanising her ... impossible'. When the first prisoner returned from Port Moresby she was released to accompany him back to Olipai. She had learnt very little Motu and had never settled to the routine of life among the wives of the police and their children. She was not seen again by government officers.

While returning to Nepa from a patrol which had taken him to Morobe on the north coast of New Guinea, Humphries saw the two prisoners who remained in Port Moresby. They had maintained a stony indifference to all efforts to gain their interest or teach them anything; but when Humphries spoke a word or two in their own language one man answered with a torrent of words Humphries could not understand. For the rest of Humphries's stay in Port Moresby the two men clung to him, not because they felt any affection for him but because they saw him as a link with their home. Humphries took one man, Iadu from Arabi, back to Nepa. Humphries believed that had Iadu been kept in Port Moresby any longer he would have died, and in any case he had now served his six months' sentence for the theft of His Majesty's tools. At the landing and at the station Humphries had to restrain Iadu from setting off immediately for his home. Having given him a meal and a bag of food Humphries released him the next morning, watched him climb the ridge overlooking the station, fling the bag of stores into the bush, and disappear.

The other prisoner sent to Port Moresby, Didiam of Olipai, returned to Nepa after nearly eighteen months' absence. He had put on weight and allowed his hair to grow; Muscutt seemed disappointed to find that, dressed in a singlet and *rami*, he looked no different from other Papuans. Muscutt escorted Didiam to within hailing distance of his home village and stood with him exposed on a rock while he called his people. To Muscutt's astonishment he first called in Motu, and then when he tried to use his own language he had lost his fluency. The Olipai men seized weapons and women snatched up their children and prepared for flight. Didiam went forward while Muscutt and the police watched:

> It was the most casual [meeting] imaginable. He went up within a few feet of them and started talking. There was not the slightest sign of any affectionate greeting by either side. It was just as if he was a stranger who had gone up to ask which way the road went.

Didiam was able to persuade six men, including one of the other men who had been to Port Moresby, to meet the patrol. Muscutt found it difficult to understand Didiam's behaviour: he was either reluctant or unable to interpret Muscutt's instruction that the Olipai were to refrain from shooting arrows at other people, and he seemed to prefer the company of the police to his own people. Before leaving, Muscutt asked Didiam to come to Nepa after one month. He did not go, and five months later when Muscutt took a patrol to Olipai to re-establish contact, he found that Didiam, having recovered his position in the village, no longer wanted any association with the foreigners. Muscutt regretted that Didiam

> had reverted back to the same state as the other natives in dress and ways. He had lost a lot of his previous plumpness, was now dirty and hardly recognizable with his hair trimmed the same fashion as the others, bone through the septum of his nose, and wearing a grass rami. He had not forgotten all his motu, still we could not get him to talk much.

About fifty Olipai, including some women, came in to meet the patrol, but they were reluctant to trade and kept making signs that the government party should take the food and move on. Didiam refused to go to Nepa or show them the track to Fish Creek. That night, as the patrol camped in the rain after a day scrambling through dense scrub, an arrow was fired from close range at Constable Pangari. Muscutt found the action of the Kukukuku most unreasonable: 'No cause whatsoever had been given to warrant any natives shooting arrows at my party.' Murray's only comment on the incident was that Constable Pangari should have had his rifle with him at the time. On the next patrol to Olipai the police contacted two men who acted as guides for a short time but in spite of the fact that Muscutt made a 'bit of fuss' over them and got them to shake hands with some Milne Bay and Orokolo labourers, they disappeared into the bush before the patrol reached the village. The village itself was deserted, and as the nearby creek was dry Muscutt thought the people might have shifted closer to water. The patrol returned to Nepa without seeing any more Kukukuku. Government officers did not speak to Didiam again.

Iadu of Arabi visited Tiveri and Nepa but at times when Muscutt was away. The two men never met again. The attempt to communicate with the Kukukuku by using men who had been to Port Moresby had failed. Just before Nepa station was closed government officers saw the Kukukuku only when men from nearby villages went infrequently to Nepa to try to get steel. The store was protected with barbed wire, and contrary to the normal rules on a station, the police were issued with ammunition in case of a sudden raid. When ten men arrived at the station in September 1919 Muscutt gave them an enthusiastic welcome, fed them and gave them a 'choice selection' on the gramophone. Although they kept together, holding each other by the wrist, Muscutt thought that they were unafraid and unimpressed. One advance was that Moveave canoemen and Kukukuku had met at Nepa. Thirty years before they had fought one another and perhaps at times had had other sorts of relationships. Now there was the possibility of a new association with the coastal peoples.

The Kukukuku had behaved differently from all other peoples encountered by government officers and miners in Papua. The officers thought this aberrant behaviour stemmed from some deficiency inherent in all Kukukuku. Humphries, while praising his own restraint, thought 'pacification' would take many years: 'Knock-kneed and splay-footed as most of them are, with great muscular legs and long arms, they are not unlike gorillas — with this exception in favour of the gorilla, he doesn't eat man.' Muscutt preferred another explanation which also owed a distant debt to nineteenth-century arguments about the evolution of man:

> They being so childlike in their ways, it was quite likely that they were somewhat puzzled to grasp the reasons why we should come up here, give them presents and then try and coax some of them to go back to Nepa ... it's quite likely that he [Didiam] found great difficulty to explain matters to them which laid [sic] outside of their small world

and beyond their childlike intelligence, and they probably put him down as being a first-rate liar.

But the Kukukuku may have acted according to values which the patrol officers could have understood. They had looked at the actions and possessions of the foreigners and there was only one thing which was of obvious value. That was steel, and they took it whenever they had the chance.

Two factors beyond the control of men influenced relations between the Kukukuku and the foreigners who came up the Lakekamu in 1909. Firstly, the south-eastern Kukukuku were among the most isolated people in Papua. They knew no neighbours who could explain who the foreigners were or what they wanted. Vavasua of Moveave, who had gone with Murray in 1910, was an exception, but he was unable to persuade the Kukukuku to talk with any members of the patrol. Secondly, the foreigners on the Lakekamu were more independent of the local population than they were on any other goldfield. The river took them and their stores to within a day's walk of the field. The miners and officials did not need long carrier lines constantly bringing goods from the coast and they never had to buy food from nearby villages or starve. But labourers taken to the Lakekamu were dependent on the miners for a safe passage home. Because the miners knew the Kukukuku as fierce warriors and they did not need them as either workers or producers of food, they may have been more inclined to shoot on sight than they were in other areas. Many years after, the Kukukuku told the Reverend Herbert Brown that when they visited miners' camps to trade they had been shot at and the dogs had been set after them. Their apparent rejection of the government's advances may be partly explained by the fact that before and after the government began its conciliatory patrols, miners were reinforcing the pattern of violence which the Kukukuku knew well.

In 1930 the Kukukuku robbed Bethune's camp on Twisty Creek. Hides, sent to show that 'they could not steal, with impunity, the property of a white man', followed the tracks of the looters up the Olipai and across the ranges to the Kapau. In a village he found a novel, a china cup, pieces of cloth and other objects taken from Bethune's camp. Men had been working on shovel blades to cut them into strips which could be fitted to Kukukuku adzes. Hides believed he could distinguish four types of Kukukuku and the people on the Kapau, he thought, were pygmies, 'the nearest approach to the original inhabitants of New Guinea'. After a violent struggle Hides arrested some men, but the patrol was ambushed on the way back to Tiveri, a constable was badly wounded, and the handcuffed prisoners escaped. On a second patrol Hides was attacked frequently, and two carriers were wounded: he had failed to carry out his instructions 'to establish friendly relations with the Kukukuku'. But by this time the most frequent intrusions into Kukukuku country were coming from the north as miners and government officials pushed out from Wau. In the early 1930s the Kukukuku killed three prospectors and many labourers from the Morobe goldfield, and attacked government patrols which attempted to investigate the killings. They earned a reputation among New Guinea officers as the 'most bloodthirsty and vicious' warriors in New Guinea.

It was not until the 1960s that government patrols could move peacefully through the country of the Anga peoples, and few of those patrols passed through the thinly populated country disturbed fifty years earlier by the miners fanning out from Tiveri. Some miners and prospectors returned after the war, but probably the people who have had most influence on the Kukukuku of the upper Lakekamu were other Papuans. Some Kunimaipa and Moveave men worked gold, began cattle and rubber projects near Bulldog, and cut timber. Papuan pastors, supported by the London Missionary Society churches on the coast, began work in the area. Some Kukukuku began to move south. In the early 1960s they established the village of Keremahaua on the Olipai just above the junction with the Lakekamu. Their culture is now strongly influenced by the coast: they are canoemen in regular contact with the outside world.

Brown 1956, 1973 and Ryan 1965 provided basic information about the Toaripi and Moveave peoples. I am particularly indebted to the Reverend Herbert Brown who talked to me about the history of the Gulf. Saunders 1965 wrote a popular biography of Brown. Mekere and Louise Morauta, students and residents of the Gulf, gave additional information. The two earliest European residents of the area wrote about the Gulf and its people: Chalmers 1898 and Edelfelt 1887, 1893. Lovett 1903 and Langmore 1974 examined Chalmers's work in the Gulf. MacGregor's account of his voyage up the Tauri and Lakekamu was printed in the *Annual Report 1892/93*, pp. 24-5. The information concerning the development of London Missionary Society schools and relationships between government officers and villagers was taken from *Annual Reports* and Kerema station papers. The work of the Moveave on the river is noted in the Nepa station papers. Brown 1956, p. 163 wrote of his conversations with Moveave men on the goldfields. Williams 1923, 1934 commented on the Vailala madness.

Neville Robinson, M.A. student, U.P.N.G., has collected material on the wartime experiences of the Toaripi. The building of the Bulldog track is described in Reinhold 1946 and McCarthy 1959. Wilson and Garnaut 1968, Brown and Ryan have recorded the attempts by the Moveave-Toaripi to change their way of life in the post-war period.

The Kovio have also been called the 'Bush Mekeo', 'West Mekeo' and 'North Mekeo'. Brown 1956, 1973 and in conversation provided much of the material used here. Other information was collected on a visit to Urulau in 1972.

Lloyd 1973, Appendix A, discusses the origin of the term, 'Kukukuku'. Gajdusek and others 1972 provide a comprehensive guide to printed material on the Anga. The references to early contact between the Kukukuku and government officers is taken from *Annual Reports*, Kerema station papers, Griffin 1925 and Monckton 1921. C.B.

Higginson described his encounter with the Kukukuku and summed up the government officers' knowledge of the Kukukuku in the *Annual Report 1907/08*, pp. 50-4. Murray's statement about the ease of civilising 'this type of native' is from the same *Report*, p. 17. The prospectors commented on the Kukukuku in their reports to Murray and in their letters: Frank to Dan Pryke 11 February 1910 and Jim to Mark Pryke, no date but late 1909 or early 1910. Murray wrote of his attempt to meet the Kukukuku in 1910 in the *Annual Report 1909/10*, pp. 12-14. The largely futile efforts of government officers to establish friendly relations with the Kukukuku are recorded in the Nepa station papers. Humphries 1923 retold the story of the men taken to Port Moresby and Murray commented on them in his introduction to the book. Hides 1935 published his account of his encounter with the Kukukuku and most of his patrol report survives in the Australian Archives, AS 13/26 item C 918 /3. Blackwood 1939a, 1939b, 1950 carried out fieldwork among the northern Kukukuku in 1936-7 not long after they had clashed with miners moving out from Wau. Sinclair 1966 summarised early relations between foreigners and Kukukuku and provided a lucid account of patrolling among the Kukukuku in the 1950s. Sinclair 1969 wrote on Hides's work on the upper Lakekamu. See McCarthy 1963 for a New Guinea officer's experiences with the Kukukuku.

EDIE CREEK

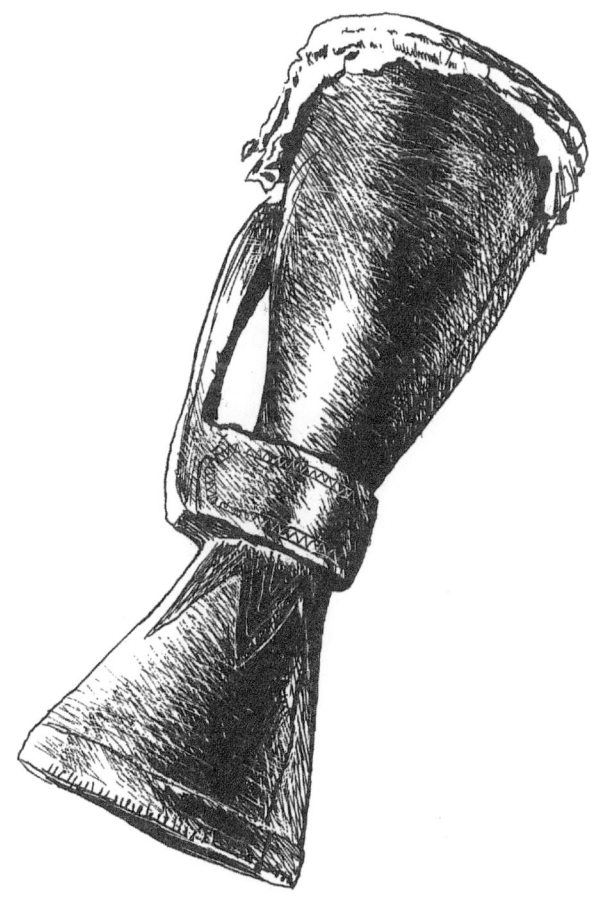

Drum, Morobe District, National Museum of Victoria 1932

14

On Gold

*a quiet whisper …
up on the Bulolo old Shark-eye's
getting gold*

In 1909 most miners in the Northern Division thought that the government prospecting expedition should use the rivers of the eastern Gulf to enter the ranges from the south; and Matt Crowe had agreed with them. But had the land north of the eighth parallel not been part of the German Empire the miners might have suggested going beyond the Waria and trying the rivers flowing into the Huon Gulf. Before Crowe asked him to join the government expedition, Frank Pryke had wondered whether or not it would be worthwhile turning 'square-head' and prospecting in German territory.

By the end of 1909 the Australian miners knew that the land to their north carried signs of gold; but they also knew that defining and working a field in that country would be difficult and dangerous. In August 1909 two German prospectors, Wilhelm Dammköhler and Rudolph Oldörp, went up the broad flat valley of the Markham and then turn ed south to follow the Watut. Dammköhler, who had spent many years prospecting, trading, farming and pearling in Australia and New Guinea, was well known in Papua. Not long after Dammköhler and Oldörp had sent their carriers back to their villages on the coast a shower of arrows ripped into the camp on the Watut and thirty yelling warriors rushed forward. Within a few minutes many of the warriors were shot dead and the prospectors were slashed and spiked with arrows. Dammköhler bled to death. Oldörp built a raft and drifted for five days to the Markham mouth. While recovering at the Lutheran mission station at Cape Arkona he chose to be remembered by writing in the guestbook a verse from Psalm 104:

> O Lord, how manifold are thy works!
> In wisdom has thou made them all:
> The earth is full of thy riches.

Having found a few grains of gold and much quartz, Oldörp was determined to go back up the Watut. But on his return to the Markham mouth in 1910 a sudden storm caught his schooner, and he was drowned.

Map 15 Morobe Goldfield

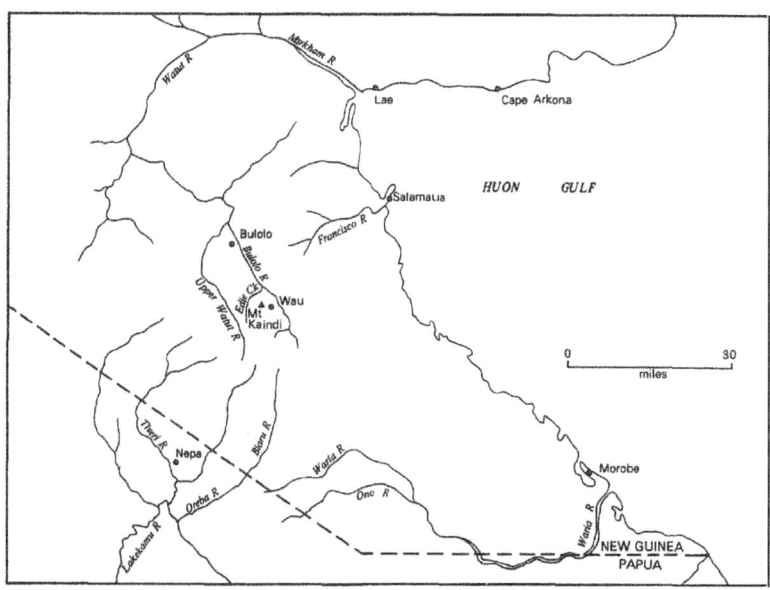

Arthur Darling had also looked north. Late in 1909 he recruited carriers from the Northern Division and went up the Markham and the Watut. Sick with fever and short of food he too was attacked and wounded. His labourers carried him back to the river and they rafted down to where they had left a whaleboat; but it was gone. They went on to the Markham mouth where Darling was nursed at the mission station. Picked up by Les Joubert, Darling rested at Buna before joining the rush to the Lakekamu. Darling said that he had found gold not far across the ranges from the Lakekamu field; but he did not have a chance to build up a reserve of cash to allow him to lead another prospecting expedition. Never having fully recovered from lack of food, exposure and spear wounds, he died early in 1911. 'Buna' Darling was born in Canada, and grew up and worked in Queensland before going to British New Guinea in about 1900. He was forty-two when he died.

In 1912 when gold was becoming hard to find on the Lakekamu Matt Crowe, William (Sharkeye) Park, James Preston and Edward Auerbach tried the Markham again. They confirmed that the villagers were determined to contest the right of any foreigners to enter their lands, but they were unable to add to the elusive stories of being close to payable gold. In 1940 Teddy Auerbach wrote that sometime before 1914 Jimmy Preston went back and found gold on Koranga Creek.

Although there is no evidence to prove Auerbach's claim it could be true. Auerbach is reasonably accurate when he writes of other incidents, and Preston made an official request to prospect in German New Guinea in 1913. If Preston did find gold he had good reason to keep the discovery quiet: he did not want other diggers crowding into the area before he could select and begin work on the richest ground, his request to enter the area had been refused, and German syndicates claimed the sole right to prospect in the area between the Waria and the Ramu. Whether or not Darling or Preston had made a strike, the stories of gold on the upper Watut were made a little easier to believe by the German New Guinea Annual Report for 1912-13 which assumed that the ranges running north-west from Papua released gold into the streams draining into the Huon Gulf.

Table 14
Principal goldmining laws
New Guinea

The Mining Ordinance 1922
    came into force on 1 January 1923. It set the fee for a miner's right at one pound a year, but generally the legislation was similar to that in force in Queensland and Papua. The disputes among the early white miners on Edie Creek arose because the ordinance seemed to allow individual alluvial miners to hold large areas of rich ground under dredging or sluicing leases. Such leases were normally granted only to those about to install machinery for large-scale operations.

The Mining Ordinance 1928
    repealed the ordinance of 1922 which had been amended every year since its introduction. The 1928 ordinance was also amended frequently; by 1940 the basic ordinance was known as the Mining Ordinance 1928-40. The provisions of the Papuan ordinance of 1907 to protect and give rights to 'native' landowners and miners were included in the 1922 and 1928 ordinances.

The people who confronted the prospectors once they left the coast were the Laewomba, a strong community expanding their territory down the Markham Valley and driving out the previous inhabitants. On the coast scattered groups, fearing another raid on their homes and gardens, had neither the power nor confidence to fight the foreigners. Like the Sudest in 1888 they looked to the foreigners for protection. They welcomed the Lutheran missionaries who established a station at Cape Arkona in 1906 and they assisted the private and government expeditions which passed through their lands to fight the Laewomba. In 1921 when late German New Guinea became the Australian Mandated Territory of New Guinea and Australian civilians took over it's administration from Australian soldiers, the Laewomba of the Markham Valley met government officers and missionaries in peace, but higher up the Watut and the Bulolo the villagers still felt able and obliged to attack foreigners. The new government in Rabaul passed a mining ordinance, similar to the Queensland and Papuan legislation, to come into effect in 1923: the country north of the eighth parallel was then Australian and open to Australian prospectors.

Two were already there; and on gold. Sharkeye Park and Jack Nettleton found gold on Koranga Creek in 1922. Park, born in England, had, like Crowe, mined in Western Australia and the Yukon before going to the Northern Division and the Lakekamu. Perhaps he had used information obtained from Darling or Preston, but it is more likely that he wanted to go beyond the country he had seen ten years before and prospect the general area said to be gold-bearing. In addition to the knowledge built up by the miners, Park could use a map published in 1915 by the missionary Georg Pilhofer of his journey from the upper Waria to the Bulolo and the Markham, and he had talked with George Ellis, who had taken a government patrol into the Bulolo Valley early in 1922. Nettleton, a member of the Australian Expeditionary Force which captured German New Guinea in 1914, returned as a member of the administering force, and stayed on after his discharge to open a trade store east of the Markham on the Huon Gulf. He left the store to team with Park on his successful prospecting trip of 1922.

Table 15
Morobe Goldfield
Men employed in mining

	Year	European	New Guinean
November	1926	219	1324
June	1932	419	3271
June	1938	700	6218

Park and Nettleton worked on Koranga for a year before outsiders knew of the strike. Then there was no rush. The extent and richness of the gold-bearing country were unknown, and Koranga was difficult to work. By the time a miner and his labourers reached the mining area their stores were almost exhausted, little food was available locally, and unless a miner made his own arrangements he could not obtain stores at the beach. It was like MacLaughlins Creek in the 1890s without the Beda gardens. In 1925 only five men attended the Christmas dinner provided for all the diggers by Doris and Charles Booth at their claim on the Bulolo. Three did not go: Park was sick and Arthur Chisholm and William Royal were too poor. Yet during the year there had been rumours that some of the Koranga and Bulolo miners were making a lot of money: according to one report in the *Papuan Courier*, two experienced miners thought it possible to win up to 80 ounces a day on

Koranga. It was true that Park and Nettleton were having trouble finding enough containers for all their gold; but that was certainly not a problem for all the diggers on the field.

Table 16
Morobe goldfield
production
(*ounces*)

1923/24	6617
1924/25	7417
1925/26	10,067
1926/27	84,768
1927/28	113,874
1928/29	87,542
1929/30	42,819
1930/31	57,874
1931/32	108,647
1932/33	196,823
1933/34	257,511*
1934/35	298,634
1935/36	300,735
1936/37	359,917
1937/38	403,652
1938/39	229,212
1939/40	271,574

* Until 1933/34 the figures are for the gold exported from the Territory; after that year they are estimates for the production on Morobe.

In January 1926 Bill Royal and one labourer spent five days cutting a track through heavy timber and scaling cliff faces to get to Upper, or Top, Edie Creek. Rising high on Mount Kaindi, Upper Edie plunged over a massive rock bar to cascade into Lower Edie. In only 4 miles from Upper Edie to the Bulolo the creek fell 4000 feet. Upper Edie was over 7000 feet above sea level. Moss covered tree trunks, roots and rotting logs; each night it rained; and often during the day the clouds piled up against the mountains, covering the mining area in more rain or dense fog. Royal and his partner, Chisholm, were in debt and almost out of stores; but in Upper Edie Royal found that he could wash up to 7 pennyweights in a dish. He left to report his strike and register claims: he had found Papua New Guinea's richest alluvial field. Upper Edie, slowed by the rock bar, had been dropping gold in 'free-wash' which was up to 8 feet deep along several miles of the main creek and its short tributaries. Where successful miners on Papuan fields took 300 ounces in a year, enough to pay their expenses and give them a spell in Samarai and Australia, on Edie they washed-up thousands of ounces and hoped to join the carefree rich, a goal as elusive as the big strike. On Papuan fields nuggets were rare; on Edie the labourers picking out stones from the head of the sluice often cried, 'Golston i stap'. They recognised the nuggets (or 'specimens') by their weight rather than by their colour.

Boats began calling at Salamaua, the *saksak* (sago leaf) town on the narrow spit jutting into Huon Gulf. It grew quickly as a port and the beginning of the track into Edie. By September 1926 there were ninety miners and 540 labourers on the goldfield, by October there were 158 miners and 790 labourers and by November there were 219 miners and 1324 labourers. Joe Sloane and Dick Glasson took 240 ounces in one day from a 15-inch box on Midas Creek. During 1927 the miners declared over 110,000 ounces at Edie and Salamaua.

Among the first to arrive at Edie were some of the men who had battled around the Papuan fields for more than twenty years: Frank Pryke, Les Joubert, Charles Ericksen, Dave Davies, Andy Gillespie, and Ned Ryan. Gordon Robertson, Peter 'Bourke' (Bjorquist), George Arnold and Joe Sloane were already on Bulolo when Royal climbed to Upper Edie. Lucky Joe Sloane became one of the 'Big Six', the group who first pegged much of Upper Edie. John Henry Sloane, a Queenslander, went to south-eastern New Guinea with James Hurley's prospecting expedition of 1894. When news of Lobb and Ede's strike on Woodlark reached north Queensland he went back to mine on the islands and then shifted to the Gira. After going to South Africa too late to fight the Boers he returned to follow the diggers to the Yodda, the Waria, the Lakekamu and then back to the Gira to mine osmiridium when it suddenly and briefly increased in value at the end of World War 1. Before he went to look at Sharkeye Park's prospect he was already known as Lucky Joe. He had been wounded on the Yodda, shot men on the Aikora, and took a woman from the Waria.

Table 17
Sepik Goldfield

From 1936/37 to 1939/40 over 6000 ounces were taken each year from the Sepik, the second most productive field in the Mandated Territory. In the year of greatest production, 1937/38, the alluvial miners obtained 11,012 ounces. In the next year there were thirty white miners and over 500 labourers working gold on the Sepik.

Young men fresh from Australia and proud of their physical strength rested exhausted on Komiatum Hill, the first tough climb out of Salamaua, and wondered how the old diggers, many still suffering from over-indulgence at the store, completed the six-day walk into Edie. It was, Frank Pryke said, the worst track he had seen used to supply a goldfield; and at the end of it Edie was 'cold and wet — a miserable kind of place'. Matt Crowe was not there: he had died in Samarai in

1925. Having spent his Lakekamu gold he chased pennyweights at Keveri and on the Awala River, one of the tributaries of the Musa. He had lived long enough to hear the talk of the gold on Koranga, but not long enough to take a share in the 'big one'. Frank Rochfort, who had instructed the Royal Commissioners in 1906 on the proper way to handle a team, stayed on the Murua Field. Andy Doyle had retired to a shack on the Laloki. His visitors listened to stories of old goldfields and watched his chickens hop on his knee and hat, roost in the rafters, and drop dung indiscriminately. Billy Ivory had died in the Northern Division with a blessed candle in his hand. Among his effects were a prayer book, rosary beads, a crucifix, three packs of cards and two loaded dice. Sir Hubert Murray thought that a theologian would say that he had faith but lacked obedience. Sharkeye Park did not shift onto Upper Edie. After four years' mining on lower creeks he sold his leases and left New Guinea in 1926 with a 'healthy five figure fortune'. White men had been searching constantly for gold in New Guinea for fifty years and mining it for thirty-eight. Sharkeye was the first alluvial digger to get rich. He married in Sydney and retired to Vancouver, Canada, where he died in 1940.

The Gosiagos were not there. The law compelled miners to recruit their labour from the Mandated Territory of New Guinea. The new labourers came from the Markham, the Huon Peninsula and the Waria, but as the field developed more and more men came from the Sepik. By 1940 over 3500 Sepik men were working in the Morobe District.

Table 18
Indentured labourers employed in mining
New Guinea

30 June	Employed in mining	Total number of indentured labourers
1931	1900	27,765
1933	3875	28,242
1936	6816	36,927
1939	7162	41,849

In 1931 nearly two-thirds of all indentured labourers worked on plantations; later in the 1930s about half of all indentured labourers worked on plantations.

Papua

Each year from 1888 to 1920 an average of about 1000 Papuans signed-on to work for miners. In 1900 over half of all indentured labourers worked for miners; in 1920 about one-tenth worked for miners.

Early miners were desperate for labourers. One man wrote to his sister in Cairns, 'we have just got to get those coons or come home'. A miner needed twenty-four labourers to work a claim efficiently; eight men to sluice and sixteen to lump stores from Salamaua. The carriers could complete a round trip in three weeks. By the end of 1926 recruiters, who had been charging £5 to £7 for a 'three year boy', were asking £20 for a man willing to sign-on for one year, and there were stories of miners paying £30 and offering more. The labourers benefited little from their scarcity. Most of them were paid 10s. a month while the standard wage for a plantation worker remained at 6s. Recruiters travelled frequently through the known and populous areas of the Morobe District, but after the early recruits had talked about the hardships of life at 'Kaindi' men were reluctant to sign-on. The recruiters became more aggressive: they 'pulled boys' with lies and threats. Some government officers needing carriers to take supplies to the new station on the field were also guilty of abuses. The people of Binamarien, a 'Gazup' village near the head of the Markham, were unlucky. The *luluai* (government-appointed headman) of a neighbouring village said that raiders from Binamarien had killed some of his people. A government party arrested twenty-six 'Gazup' men and brought them back to Salamaua. They admitted taking part in the raid, explaining that it was in retaliation for a previous attack on them. Those 'Gazup' prisoners passed fit were made carry packs to Wau; some died on the track and some in the Salamaua hospital. Twelve men survived to return to Binamarien. Shortly after their arrival, they were attacked and another man was killed. The Binamarien men, from a Tairora-speaking village within the Gadsup area east of Kainantu, were among the first Highlanders to go to *nambis* (the beach) and work for white men. Batong of Gensiko village on the Huon Peninsula went to work as a carrier because he believed that once his name was written down he had no choice: how could he '*sakim tok bilong Mast a Pits*' (contradict the Australian police officer)? Umauna of Uluo volunteered. The government officer obtaining men to carry to Wau told Raikupa that he must go, but he had two wives and three children to look after so Umauna, a single man, took his place. Endong, a senior man of Koromanau, went unwillingly. His *luluai* was punched about the head and threatened with gaol, and others who protested were abused and slapped. The men of Koromanau were not asked whether they wanted to go, and they saw that resistance was futile and dangerous.

The actions of the *kampani masta* (private recruiters) were rarely subjected to the same scrutiny as those of government officers, but there were many stories of recruits being tricked or intimidated. Most private recruiters were probably guilty of crimes that could have earned them spells in Townsville's Stuart Creek gaol. They excused their behaviour by saying that the 'coons' treated one another a lot worse in the bush, that the returned labourer was a bigger, fitter and more knowledgeable man, and that without labourers there would be no economic development. Errol Flynn wrote that he obtained recruits by deceit; but he was as skilful at misleading readers as he was at creating an illusion on film. It is uncertain how he came by the men who worked for him at Edie in 1933. Bumbu of Busama village north of Salamaua on the Huon Gulf was given his *luluai's* cap in 1926. He increased his wealth and power by supplying recruiters with men. By 1940 he was a village despot, exploiting the villagers and hoodwinking the Australian government officers.

Miners' camp near the junction of Edie and Merri Creeks, 'one of the picked spots'. Late 1920s
PHOTOGRAPH: FRANK PRYKE

Frank Pryke's hut on Edie Creek
PHOTOGRAPH: FRANK PRYKE

Bishop Mambu of the Lutheran church was fifteen in 1938 when a *kampani* and one of Mambu's *wantoks* came to Bukaua village on the Huon Gulf. Mambu had spent some time at the village and mission station schools, but there was no place for him at the training school for mission teachers. He offered to go away with the recruiter who then gave him a pound and quietened his mother's protests with sweet talk. Flown to Wau, he was told to work in the hotel, but the lawlessness of the men of the town frightened him. Although his *masta* told him that he was too young to work as a miner on the Watut, he asked to be allowed to go. His *masta* warned him, '*sapos yu no wok strong long karim ston bai mi paitim yu*'. Mambu agreed that if he did not work hard he would be beaten. On the Watut he carried stones, dammed creeks, dug races, directed the hoses while sluicing, and amalgamated gold. He worked alongside men from the Markham, the Sepik, Bougainville and New Britain. They settled disputes in games of football: a practice he called *pilai kik*. He learnt of a sharp division between *masta* and *boi*:

> *Mipela i no save bung wantaim ol wait skin na sindaun na tok tok wantaim na kaikai wantaim, nogat. Masta bilong mi ino save mekim planti tok gris wantaim mi, nogat. Olsem mi no pilim wait skin i no sindaun klostu long mi na toktok wantaim mi, nogat. Olsem birua. Ol i bosim Nu Gini bilong wok tasol, i no gat skul i tok long wok. Lainim long han wantaim pait tasol.*

Mambu knew no times of friendly talk between himself and his masta. He was given food, a bush house to sleep in, instructions and a cuff when he was slow to act or learn.

Salamaua's jovial billiard saloon keeper, Bill Cameron with a few of his patrons. June 1929
PHOTOGRAPH: H.L. DOWNING

Some areas suffered from heavy recruiting. F.B. (Monty) Phillips who conducted an inquiry into recruiting in 1927 said that some accessible

Labourers sluicing with a monitor on Koranga Creek
PHOTOGRAPH: FROM THE ALBUM 'ON SILVER WINGS TO THE GOLDFIELDS OF NEW GUINEA'. NATIONAL LIBRARY OF AUSTRALIA

Labourers carrying ore to the tram-line that serves the crusher, 1938, Kupei. Bougainville, now the site of a giant copper mine
PHOTOGRAPH: COURTESY MRS M. SCOTT

villages had lost nearly all their able-bodied men. At Sialum village on the north coast of the Huon Peninsula fifty-three adult males were away. This left eight fit adult men in a village of 149 females of all ages, eight old men and eighty-six male children. Other villages, Phillips reported, had suffered in the same way. The District Officer at Salamaua estimated that by February 1927 over 3000 New Guineans were employed in connection with mining, and that the government needed about 450 men to build roads and supply its staff and hospital on the field. Many groups from villages and labouring camps dressed and danced at the 1927 goldfields Christmas *singsing*.

Much that the young men from Australia found strange, violent and appalling was familiar to the old diggers. The people of Kaisenik village on the upper Bulolo killed some carriers, and government officers, miners and labourers combined to shoot some of them. Few New Guineans lived near Edie Creek, but once the miners moved further west and south they entered the country of the Anga peoples, the Kukukuku, with whom they traded and fought. A dysentery epidemic broke out at the end of 1926 and nineteen carriers died on the Gadugadu track between Salamaua and Edie. Others died on Edie, but generally the death and desertion rates were much lower than on the Gira and Lakekamu. One digger arrived at Salamaua with a team of mules, and turned them loose rather than carry them to Edie.

Sharkeye Park, staying at Petty's Hotel in Sydney, read lurid accounts of Kaisenik cannibals, kanakas and raids; and he wrote to the Minister. He said that he knew all the villagers well, they could not muster nearly as many fighting men as the reports claimed, and that before the violence they had suffered greatly from miners and carriers who damaged their gardens and stole their property.

Sir George Pearce as Minister for Home and Territories in the Australian Government received the first official reports of the find at Edie. In 1926 Pearce had been Minister for five years and he served another five years before 1940. In the 1890s Pearce had briefly given away his trade as a carpenter to look for gold at Kurnalpi in Western Australia. The Aborigines killed two prospectors in the Kurnalpi area, and Pearce himself used his revolver twice against Aborigines. He fired at unseen men who threw a spear into his camp at night, and he shot to frighten a man who he thought was about to steal from his tent. Pearce was not excited by the stories of gold and violence in New Guinea. He and other Ministers were mainly concerned with warning diggers of the hardships that they faced in New Guinea and denying reports that labourers were recruited by force and treated cruelly by employers.

In April 1927 Pard Mustar made the first flight from Lae to 'the Wau'. Soon more freight was being flown from Lae than from any other airport in the world. Miners paid £35 for themselves and 1s. a pound for their stores and 'boys' for a flight to Wau. By the end of 1927 more planes were operating and the rates fell. The pilots flew up the Markham, turned south along the Wampit to its headwaters, crossed into the Watut Valley and from there on a good day they could see the patch of kunai grass at the head of the Bulolo and other goldfield landmarks. They landed uphill on the sloping Wau strip. The flight took fifty minutes.

Putting the alluvial of Upper Edie through the boxes was only the beginning of the development of the Morobe goldfield. Morobe followed the pattern dreamt about by those optimistic diggers on every Papuan field who planned companies to exploit lodes and dredge. Companies worked rich reefs and Bulolo Gold Dredging flew in giant dredges piece by piece, assembled them at Bulwa, and set them churning in the broad valley of the Bulolo. In 1934 the Morobe Goldfield produced over one million pounds worth of gold, and in 1940 it returned three million.

The alluvial miners moved on. They worked gold on the upper Ramu, the Sepik and Bougainville. They prospected across the central highlands to the border with Dutch New Guinea and found a little gold. Local communities evaded them, worked for them, traded with them, fought them, and slept with them. Their behaviour towards the miners was influenced by their previous meetings with foreigners, their beliefs about how men ought to behave towards outsiders, their relationships with other New Guineans of the area who could be seen as possible enemies or allies of the miners, and accident.

Table 19
Gold exports Papua

From 1895/96 until 1915/16 gold was the most valuable export from the Territory of Papua; but the total value of all exports was low.

Year	Gold exports £ A 000	Total exports £ A 000
1900/01	33	50
1905/06	58	80
1910/11	67	117
1915/16	43	125

From 1916/17 until 1937/38 copra and dessicated coconut were more important than gold. In 1938/39 gold was again the most valuable export, and the next year rubber exceeded the value of both gold and copra products.

1920/21	11	172
1930/31	23	274
1940/41	133	493

Australian newspaper reports of the Morobe gold strikes
PHOTOGRAPH: NATIONAL LIBRARY OF AUSTRALIA

Jack O'Neill, who prospected and mined on Morobe, the Waria and the eastern Highlands in the 1930s, set down his memories of the miners' relationships with New Guineans:

> We white men treated our boys as inferiors; some with tolerance, some with unadmitted affection, some few with hatred bred of fear, but it was impossible to harbour such attitudes towards these wild free people [of the Dunantina valley on the eastern edge of the Highlands], They were just men; most very likable, a few nasty bastards; just like our own kind. A few were very real friends. While prospecting the Agutina creek one day I sat down on the bank for a smoke. An oldish looking man sat down beside me and produced his pipe also. I gave him a light and then presented him with the few remaining matches in the box. These he accepted, but immediately dipped into his string bag and gave me two cucumbers in return. That is the attitude we found; I hope it has not been destroyed. Often while travelling, a string of men would come streaming down a mountain spur and run up and greet us. Invariably they would produce cucumbers — just like our "Apple" cucumbers — and hand them around. These were most welcome on a dry track between rivers; never did think much of cucumbers as a food, but these proved a wonderful thirst quencher, and were always cool. Looking back, I fear we gave these people much less than they gave us; the civilisation we imposed on them was just another Trojan Horse.

Table 20
Gold exports
New Guinea

Year	Gold exports £ A 000	Total exports £ A 000
1923/24	17	719
1925/26	25	1105
1927/28	256	1471
1929/30	96	997
1931/32	399	1109
1933/34	1368	1766
1935/36	1704	2573
1937/38	2029	2980
1939/40	3022	3674

In 1922/23 copra made up 98 per cent of all exports from the Territory of New Guinea; in 1939/40 gold made up more than 80 per cent of all exports.

Some communities certainly suffered from the coming of the miners. Some Muruans must have lamented their presence, and so must some villagers on the Yodda, Gira and Waria (although in the Northern Division the miners came as a battering ram rather than as a Trojan Horse). The Keveri Valley, once the home of a 'friendly lot of coloured men' is now almost deserted, but perhaps the people would have left had the miners never worked the rocky creeks below their garden lands. But not all communities suffered. The Misima shared a small island with alluvial miners and reefers; their numbers have increased and they still own their own land. They are freer and better fed than most people in the world. Most southern Kukukuku hamlets survived the mining on the Lakekamu. Apart from loosing arrows and taking steel they chose to have nothing to do with the foreigners: they did not fall for the Trojan Horse trick.

The search for gold brought many strangers face to face. It was the reason why many Australians went to Papua New Guinea, and until the invasion by the Japanese in 1942 it was the main reason why most of them went beyond the beaches. To say that the encounters were exciting, complex, varied and dynamic is to say the expected; but the drama, complexity, variety and events which flowed from the meetings have rarely been shown in detail. Only a close examination can trace the shifts between exploitation, interdependence and manipulation. The Orokaiva lined up behind their shields and charged the rifles of the government officers, police and miners. Yet they did not blindly fight the foreigners; they strove to divert them and exploit their power, they sought to select from the new order and not be overwhelmed by it. Nor did the meetings absorb all the energy of the communities which faced the men who came to mine and govern. A Binandere clansman in 1897 may not have thought that the most important fight he was involved in was the clash with the foreigners. In fact his decision to fight, evade or compromise with the foreigners was largely determined by his relations with other Binandere clans and neighbouring peoples. The most fundamental change experienced by the Beda was their shift from being a Koiari to a Fuyuge people; a process which seems to have gone on during the most varied and violent of encounters with foreigners. All communities had many meetings with different sorts of strangers. Miners, planters, missionaries and government officers of course wanted different things from Papua New Guinea and its peoples. But the most numerous outsiders encountered by nearly all Papua New Guinea communities have been other Papua New Guineans. Perhaps the most important effect of the goldmining industry on the history of Papua New Guinea is that it caused thousands of Papua New Guineans to travel beyond the lands that they had known.

This book is a study of meetings between peoples. Those meetings were important in the history of the people of Papua New Guinea. It does not say that the people have a history only when they have been in contact with foreigners. On the contrary, while concentrating on one encounter, the writer has tried to show that Papua New Guineans have had many meetings, and the meetings have changed them, but not blotted out their past.

Expressing his readiness to become a resident of German New Guinea, Frank Pryke actually wrote '☐Head' in a letter to Dan Pryke, 9 February 1909.

Dammköhler and Oldörp 1909 and Oldörp 1909 wrote of their prospecting expeditions. Flierl 1932, Healy 1965 and 1967, Holtzknecht 1973-4, Parr 1974, Souter 1964, and Willis 1972 and 1973a refer to the two German prospectors.

Arthur Darling's last trip is omitted from some accounts or wrongly dated. Two accounts which at least fix the date are Murray's diary, 28 December 1909, 6 April and 15 June 1910, and Holtzknecht 1973-4 who used the mission records. Murray said that Darling claimed to have found gold. Auerbach wrote about his expedition of 1912 and Preston's later trip in *Pacific Islands Monthly*, March, June and July 1940.

The early history of the peoples of the Markham Valley is taken from Willis 1972 and Holtzknecht 1973-4. McElhanon 1970 and Hooley and McElhanon 1970 provide a guide to research into languages in the area and map the distribution of languages.

There are many accounts of the early strikes on Koranga and Edie; few of them agree. Clune 1951, Demaitre 1936, Idriess 1933, Rhys 1942, and Taylour and Morley 1933. Healy 1967 and Parr 1974 have examined this literature. Booth 1929, Leahy and Crane 1937, and Struben 1961 have published accounts of their experiences on the Morobe goldfield.

The story of Royal's find is based on his evidence given to the Royal Commission, 'Report of the Royal Commission on the Edie Creek (New Guinea) Leases', 1927, Transcript of Evidence.

The biographical notes on Sloane are from Clune 1951 and *Pacific Islands Monthly*, April 1946, pp. 48-9. Sloane sailed for South Africa with the 7th Australian Commonwealth Horse in 1902 and arrived after peace was declared (P. Murray 1911). Matt Crowe's death is recorded in *Papuan Courier*, 14 August 1925. Frank Pryke made his comments on the track and Edie to the Royal Commission of 1927.

Although many outstation records from the Mandated Territory were destroyed during World War II some, sent to Canberra, survived. The most important mining records are in the Australian Archives, series A 518, and include: 'New Guinea Mining, Wardens' Reports, 1927-41', AA 834/2, parts 1-6; 'New Guinea, Discoveries of Gold, 1927-34', F 834/2 (also refers to fields outside the Morobe District); 'Morobe Goldfields, Press Cuttings', AB 834/2, parts 1 and 2; 'Commission of Inquiry re Recruiting in Morobe District New Guinea 1927' (by F.B. Phillips), AD 840/1/3. Two other useful files are 'W.M. Park Re Treatment of Natives 1927', CRS A1 27/2571; and 'Salamaua 1927' (Kaisenik attack), CRS A1 27/728. The cases of recruiting in the Markham Valley and Huon Peninsula are from Phillips's report. Surviving Patrol Reports are listed in *Pambu*, No. 35, April-June 1974. Flynn 1960 wrote about his experiences in New Guinea. Moore 1975 checked his accuracy.

The history of the later development of Bulolo has been written by Healy 1967.

The work of the alluvial miners away from Edie is recorded in *Annual Reports*, Beazley, Fox, Leahy and Crane 1937, Radford 1972, Tudor 1966, Willis 1969a and 1969b, and Fulton.

The quotation from O'Neill is from his manuscript, A Prospector's Diary: New Guinea 1931-1937.

Some of the activity on Edie and Bulolo was recorded on film (see list in bibliography).

Bibliography

1. MANUSCRIPTS

Government Records. Much government material is held in the National Archives of Papua New Guinea, Port Moresby and on microfilm in the National Archives, Canberra. The location of documents which may exist in one place only is given.

Angau War Diary, records of the war of 1939–45, Australian War Memorial Library.

New Guinea Campaign Records, War of 1914–18, Australian War Memorial Library.

Papers Relating to U.S.A. forces in Woodlark Island, C.A.O. 609/7/9, 616/4/1–5. Australian War Memorial Library.

Trials held by Angau for various civil offences committed by natives 1943–4, File 506/1/4, records of the war of 1939–45, Australian War Memorial Library.

Barry, J.V. 1945. Commission of Inquiry ... into the circumstances relating to the suspension of the Civil Administration of the Territory of Papua, February 1942, Australian Archives, A518 W 800/1/5.

Correspondence of the Lieutenant-Governor or Administrator of British New Guinea and Papua to the Governor of Queensland and later the Governor-General of Australia.

Delimitation of the Boundary between Papua and New Guinea, Australian Archives. C.A.O., A1 14/4329.

Morobe Goldfields, Wardens' Reports, 1927–41, Australian Archives, A318, AA 834/2, Parts 1–6.

Murray, J.H.P. 1905–40. Notebooks of cases heard as a judge of the Central Court, Australian Archives, C.R.S. G 178, G 179.

New Guinea, Discoveries of Gold, 1927–34, Australian Archives, A518, F 834/2.

New Guinea, Labour Conditions on the Goldfields, Natives 1930–41, Australian Archives, A518, O834/2, Part 1.

New Guinea, Goldfields Maps, 1930–1, Australian Archives, A518, N 834/2.

New Guinea, Morobe Goldfields — Press Cuttings, Australian Archives, A518, AB 834/2, Parts 1 and 2.

Papua, Mines Papers, 1907–27, National Archives of Papua New Guinea G 70.

Patrol Reports and Journals from outstations: Abau, Bogi, Buna, Bwagaoia, Cape Nelson, Ioma, Kerema, Kokoda, Kulumadau, Nepa, Nivani, Papaki, Samarai, Tamata.

Phillips, F.B. 1927. Commission of Inquiry re Recruiting in the Morobe District of New Guinea ... Australian Archives, A518, AD 840/1/3.

Register of Criminal Cases, Central Court, British New Guinea, National Archives of Papua New Guinea.

Resident Magistrate, South-Eastern Division, General Correspondence 1919–22, National Archives of Papua New Guinea, G 180.

Rex v. Boutillier and Downey, Samarai, 1948, Supreme Court Building, Port Moresby.

Robinson, C.S. 1903–4. Diary, May to May, National Archives of Papua New Guinea.

Royal Navy, Australian Station, New Guinea, 1884–8, Vol. 3, Microfilm, National Library of Australia.

Private Papers

Anglican Mission Papers, New Guinea Collection, University of Papua New Guinea.

Bardsley, George H. 1891–2. Diary, May to January, Pacific Manuscripts Bureau.

Beazley, R.A. (Marta Bele). New Guinea Adventure, manuscript, and some letters, diaries and photographs, Fryer Library, University of Queensland.

Bensted, J.T. Letters, articles and photographs, Papua 1900–30, National Library of Australia.

Bramell, B.W. Papers (including many news-cuttings, Papua and New Guinea, 1897–1940), National Library of Australia.

Brierly, Oswald. Paintings and sketches, Mitchell Library and National Library of Australia.

Carne, Joseph E. 1911–12. Father's New Guinea Diary, 30 December to 20 September, New Guinea Collection, University of Papua New Guinea.

Chinnery, E.W.P. Stone work and Goldfields in British New Guinea, typescript of paper read before the British Association for the Advancement of Science, 1919, National Library of Australia.

Corbould, William H. The Life of Alias Jimmy, manuscript, Mitchell Library.

Deland, Anne. Letters, National Library of Australia.

Deland, C.C. Bougainville — New Guinea, manuscript, National Library of Australia.

Dexter, Henry. Reminiscences of a 'Gin-soaked' Trader, manuscript, Pacific Manuscripts Bureau.

Downing, Henry L. Photographs of the Morobe Goldfields and elsewhere in Papua New Guinea, Mitchell Library.

English, A.C. Letters, news-cuttings, National Library of Australia.

Fellows, S.B. 1891–3. Diary, July to October, Panaeate, National Gallery of Australia.

Fox, John R. 1934–5. Diary of prospecting trip from Mount Hagen to Dutch New Guinea border, reminiscences, and tape-recorded interview with Roger Southern, New Guinea Collection, University of Papua New Guinea.

Fulton, Edward. 1936–48. Papers (letters, diaries, gold returns, accounts), Pacific Manuscripts Bureau.

Green, John. 1892–6. Letters, Pacific Manuscripts Bureau.

Hunt, Atlee. Papers (includes correspondence with Abel, Campbell, Griffin, Monckton, Murray, Sabine and Smith), National Library of Australia.

Imlay, N.G. Photographs of Port Moresby and Samarai about 1910, National Library of Australia.

Johns, J.H.W. Letters, Archives, University of Melbourne.

Lett, Lewis. Papers, National Library of Australia.

London Missionary Society Archives, microfilm, National Library of Australia.

MacGregor, William. 1890–2. Diary, November to October, National Library of Australia.

Marshall, Charles W. Diaries of oil and gold prospecting expeditions in Papua and the Highlands of New Guinea, 1927, 1928, and 1933, and photographs, The Australian Museum.

Miller, James. 1928–32. Diaries (not complete), Pacific Manuscripts Bureau.

Murray, J.H.P. Draft of Book of Reminiscences and Diary, 1904–17, Mitchell Library.

Murray Family Papers (in particular the letters of Hubert to Gilbert and Rosalind Murray), National Library of Australia.

Musgrave, A. Files of news-cuttings compiled for official use, Mitchell Library.

Nixonwestwood, J.E. Memories of British New Guinea (Papua): being the thoughts of a resident of that territory from 1908–15 and put into book form for all interested in the earlier history, exploration and other activities of what is known in Australia as the Territory of Papua, by Toaguba (J.E. Nixonwestwood), manuscript, New Guinea Collection, University of Papua New Guinea. Letters and notes on the white residents of Papua, National Library of Australia. Photographs, Mitchell Library.

Nurton, A. Papers, photographs, maps and patrol reports (including some from Upper Ramu), National Library of Australia.

O'Neill, Jack. A Prospector's Diary: New Guinea 1931–37, manuscript, National Library of Australia.

Paterson, A.B. ('Banjo') 1935. 'The Land of Adventure', Number 2, 'New Guinea Gold', script of ABC radio program.

Pryke Brothers. Papers of Dan, Frank and Jim Pryke, diaries, letters, etc. National Library of Australia, Mitchell Library, and personal possession of Pryke family (Mr Frank Pryke, Sydney, and Mrs Leonie Christopherson, Canberra).

Rentoul, A.C. Papuan Adventure, manuscript, National Library of Australia.

Sharp, George. Elusive Fortune: The Adventures of George Sharp while Prospecting in the Kimberleys, New Guinea and Coolgardie, 1885-1907, manuscript edited by H.J. Gibbney, and in his possession.

Smith, M.C. Staniforth. Diaries, letters, newspaper cuttings and other papers, National Library of Australia. (Although Smith was Director of Mines in Papua the papers contain little about goldmining.)

Souter, Gavin. Letters and notes related to *New Guinea: the Last Unknown*, National Library of Australia.

Stanley, Owen. 1849. Journal, 3 June to 2 October; xerox copy, New Guinea Collection, University of Papua New Guinea.

Stone-Wigg, Bishop Montague. 1898–1908. Diary, 7 vols., New Guinea Collection, University of Papua New Guinea.

United Church Papers, Minute Book of the Panaeate Station and Circuit, Minutes and Journals of the Annual District Synod, South-East Papua.

Wells, Charles. News-cuttings (comprehensive and identified on gold and airways in Papua and New Guinea, 1926–41); business papers of Guinea Gold and Guinea Airways; Wells's diaries of trips to New Guinea, 1932–6; Cecil Levien's diary of trip to Upper Ramu, 1929; letters; other papers; and photographs; National Library of Australia.

Wilcox, J.F. 1848–9. Journal, January–July, Mitchell Library.

2. GOVERNMENT PUBLICATIONS

Allied Forces, Southwest Pacific Area, Allied Geographical Section, *Terrain Study, No. 28, Main Routes, Across New Guinea*. — *Terrain Study, No. 34, Area Study of Louisiade Archipelago*. — *Terrain Study, No. 33, Area Study of Woodlark Island*. ...

Annual Reports British New Guinea, 1885–1906; Papua, 1906–40; New Guinea, 1914–40; German New Guinea (translated), 1900/01–1904/05, 1906/07–1907/08, 1909/10–1912/13, typescript, National Library of Australia.

Commonwealth Parliamentary Debates

Commonwealth Parliamentary Papers, 'Papers Relating to the Administration of Justice in the case of Mr O'Brien, British New Guinea', Vol. 2, 1906. — 'Papua (British New Guinea). Spirituous Liquors — Correspondence Respecting the Suggested Prohibition Against their Introduction, Manufacture and Sale', Vol. 2, 1904. — 'Report of the Royal Commission on the Edie Creek (New Guinea) Leases', Vol. 4, 1926/27/28; Transcript of Evidence, Australian Archives, CP 660 Series 25 4 Volumes. — 'Report of the Royal Commission of Inquiry into the Present Conditions, including the method of Government, of the Territory of Papua, and the best method for their improvement', 1907.

Fry, T.B. (ed.). 1947. *The Laws of the Territory of New Guinea 1921-1945* (annotated), 5 vols., Sydney.

— 1949. *The Laws of the Territory of Papua 1888-1945* (annotated), 5 vols., Sydney.

Government Gazettes: *British New Guinea Government Gazette, New Guinea Gazette, Territory of Papua Government Gazette*.

Great Britain Parliamentary Papers, 'Further Correspondence respecting New Guinea', 1883, Vol. 47. — 'Further Correspondence respecting New Guinea and other islands, in the Western Pacific Ocean', 1884–5, Vol. 54.

Lett, Lewis (compiler). 1938. *The Official Handbook of Papua*, Port Moresby.

Manser, W. and Freeman, C. 1970. *Commonwealth of Australia, Department of National Development, Bureau of Mineral Resources, Geology and Geophysics Report No. 141, (Report PNG 5), Bibliography of the Geology of Eastern New Guinea (Papua New Guinea)*, Canberra.

Murray, P.L. 1911. *Official Records of the Australian Military Contingents to the War in South Africa*, Melbourne.

Naval Intelligence Division, *Geographical Handbook Series, Pacific Islands, Volume IV, Western Pacific (New Guinea and Islands Northward)*, 1945.

Official Handbook of the Territory of New Guinea Administered by the Commonwealth of Australia ... , Canberra. 1937.

Papua New Guinea, Village Directory 1973, Port Moresby, 1973.

Queensland Parliamentary Papers, 'Seizure of the "Forest King"' by PLM. Gunboat "Swinger" and subsequent Proceedings in the Vice-Admiralty Court in connection therewith, Vol. 2, 1884, pp. 853–915. — 'Report with minutes of evidence taken before the Royal Commission appointed to inquire into the circumstances under which labourers have been introduced into Queensland from New Guinea and other islands etc.', Vol. 2, 1885, pp. 797–988. — 'Correspondence respecting the return of the New Guinea Islanders', Vol. 2, 1885, pp. 1053–74. — 'Massacres in British New Guinea. (Correspondence respecting, and reports of Special Commission upon)', Vol. 3, 1887, pp. 719–726. — 'Return of Louisiade Islanders to their Native Islands', Vol. 3, 1888, pp. 611–19. — A.G. Maitland, 'Geological Observations in British New Guinea in 1891', Vol. 2, 1893, pp. 695–728.

Smith., M.C. Staniforth (compiler). 1927. *Handbook of the Territory of Papua*, Melbourne, 1907, 1909, 1912 and Canberra, 1927.

Stanley, Evan. 1912. 'Report on the Geology of Woodlark Island (Murua)', *Annual Report, 1911/12*, pp. 189–208.

— 1915. Report on the Geology of Misima, (St Aignan), Louisiade Gold-Field', *Bulletin of Territory of Papua*, No. 3, Melbourne.

Papua New Guinea, Lands Department, Woodlark Island (Special Report), Port Moresby, 1972, New Guinea Collection, University of Papua New Guinea.

3. NEWSPAPERS AND PERIODICALS

Brisbane Courier

Cairns Argus

Cairns Post

Cooktown Courier

Cooktown Independent

Grey River Argus

Healesville Guardian

Historical Studies

Journal of the Morobe District Historical Society

The Journal of Pacific History

The Journal of the Papua and New Guinea Society

Labour History

Lone Hand

Missionary Review

Morobe News

Mount Alexander Mail
New Guinea and Australia, the Pacific and South-East Asia
Oral History
Pacific Islands Monthly
Pambu (Newsletter of Pacific Manuscripts Bureau)
The Papuan (Anglican Mission paper)
Papuan Courier
Papuan Times
The Pilot (Thursday Island)
Sydney Mail
Sydney Morning Herald
Town and Country Journal

4. BOOKS. ARTICLES AND THESES

Abel, Charles. 1902. *Savage Life in New Guinea: The Papuan in Many Moods*, London.

Abel, Russell. 1934. *Charles W. Abel of Kwato: Forty Years in Dark Papua*, New York.

Affleck, Arthur H. 1964. *The Wandering Years*, Melbourne.

Affleck, D.A. 1971. Murua or Woodlark Island: A Study of European-Muruan Contact to 1942, B.A. hons thesis, Australian National University.

Armstrong, W.E. 1928. *Rossel Island: An Ethnological Study*, Cambridge.

Austin, A.R. 1972. The History of Technical Education in Papua 1874-1941, M.Ed. thesis, U.P.N.G.

Baker, Liese. 1971. The Beginning of the Indentured Labour System in Papua 1888-1908, B.A. hons thesis, Australian National University.

Barereba (Tago), Stephen. 1964. 'How My Grandfather Killed Mr J. Green', *Australian Territories*, Vol. 4, No. 3, pp. 15–18.

Beautemps-Beaupré, C.F. 1807. *Atlas du Voyage de Bruny-Dentrecasteaux*, Paris.

Beier, U. (ed.). 1973. *Black Writing from New Guinea*, Brisbane.

Belshaw, C.S. 1955. 'In Search of Wealth: A Study of the Emergence of Commercial Operations in the Melanesian Society of Southeastern Papua', *American Anthropologist*, Vol. 57, No. 1, Pt 2, Memoir No. 80.

Bengo, Paul. 1973. 'My Father on the Goldfields and at the Outbreak of the Second World War', *Oral History*, Vol. 1, No. 3, pp. 2–8.

Benson, James. 1957. *Prisoner's Base and Home Again: The Story of a Missionary P.O.W.*, London.

Bevan, Theodore. 1890. *British New Guinea from the Protectorate to the Sovereignty, 1884-1888: Toil, Travel and Discovery in British New Guinea*, London.

Binnie, J.H. 1944. *My Life on a Tropic Goldfield*, Melbourne.

Biskup, Peter (ed.). 1968. 'Herman Detzner: New Guinea's First Coast Watcher', *Journal of the Papua and New Guinea Society*, Vol. 2, No. 1, pp. 5–21.

— 1974. *The New Guinea Memoirs of Jean Baptiste Octave Mouton*, Canberra.

Black, R.H. 1957. 'Dr Bellamy of Papua', *Medical Journal of Australia*, 10, 17 and 24 August.

Blackwood, Beatrice. 1939a. 'Folk-Stories of a Stone Age People in New Guinea', *Folk-Lore*, Vol. 50, No. 3, pp. 209–42.

— 1939b. 'Life on the Upper Watut, New Guinea', *Geographical Journal*, Vol. 94, No. 1, pp. 11–28.

— 1950. *The Technology of a Modern Stone Age People in New Guinea*, Pitt Rivers Museum, Occasional Papers on Technology, 3, Oxford.

Blainey, Geoffrey. 1963. *The Rush that Never Ended: A History of Australian Mining*, Melbourne.

Blum, Hans. 1900. *Neu-Guinea und der Bismarckarchipel: Eine Wirtschaftliche Studie*, Berlin.
Bolton, G.G. 1963. *A Thousand Miles Away: A History of North Queensland to 1920*, Brisbane.
Booth, Doris. 1929. *Mountains, Gold and Cannibals*, Sydney.
de Bougainville, Louis. 1967. *A Voyage Round the World*, Amsterdam.
Brass, L.J. 1959. 'Results of the Archbold Expeditions, No. 79, Summary of the Fifth Archbold Expedition to New Guinea (1956/57)', *Bulletin of the American Museum of Natural History*, Vol. 118, Article 1, New York.
Brown, H.A. 1956. The Eastern Elema, Diploma of Anthropology thesis, University of London.
— 1973. 'The Eleman Language Family', in Karl Franklin (ed.). *The Linguistic Situation in the Gulf District and Adjacent Areas*, Papua New Guinea, Pacific Linguistics, Series C, No. 26, Canberra.
Browne, R. Spencer. 1927. *A Journalist's Memories*, Brisbane.
Bushell, Keith. 1936. *Papuan Epic*, London.
Carey, C.N. 1892. *Australian Miners' Guide*, Sydney.
Chalmers, James. 1898. 'Toaripi', *Journal of the Anthropological Institute of Great Britain and Ireland*, Vol. 27, pp. 326–34.
Cheesman, Lucy Evelyn. 1935. *The Two Roads of Papua*, London.
— 1957. *Things Worth While*, London.
Cheyne, Andrew. 1852. *A Description of Islands in the Western Pacific Ocean North and South of the Equator ...* , London.
Chignell, A.K. 1911. *An Outpost in Papua*, London.
— 1913. *Twenty-one Years in Papua: A History of the English Church Mission in New Guinea, 1891-1912*, London.
Chinnery, E.W.P. 1931. 'Natives of the Waria, Williams and Bialolo Watersheds', *Territory of New Guinea, Anthropological Report*, No. 4.
— 1934. 'The Central Ranges of the Mandated Territory of New Guinea from Mount Chapman to Mount Hagen', *Geographical Journal*, Vol. 84, pp. 398–412.
— and Beaver, W.N. 1917. 'The Movements of the Tribes of the Mambare Division of Northern Papua, *Annual Report 1914/15*, pp. 158–61.
Clune, Frank. 1942. *Prowling through Papua with Frank Clune*, Sydney.
— 1951. *Somewhere in New Guinea: A Companion to Prowling through Papua*, Sydney.
Collinson, J.W. 1941. *Tropic Coasts and Tablelands*, Brisbane.
Corfield, W.H. 1921. *Reminiscences of Queensland 1862-1899*, Brisbane.
Corris, Peter. 1968. '"Blackbirding" in New Guinea Waters, 1883-84: An Episode in the Queensland Labour Trade', *The Journal of Pacific History*, Vol. 3, pp. 85–105.
— 1973. *Passage, Port and Plantation: A History of Solomon Islands Labour Migration 1870-1914*, Melbourne.
Crocombe, Marjorie. 1972. 'Ruatoka: A Polynesian in New Guinea History', *Pacific Islands Monthly*, November, pp. 69–75 and December, pp. 69–76.
Dammköhler, W. and Oldörp, R. 1909. 'Bericht der Herren Dammköhler und Oldörp über eine Reise in Neuguinea 1908-09', *Amtsblatt für das Schutzgebiet Deutsch- Neuguinea*, Vol. 1, No. 17, pp. 135–6.
Davidson, J.W. and Scarr, D. (eds.). 1970. *Pacific Islands Portraits*, Canberra.
Demaitre, E. 1936. *New Guinea Gold: Cannibals and Gold-Seekers in New Guinea*, London.
Detzner, Capitaine H. 1935. *Moeurs et Coutumes des Papous: quatre ans chez les cannibales de Nouvelle-Guinée (1914-1918)*, Paris.
Dexter, David. 1961. *Australia in the War of 1939-1945: The New Guinea Offensives*, Canberra.
Dickson, Diane and Dossor, Carol (compilers). 1970. *World Catalogue of Theses on the Pacific Islands*, Canberra.

Douglas. John. 1888. 'Notes on a Recent Cruise through the Louisiade Group of Islands', *Transactions and Proceedings of the Royal Geographical Society of Australasia*, Victoria Branch, Vol. 5, pp. 46–59.

— 1890. 'Sudest and the Louisiade Archipelago', *Proceedings and Transactions of the Queensland Branch of the Royal Geographical Society of Australasia*, Vol. 4, 1888/89, pp. 2–16.

Du Toit, Brian M. 1975. *Akuna: A New Guinea Village Community*, Rotterdam.

Dutton, T.E. 1969. *The Peopling of Central Papua: Some Preliminary Observations*, Pacific Linguistics, Series B, No. 9, Canberra.

— 1973. *A Checklist of Languages and Present-day Villages of Central and South-East Mainland Papua*, Pacific Linguistics, Series B, No. 24, Canberra.

— 1971. 'Languages of South-East Papua: A Preliminary Report', *Papers in New Guinea Linguistics*, No. 14, Canberra.

D'Urville, Dumont. 1832. *Voyage de la Corvette L'Astrolabe* ... , Vol. 4, Paris.

— 1846. *Voyage au Pole Sud et dans L'Oceanie sur les Corvettes L'Astrolabe et la Zelee* .... , Vol. 9, Paris.

Edelfelt, E.G. 1887. 'Motu-Motu, and Customs of the People' in J.W. Lindt, *Picturesque New Guinea*, London.

— 1893. 'Customs and Superstitions of New Guinea Natives', *Proceedings and Transactions of the Queensland Branch of the Royal Geographical Society of Australasia*, 1890/91 , Vol. 3, Pt 1, pp. 9–28.

— Egloff, Brian. 1971. Collingwood Bay and the Trobriand Islands in Recent Prehistory: Settlement and Interaction in Coastal and Island Papua, Ph.D. thesis, Australian National University.

An Ethnographic Bibliography of New Guinea. 1968. Department of Anthropology and Sociology, Australian National University, 3 vols., Canberra.

Evans-Pritchard, E.E. and others (eds.). 1934. *Essays Presented to C.G. Seligman*, London.

Flierl, Johannes. 1927. *Forty Years in New Guinea*, Chicago.

— 1932. *Christ in New Guinea: Former Cannibals become Evangelists by the Marvellous Grace of God*, Tanunda, S.A.

Flynn, Errol. 1960. *My Wicked, Wicked Ways*, London.

Fortune, R.F. 1932. *Sorcerers of Dobu: The Social Anthropology of the Dobu Islanders of the Western Pacific*, London.

Franklin, Karl (ed.). 1973. *The Linguistic Situation in the Gulf District and Adjacent Areas, Papua New Guinea*, Pacific Linguistics, Series C, No. 26, Canberra.

Gajdusek, D.C., Fetchko, P., Van Wyk, N.J. and Ono, S.G. 1972. *Annotated Anga (Kukukuku) Bibliography*, Bethesda, Maryland.

Gibbney, H.J. 1972. 'The New Guinea Gold Rush of 1878', *Journal of the Royal Australian Historical Society*, Vol. 58, Pt 4, pp. 284–96.

Gill, G.H. 1968. *Royal Australian Navy, 1942-45*, Canberra.

Gillison, Douglas. 1962. *Royal Australian Air Force, 1939-42*, Canberra.

Griffin, H.L. 1925. *An Official in British New Guinea, with earlier Reminiscences of Harrow and the Royal Artillery*, London.

Grimshaw, Beatrice. 1911. *The New New Guinea*, London.

— 1930. *Isles of Adventure*, London.

— [n.d.] *Guinea Gold* (a novel), London.

Haddon, A.C. 1894. *Royal Irish Academy, 'Cunningham Memoirs' — No. 10, The Decorative Art of British New Guinea: A Study in Papuan Ethnography*, Dublin.

Haddon, A.C. and Hornell, J. 1937. *Canoes of Oceania*, Vol. 2, *The Canoes of Melanesia, Queensland, and New Guinea* by A.C. Haddon, Bernice P. Bishop Museum Special Publication, No. 28, Honolulu.

Hamy, E.T. 1889. 'Etude sur les Papouas de la Mer D'Entrecasteaux', *Revue D'Ethnographie*, Vol. 7, Paris, pp. 503–19.

Harding, Thomas G. 1967. *Voyagers of the Vitiaz Strait: A Study of a New Guinea Trade System*, Seattle.

Healy, A.M. 1965. 'Ophir to Bulolo: The History of the Gold Search in New Guinea', *Historical Studies Australia and New Zealand*, Vol. 12, No. 45, pp. 103–18.

— 1967. *Bulolo: A History of the Development of the Bulolo Region*, New Guinea, New Guinea Research Bulletin No. 15, Canberra.

Heydon, Peter. 1965. *Quiet Decision: A Study of George Foster Pearce*, Melbourne.

Hides, Jack. 1935. *Through Wildest Papua*, London.

— 1938. *Savages in Serge*, Sydney.

Higginson, J.B. 1908. 'Memories of Papua', *Lone Hand*, 1 June, pp. 117–23.

Hill, W.R.O. 1907. *Forty-Five Years' Experiences in North Queensland, 1861 to 1905* ... Brisbane.

Hogbin, H. Ian. 1951. *Transformation Scene: The Changing Culture of a New Guinea Village*, London.

Holtzknecht, H. 1973–4. 'The Exploration of the Markham Valley, 1st part, up to 1910', *Journal of the Morobe District Historical Society*, Vol. 1, No. 3, December 1973, pp. 33–52; Vol. 2, No. 1, May 1974, pp. 20–33.

Hooley, B.A. and McElhanon, K.A. 1970. 'Languages of the Morobe District — New Guinea', in S.A. Wurm and D C. Laycock, eds., *Pacific Linguistic Studies in Honour of Arthur Capell*, Pacific Linguistics, Series C, No. 13, Canberra.

Humphries, W.R. 1923. *Patrolling in Papua: With an Introduction by J.H.P. Murray* ... , London.

Huxley, Leonard. 1900. *Life and Letters of T.H. Huxley*, 2 vols., London.

Huxley, T.H. (J. Huxley, ed.). 1935. *T.H. Huxley's Diary of the Voyage of H.M.S. Rattlesnake*, London.

Idriess, I.L. 1933. *Gold-Dust and Ashes: The Romantic Story of the New Guinea Gold-fields*, Sydney.

Jack, R.L. 1921. *Northmost Australia: Three Centuries of Exploration, Discovery, and Adventure in and around the Cape York Peninsula, Queensland* ... , London.

Jenness, D. and Ballantyne, A. 1920. *The Northern D'Entrecasteaux*, Oxford.

Jinks, B., Biskup, P. and Nelson, H. (eds.). 1973. *Readings in New Guinea History*, Sydney.

Jones, R. Fleming. 1912. 'Tropical Diseases in British New Guinea', *Transactions of the Royal Society of Tropical Medicine*, pp. 93–105.

Kekeao, T.H. 1973. 'Vailala Madness', *Oral History*, Vol. 1, No. 7, pp. 1–8.

King, J. 1909. *W.G. Lawes of Savage Island and New Guinea*, London.

Kruger, Walter. 1953. *From Down Under to Nippon: The Story of Sixth Army in World War 2*, Washington.

Labillardiére, J.J.H. de. 1800. *Voyage in Search of La Pérouse....* translated, 2 vols., London.

Langmore, Diane. 1974. *Tamate — a King: James Chalmers in New Guinea 1877-1901*, Melbourne.

Laracy, H.M. 1969. Catholic Missions in the Solomon Islands, 1845-1966, Ph.D. thesis, Australian National University.

— 1970. 'Xavier Montrouzier: A Missionary in Melanesia', in J.W. Davidson and D. Scarr, eds., *Pacific Islands Portraits*, Canberra.

Lauer, P.K. 1970. Pottery Traditions in the D'Entrecasteaux Islands of Papua, Ph.D. thesis, Australian National University.

Leahy, M. and Crain, M. 1937. *The Land that Time Forgot, Adventures and Discoveries in New Guinea*, London.

Lee, Ida. 1912. *Commodore Sir John Hayes: His Voyage and Life with Some Account of Admiral D'Entrecasteaux's Voyage of 1792-3*, London.

Legge, John. 1956. *Australian Colonial Policy*, Sydney.

Lett, Lewis. 1935. *Knights Errant of Papua*, London.
— 1943. *Papuan Gold: The Story of the Early Gold Seekers*, Sydney.
Lindt, J.W. 1887. *Picturesque New Guinea*, London.
Lloyd, Richard G. 1973. 'The Angan Langnage Family', in Karl Franklin (ed), *The Linguistic Situation in the Gulf District and Adjacent Areas*, Papua New Guinea, Pacific Linguistics, Series C, No. 26, Canberra.
Lovett, R. 1903. *James Chalmers: His Autobiography and Letters*, London.
Lutton, Nancy. 1972. C.A.W. Monckton's Trilogy of his Adventures in New Guinea: Fact or Fiction? B.A. hons., thesis, U.P.N.G.
— (compiler) 1974. *Preliminary List of Manuscripts held in the New Guinea Collection of the University of Papua New Guinea Library*, Port Moresby.
Lyne, C. 1885. *New Guinea: An Account of the Establishment of the British Protectorate over the Southern Shores of New Guinea*, London.
McArthur, A.M. 1961. The Kunimaipa: The Social Structure of a Papuan People, Ph.D. thesis, Australian National University.
McCarthy, D. 1959. *South-West Pacific Area — First Year: Kokoda to Wau*, Canberra.
McCarthy, J.K. 1963. *Patrol into Yesterday: My New Guinea Years*, Melbourne.
MacDonald, Alexander. 1907. *In the Land of Pearl and Gold: A Pioneer's Wanderings in the Backblocks and Pearling Grounds of Australia and New Guinea*, London.
McElhanon, K.A. 1970. 'A History of Linguistic Research in the Huon Peninsula, New Guinea', in S.A. Wurm and D.C. Laycock (eds.), *Pacific Linguistic Studies in Honour of Arthur Capell*, Pacific Linguistics, Series C, No. 13, Canberra.
MacFarlane, S. 1888. *Among the Cannibals of New Guinea: Being the Story of the New Guinea Mission of the London Missionary Society*, London.
MacGillivray, J. 1852. *Narrative of the Voyage of H.M.S. 'Rattlesnake'; Including Discoveries and Surveys in New Guinea, the Louisiade Archipelago....* London, 2 vols.
MacGregor, William. 1897. *British New Guinea: Country and People*, London.
MacKay, Kenneth. 1909. *Across Papua: Being an Account of a Voyage Round, and a March Across the Territory of Papua, with the Royal Commission*, London.
MacKenzie, S.S. 1927. *The Official History of Australia in the War of 1914-1918, Vol. 10, The Australians at Rabaul: The Capture and Administration of the German Possessions in the Southern Pacific*, Sydney.
McRae, Keith. 1974. 'Kiaps, Missionaries and Highlanders', *New Guinea and Australia, the Pacific and South-East Asia*, Vol. 9, No. 1, pp. 16–28.
Maguire, H.R. 1903. 'Impressions of a Year's Sojourn in British New Guinea', *Proceedings and Transactions of the Royal Geographical Society of Australasia, Queensland Branch*, Vol. 17, 1901–2, pp. 117–43.
Mair. L.P. 1948. *Australia in New Guinea*, London.
Malinowski, B. 1922. *Argonauts of the Western Pacific: An Account of Native Enterprise and Adventure in the Archipelago of Melanesian New Guinea*, London.
— 1934. 'Stone Implements in Eastern New Guinea', E.E. Evans Pritchard and others, (eds.), *Essays Presented to C.G. Seligman*, pp. 189–96, London.
Mambu, Bishop. 1975. 'Bishop Mambu I Tingting Bek', *Journal of the Morobe District Historical Society*, Vol. 2, No. 3, pp. 17–27.
'The Man Who Named Kokoda Trail'. 1971. *Leagues' Club Journal*, May, p. 5.
Marshall, A.J. 1970. *Darwin and Huxley in Australia*, Sydney.
Mayo, John. 1972. Oddity of Empire: British New Guinea 1884-1888, M.A. thesis, U.P.N.G.
— 1973. 'A Punitive Expedition in British New Guinea, 1886', The Journal of Pacific History, Vol. 8, pp. 89–99.
Meek, A.S. 1913. *A Naturalist in Cannibal Land*, London.

Miller, John Jr. 1959. *United States Army in World War II. The War in the Pacific. Cartwheel: The Reduction of Rabaul*, Washington.

Monckton, C.A.W. 1921. *Some Experiences of a New Guinea Resident Magistrate*, London.

— 1922. *Last Days in New Guinea: Being Further Experiences of a New Guinea Resident Magistrate*, London.

— 1934. *New Guinea Recollections*, London.

Moore, John Hammond. 1975. *The Young Errol: Flynn before Hollywood*, Sydney.

Moresby, John. 1876. *New Guinea and Polynesia: Discoveries and Surveys in New Guinea and the D'Entrecasteaux Islands* ... London.

Morison, S.E. 1950. *History of United States Naval Operations in World War II, Volume VI, Breaking the Bismarcks Barrier 22 July 1942-1 May 1944*. Boston.

Morrell, W.P. 1960. *Britain in the Pacific Islands*, Oxford.

Murray, J.H.P. 1912. *Papua or British New Guinea*, London.

— 1923. *Recent Exploration in Papua*, Sydney.

— 1925. *Papua of Today*, London.

Newton, H. 1914. *In Far New Guinea: A Stirring Record of Work and Observation amongst the People of New Guinea, with a Description of their Manners, Customs and Religions*, London.

Nelson, H. 1973a. 'Miners, Labourers and Officials on the Lakekamu Goldfield of Papua', *Labour History*, No. 25, pp. 40–52.

— 1973b. 'Our Boys up North: The Behaviour of Australians in New Guinea', *Meanjin Quarterly*, No. 4, pp. 433–41.

Nisbet, Hume. 1891. *A Colonial Tramp: Travels and Adventures in Australia and New Guinea*, London.

Parr, E.A. 1974. Edie Creek, 1926-27: The Indian Summer of Red-Shirt Capitalism , B.A hons. thesis, U.P.N.G.

Pearce, George F. 1951. *Carpenter to Cabinet*, London.

Pearl, Cyril. 1967. *Morrison of Peking*, Sydney.

Pike, Douglas (ed.). 1966–74. *Australian Dictionary of Biography*, 5 vols., Melbourne.

Pilhofer, G. 1915. 'Eine Durchquerung Neuguineas vom Waria-zum Markam fluss' *Petermanns Geographische Mitteilungen*, Vol. 61, pp. 21–5 and 63–6.

Pryke, Frank. 1937. *Poems: New Guinea*, privately printed.

Radford, Robin. 1972. 'Missionaries, Miners and Administrators in the Eastern Highlands', *Journal of the Papua and New Guinea Society*, Vol. 6, No. 2, pp. 85–105.

Reed, S.W. 1943. *The Making of Modern New Guinea: With Special Reference to Culture Contact in the Mandated Territory*, Philadelphia.

Reinhold, W.J. 1946. *The Bulldog-Wau Road* (John Thomson Lecture for 1945), Brisbane.

Reynolds, H. (ed.). 1972. *Aborigines and Settlers: The Australian Experience 1788-1939*, Melbourne.

Reynolds, John. 1974. *Men and Mines: A History of Australian Mining*, Melbourne.

Rhys, Lloyd. 1942. *High Lights and Flights in New Guinea, Being in the Main an Account of the Discovery and Development of the Morobe Goldfields*, London.

Romilly, H.H. 1886. *The Western Pacific and New Guinea: Notes on the Natives, Christian and Cannibal, with Some Account of the Old Labour Trade*, London.

— 1889. *From My Verandah in New Guinea*, London.

— 1893. *Letters from the Western Pacific and Mashonoland 1878-1891* ... Ed. with Memoir by ... S.H. Romilly.... London.

Roe, Margriet. 1962. A History of South-East Papua to 1930, Ph.D. thesis, Australian National University.

Ross, Ronald. 1923. *Memoirs: With a Full Account of the Great Malaria Problem and its Solution*, London.
de Rossel, Elizabeth. 1808. *Voyage de Dentrecasteaux, Envoyé à la Récherche de La Pérouse....* Paris.
Rowe, John. 1974. *The Hard-Rock Men: Cornish Immigrants and the North American Mining Frontier*, Liverpool.
Rowley, C.D. 1958. *The Australians in German New Guinea 1914-1921*, Melbourne.
Ryan, Dawn. 1965. Social Change among the Toaripi, Papua. M.A. thesis, University of Sydney.
Ryan, Peter (general editor). 1972. *Encyclopaedia of Papua and New Guinea*, Melbourne, 3 vols.
Saunders, Garry. 1965. *Bert Brown of Papua*, London.
Scarr, D. 1967. *Fragments of Empire: A History of the Western Pacific High Commission 1877-1914*, Canberra.
Seligman, C.G. 1910. *The Melanesians of British New Guinea*, Cambridge.
— and Strong, W.M. 1906. 'Anthropological Investigations in British New Guinea', *Geographical Journal*, No. 3, Vol. 27, pp. 225–42, and No. 4, pp. 347–69.
Sharp, Andrew. 1960. *The Discovery of the Pacific Islands*, Oxford.
Shepherd, Ernie. 1971. 'Akmana: A New Name in the Continuing Story of New Guinea Exploration', *Pacific Islands Monthly*, April.
Sinclair, James. 1966. *Behind the Ranges: Patrolling in New Guinea*, Melbourne.
— 1969. *The Outside Man: Jack Hides of Papua*, Melbourne.
Souter, Gavin. 1964. *New Guinea: The Last Unknown*, Sydney.
Smith, M.C. Staniforth. 1903. *British New Guinea: With a Preface on Australia's Policy in the Pacific*, Melbourne (pamphlet).
Stanley, E.R. 1924. *The Geology of Papua*, Melbourne.
Stevens, H.N. (ed.). 1930. *New Light on the Discovery of Australia as Revealed by the Journal of Captain Don Diego de Prado y Tovar*, London.
Stewart's Hand Book of the Pacific Islands, Sydney. First published in 1907. Robert Langdon, Australian National University, has 1921 edition with Nixonwestwood's comments on the list of white residents of Papua.
Stone, Octavius. 1880. *A Few Months in New Guinea*, London.
Struben, Roy. 1961. *Coral and Colour of Gold*, London.
Tamanabae, Leila. 1972. 'The Mamba in Moresby', *Oral History*, Vol. 1, No. 1, pp. 3–7.
Taylour, H. and Morley, I.W. 1933. 'The Development of Gold Mining in Morobe, New Guinea', *Australasian Institute of Mining and Metallurgy Proceedings*, March, No. 89, pp. 1–81 and June, No. 90, pp. 247–53.
Tindale, N.B. and Bartlett, H.K. 1937. 'Notes on Some Clay Pots from Panaeati Island, South-East of New Guinea', *Transactions and Proceedings of the Royal Society of South Australia*, Vol. LX I, pp. 159–62.
Thomson, B.H. 1889a. 'New Guinea: Narrative of an Exploring Expedition to the Louisiade and D'Entrecasteaux Islands', *Proceedings of the Royal Geographical Society*, Vol. XI, pp. 525–42.
— 1889b. 'Narrative of an Exploring Expedition to the eastern part of New Guinea', *Scottish Geographical Magazine*, No. 5, pp. 513–27.
Thomson, J.P. 1892. *British New Guinea*, London.
Thomson, N. 1975. 'Prehistory in the East Central District', *Oral History*, Vol. 3, No. 1, pp. 2–34.
Tomlin, J.W.S. 1951. *Awakening: A History of the New Guinea Mission*, London.
Townsend, G.W.L. 1968. *District Officer: from Untamed New Guinea to Lake Success*, 1921-46, Sydney.
Tudor, Judy. 1966. *Many a Green Isle*, Sydney.

Tueting, Laura. 1935. 'Native Trade in Southeast New Guinea', *Bernice P. Bishop Museum, Occasional Papers*, No. 15, Vol. XI, Honolulu.

Vaughan, Berkeley. 1974. *Doctor In Papua*, Adelaide.

Verguet, L'Abbé L. 1861. *Histoire de la Première Mission Catholique au Vicariat de Mélanésie*, Paris.

van der Veur, P.W. 1966a. *Documents and Correspondence on New Guinea's Boundaries*, Canberra.

— 1966b. *Search for New Guinea's Boundaries. From Torres Strait to the Pacific*, Canberra.

Waiko, John. 1972. The Binandere Response to Imperialism, B.A. hons. thesis, U.P.N.G.

— 1970. 'A Payback Murder: The Green Bloodbath', *Journal of the Papua and New Guinea Society*, Vol. 4, No. 2, pp. 27–35.

Ward, R.G. (ed.). 1966-7. *American Activities in the Central Pacific 1790-1870: A History, Geography and Ethnography Pertaining to American Involvement and Americans in the Pacific Taken from Contemporary Newspapers etc.*, Ridgewood, N.J., 8 vols.

Wawn, William T. (Peter Corris, ed.). 1973. *The South Sea Islanders and the Queensland Labour Trade*, Canberra.

Weston, Bert. 1925. 'Pre-War Recruiting in New Guinea', *Journal of the Morobe District Historical Society*, Vol. 2, No. 3, pp. 39–46.

Wetherell, David. 1973. 'Monument to a Missionary: C.W. Abel and the Keveri of Papua', *The Journal of Pacific History*, Vol. 8, pp. 30–48.

White, Gilbert. 1929. *A Pioneer of Papua: Being the Life of the Reverend Copland King, M.A ...* London.

White, Osmar. 1945. *Green Armour*, Sydney.

— 1965. *Parliament of a Thousand Tribes: A Study of New Guinea*, London.

Williams, F.E. 1923. *The Vailala Madness and the Destruction of Native Ceremonies in the Gulf Division*, Territory of Papua, Anthropology Report No. 4, Port Moresby.

— 1928. *Orokaiva Magic*, Oxford.

— 1930. *Orokaiva Society*, Oxford.

— 1934. 'The Vailala Madness in Retrospect', in E.E. Evans-Pritchard and others (eds.), *Essays Presented to C.G. Seligman*, pp. 369–79, London.

— 1940. *Drama of Orokolo: The Social and Ceremonial Life of the Elema*, Oxford.

— 1944. 'Mission influence amongst the Keveri of south-east Papua', *Oceania*, Vol. 15, No. 2, pp. 89–141.

Williams, Ronald G. 1972. *The United Church in Papua, New Guinea, and the Solomon Islands*, Rabaul.

Willis, Ian. 1969a. 'Who Was First? The First White Man into the New Guinea Highlands', Journal of the Papua and New Guinea Society, Vol. 3, No. 1, pp. 32–45.

— 1969b. 'an Epic Journey: The Journey of Michael Leahy and Michael Dwyer across New Guinea in 1930', M.A. qualifying thesis, U.P.N.G.

— 1971. 'Next to Horse Stealing: The Labour Trade out of Lae Before World War 2', Seminar paper, History Department, U.P.N.G.

— 1972. *Lae ti mala'hu: Lae and Its Local Villages*, M.A. thesis, U.P.N.G.

— 1973a. 'History of the Morobe District', *Journal of the Morobe District Historical Society*, Vol. 1, No. 3, December, pp. 3–12.

— 1973b. 'Village and Town: The Changes Produced in Villages around Lae by Expatriate Settlement', *Journal of the Morobe District Historical Society*, Vol. 1, No. 3. December, pp. 14–30.

— 1974. *Lae: Village and City*, Melbourne.

Wilson, D. 1969. 'The Binandere Language Family', in A. Capell and others, *Papers in New Guinea Linguistics*, No. 9, Canberra.

Wilson, R. Kent and Garnaut, Ross. 1968. 'Moveave Co-operative Sawmill', in *New Guinea Research Bulletin Number 25, A Survey of Village Industries in Papua-New Guinea*, pp. 74–175, Canberra.

Woodhouse, M. (compiler) and R. Langdon (ed.). 1968. *Cumulative Index to the Pacific Islands Monthly Volumes 1 to 15 [August, 1930 to July, 1945]*, Sydney.

Wurm, S.A. and Laycock, D.C. (eds.). 1970. *Pacific Linguistic Studies in Honour of Arthur Capell*, Pacific Linguistics, Series C, No. 13, Canberra.

Young. M. 1971. *Fighting with Food: Leadership, Values and Social Control in a Massim Society*, Cambridge.

5. FILMS

The Air Road to Gold, Commonwealth of Australia, documentary, black and white, sound, 10 minutes, about 1933.

The Green and the Gold, The Timber Development Association of Australia, documentary, sound, colour, 30 minutes, 1954.

Guinea Gold: A Romance of Australian Enterprise, Guinea Expedition, producer Harry Gilles, documentary, black and white, sound, 20 minutes, 1932.

The Jungle Women, Stoll-Hurley, feature, black and white, silent, 60 minutes, 1926.

New Guinea 1904-1906, made by the Department of Anthropology, University of Vienna, from film taken by Dr Rudolf Poch, documentary, black and white, sound (text from lecture by Poch in 1907), 15 minutes, 1904–6.

Pioneering in New Guinea, Allan Dawes, a Herschell Production, documentary, black and white, silent, 15 minutes, about 1930.

The Unsleeping Eye: A Pioneering Tale of New Guinea, the Last Stronghold of the Savage, Alexander MacDonald, Seven Seas Screen Productions, feature, black and white, silent, 60 minutes, 1927.

With the Administration in New Guinea, Commonwealth of Australia, documentary, black and white, silent, 20 minutes, about 1930.

Index

Abau Island (Map p.184), 83, 188
Abel, Charles, 178-81, 189
Aborigines (of Australia), 163; on boats in New Guinea waters, 6; and Australian miners, 10, 19, 61, 76, 81; compared with Papuans, 77, 81; with Clark, 88, 95
Adau River (Map p.184), 183, 184
Ade, 108
Africa, 6 ,81
Age, 81, 97
Agunomi, 139
Aiga, 93
Ai-i-ia, 216
Aikora River, 138, 144, 161, 195; gold found on, 119, 120, 125; violence on, 135-6, 137; mining on, 146; miners and women, 159
Air services, 171, 237, 265
Aitcheson, J., 166
Aiv Avi River, *see* Arabi
Albatross, 53, 54
Alcohol, prohibition of, 60-1
Alexander, Bill, 87-8
Alhoga, 36
Alice Meade, 6
Alice River, 221
Amau (Map p.184), 189
Amburo, 104, 105, 106
Ambush Point, 90
Ana Creek (Map p.29), 28, 40, 42
Andersen, Thomas, 83, 148
Anderson, Neil, 85
Anga speakers, see Kukukuku
Angau (Australian New Guinea Administrative Unit), 46, 71
Anglicans, see Missionaries
Anglo-German boundary commission, 141-2
Anjiga, 108
Annie Brooks, 6
Anopheles, 130
Aposi (Map p.98), 100, 106
Arabi (Map p.200), 197; prospecting on, 219, 224; Germans on, 227; Kukukuku, 233, 242, 243, 245, 246, 248
Arbouin, Charles, 42

Ariotti, Severine ('Harry'), 223
Armed Native Constabulary, *see* Police
Armidale, 194
Armit, Lionel, 206, 245
Armit, William, 121, 142, 143, 153; biography, 163
Arnold, George, 167, 188, 189, 223, 259
Asiba Creek, 116
Auerback, Edward, 150-1, 221, 255-6
Australian investment, 59
Australian Labor Party, 181, 207
Australian Mandated Territory, 257, 260
Australian miners, attitudes, v, 10-11, 18; *see* also particular goldfields
Australian Naval and Military Expeditionary Force, 257
Australian newspapers, 56, 76, 113; *see also* particular newspapers
Australian Parliament, 163, 169, 207-9
Aviri, Constable, 224
Awala River (Map p.184), 260

Babaga, 34, 38, 46
Babila, 133
Bagalina, 66
Baibara Island, 85
Baiwa, 127, 143
Bakeke, Sergeant, 106, 133, 154, 171
Bakem, William, 20
Bamu River, 157, 210, 212
Baniara, 43
Barigi, Sergeant, 106, 167, 171
Bariji River, 129
Bartle Bay (Map p.84), 85, 86
Bartlett, Reverend Harry, 46
Barton, Francis, 156, 165, 168, 184
Barua, 187
Baseta, 143
Basilisk, 90, 178
Batchelor, Egerton, 207-9, 217
Batong, 261
Batow, 154
Bauwaki language, 183
Bêche-de-mer, 21, 53
Beaver, Wilfred, 145, 234

Beda (Neneba), 142; meet foreigners peacefully, 116; feed miners, 127, 186, 257; attacked by Elliott, 133-5, 172; change culture, 269
Belfield's Gully, 228
Bell, L.L., 136
Bellamy, Rayner, 131, 132, 150, 156; biography, 163-4, 165
Bete, 203
Bethune, Albert, 230, 249
Beya, 106
Bi (Queen Bee), 179, 181
Bia, Corporal, 106, 129, 165
Biagi, *see* Mountain Koiari
Biawaria River, 139
Billy Bong Creek, 26
Billy the Cook's, 156, 176
Binamarien, 261
Binandere, 93-4, 145; warfare, 92-4, 104, 107-9, 135, 269; attack Clark, 94-5, 144; and John Green, 95-106; and missionaries, 109-10; relations with Beda, 116; and deserters, 148; as police, 155, 161-2
Biowa, 13
Black River, 221
Blackenbury, James, 135, 168
Blayney, Joseph, 119, 183, 184, 238
Boer War, 34, 106, 162, 172
Bogi, 123, 124, 125, 129, 130, 132, 145, 153, 169; station opened, 122; Bogi-Yodda track, 127, 130, 149
Boie, 148
Boiomea, 54
Bokina, Corporal, 137
Bonagai, 58, 66
Bongata, 108
Booth, Charles, 257
Booth, Doris, 257
Bougainville, 5, 71, 263, 266
Bourke' (Bjornquist), Peter, 259
Bousimai, 94, 99, 103-8
Bowden, Norman, 199
Bowden, William, 67
Bowler, Michael, 180
Bramble, 5, 6
Brammell, Bertram, 202, 214
Brandon, W., 243
Bridge, Captain Cyprian, 7, 53
Brierly Island, 5, 6, 7
Brisbane, 39, 113, 167
Brooker Island, 6, 13, 14, 30
Brough, Dr C.A., 119
Brown, Reverend Herbert, 236, 249
Brown, Louis, 31

Brown River (Map p.77), 77
Bruce, William, 129, 154
Buchanan, W.E., 61
Buhutu (Map p.177), 43, 178, 180
Bukaua, 262
Bulega, 46-7
Bulldog, 237, 250
Bulldog, 197, 209, 211, 218, 220, 228, 229, 237, 245
Bulldog Track, 237
Bulolo (Map p.255), 142, 228, 256-8, 265, 266
Bulwa, 266
Bumbu, 262
Buna (Map p.91), 108, 130, 131, 143, 150, 153, 155, 165, 169, 171
Bundowi, 'Mary', 246
Burfitt, George, 46
Burfitt, Henry, 24
Burns, miner, 116
Burns Philp, 55, 62, 86, 114
Busai (Map p.51), 55, 56, 58, 59, 60, 65, 192
Butler, John, 137, 220
Butterworth, Archibald, 97, 99-100, 105
Bwagabwaga (Map p.29), 36, 37, 38
Bwagaoia (Map p.29): anchorage, 30, 32, 34; administrative centre, 39, 40, 41, 42, 43, 44, 46, 47

Cadigan, Johnny 'Fiji', 88
Cairns, 3, 10, 29, 176, 197, 205, 261
Cairns *Argus*, 88
Cairns prospectors, 86, 87, 88, 93, 96
Caledonian mine, 22
Calico, 144
Calvados Chain (Maps pp.4, 8), 2, 9, 35, 46, 47
Cameron, Cyril, 245
Cameron, John, 19
Campbell, Alexander, 186, 214; on Sudest, 23, 24; on Misima, 34, 37, 38, 39, 40; on Woodlark, 56, 57, 58, 63, 65, 70, 72; and recruiting, 68, 150-1; at Milne Bay, 181
Campbell, William, 22, 32
Campion, Tom, 122-3, 126, 143
Campions Beach, 119
Canada, 156, 260
Cape Arkona (Map p.255), 254, 256
Cape Nelson (Map p.91), 90, 93, 148, 150, 165
Cape Possession, 234
Cape Rodney, 185
Cape Vogel, 162
Capital punishment, 20, 34, 46, 47, 53
Cardwell, 76
Carlow, James (Jimmy the Reefer), 40
Carvey, Patrick, 24
Cashman, James, 29

INDEX

Cassowary Creek (Map p.200), 219, 220, 225, 229
Castleton, Claud, 216, 217
Catholics, *see* Missionaries
Caution Point, 90
Ceara, 8
Central Division, 223
Chalmers, Reverend James, 7, 12, 232-3
Champion, Herbert, 246
Champion, Ivan, 37
Charters Towers, 86
Chester, Henry, 80
Chester, H. Neville, 33, 34, 38, 39
China, 5
Chinese, 6, 10, 18, 21, 27n., 78
Chirima River (Map p.91), 112, 116, 118, 127, 135, 136, 173
Chisholm, Arthur, 257, 258
Chisholm, Frederick, 218, 223, 225-7, 242-5
Christie, Jack, 87
Clara Ethel, 56
Clark, George: leader Cairns prospectors, 86; Milne Bay, 87-8; on Mambare, 94-6, 104, 105, 107, 110n., 112, 123, 144, 155, 172
Clark Fort, 112
Clarke, Sir Rupert, 221
Clayton, Captain Francis, 31
Cloudy Bay (Map p.84), 85, 127, 148, 183, 184, 185
Clunas, Alex, 99, 112, 118, 119
Clunas and Clark, storekeepers, 144, 145, 148, 150, 167, 168, 193
Clunas, Hugh, 169
Clunn's Hotel, 10, 156
Clyde River (Mambare), 90
Coleman, miner, 140
Colemans Creek, 58
Collingwood Bay, 49
Collomb, Monseigneur Jean, 50-1
Combley, Nurse, 201
Companies, *see* Goldmining companies
Conde Point, 20
Cooktown, 2, 3, 7, 10, 11, 12, 29, 32, 55, 56, 77, 80, 81, 83, 85, 86, 87, 95, 176, 185, 199
Cooktown Courier, 10, 11, 20
Cooktown Creek, 40
Cooktown Independent, 101, 113
Coppard, Charles, 39
Coral Haven (Map p.8), 5
Coral Sea, Battle of the, 47
Coranderrk, 97
Cosmopolitan Hotel, 25, 156
Coutance, Louis, 5
Craig, Captain J.C., 9, 12, 13, 25

Crowe, Matt: on the Waria, 102, 138; on the Yodda, 120, 144, 169, 171; biography, 192-3, 257; on the Lakekamu, 194-8, 219, 238, 239, 240; on the Markham, 221, 254, 255; death, 259-60
Cuthbert, Freddie, 41

Dabney, Martin, 115
Dagomi, 13
Daisy, 6
D'Albertis's Attack Point, 223
Dambia, Constable, 166-7
Dammköhler, Wilhelm, 254
Dandata, 94
Darling, Arthur, 138, 140, 158, 193; biography, 255-6
Daru, 221, 223
Davies, Dave, 100-2, 113, 171, 203, 209, 223, 259
Davitt, Thomas, 143
Dawari Odari, 93
Deakin, Alfred, 61, 180
Debera, 105
Deboyne Islands (Map p.29), 31, 35, 40
Dedele (Map p.84), 83-4, 148
Degen, August, 40
Delaney, James, 135
de Moleyns, Richard, 125-6, 127
D'Entrecasteaux Islands (Map p.66), 4, 5; as source of labour recruits, 42-3, 45, 66-9, 152, 184, 185, 210; *see also* Gosiagos and particular islands
D'Entrecasteaux, J. A-R. Bruny, 29
Desertion by labourers, 69, 147-52, 211-14, 235
Detzner, Hermann, 226-7
Dexter, Henry, 182
Diamond, H.M.S., 12, 31
Didiam, 247, 248
Dikoias (Map p.51), 65, 70, 72
Dobu Island (Map p.66), 23, 69, 127, 156
Dobuduru, 132
Dogi, 93
Dogura (Map p.84), 86, 109
Domara (Map p.84), 83-5
Domata, Village Constable, 137
Donabai, Constable, 131
Dorevaide, 184, 188
Double Crossing, 144
Douglas, John, 6, 11, 13, 30-1
Dowell, Peter, 202
Doyle, Andy, 260
Dredging, 225, 228, 237, 266
Driscoll, Edward, 140
Drislane, Tom, 95, 97

Dumai, 100-6 *passim*
Dunantina River, 268
Durietz, William, 160
Dutch New Guinea, 266
D'Urville, J. Dumont, 5
Duvira, 106, 107

Eastern Highlands, 268
Eaus (Map p.29), 43
Ebora (Map p.29), 40
Ede, Richard, 54, 259
Edie Creek (Map p.255), 258-9, 262, 265, 266
Edmunds, Harry, 157, 164
Efogi, 212
Eia River (Map p.91), 91, 93, 137
Ekau-hu, Constable, 228
Elema, 236; *see also* Moveave-Toaripi
Elliott, Alexander, 163, 172; appointed to government service, 122; investigates killing of Campion and King, 122-3; at Bogi, 129, 150, 154-5; at Beda, 133-4; at Tamata, 135; on Yodda, 143, 145
Elliott, Robert, 118, 119, 171, 203, 219-20
Ellis, George, 257
Emanboga, Corporal, 154-5
Emily, 6, 9, 12
Endong, 261
English, Albert, 183, 184, 187
Eni, 149
Eoro, 189
Eraga, 93
Eruvo, 232
Ericksen, Charlie, 136, 159, 259
Ernst, Carl, 40
Erskine, Commodore James, 7
Eruwatutu (Map p.98), 95, 96, 99
Ewarupa, 214
Ewena, 45
Ewia, 9
Executive Council (Papua), 158, 204

Falke, 101
Faulkner, Joe, 158
Favourite, 52
Federation lease, 59
Fellows, Reverend Samuel, 35-6, 37-8
Fergusson Island (Map p.66), 28, 67, 68, 98, 151
Fiji, Joe, 85
Filipino traders, 25, 46
Finnegan, Patrick, 170
Finnegans Creek, 127, 134, 138, 145, 185, 194
Finnigan, Patrick, 57
Fiolini, Lucien, 19
Fish Creek (Map p.200), 219, 220, 247

Fishermans Island, 204
Fletcher, Henry, 206, 223
Flint, Leo, 188
Fly Gully, 228
Fly River, 81, 152, 221-3, 224
Flynn, Errol, 262
Foley, J., 157
Fonu, Constable, 224
Forbes, Henry, 9, 12, 13
Ford, Jerry, 216
Forrest King, 9
Four Mile, Sudest Island, 2
Four Mile, Yodda, 144
Freddy, 29
Fry, miner, 100, 113, 172
Fuyuge, 116, 135, 269

Gabagabuna, 185
Gadara, 102
Gadsup (Gazup), 261
Gadugadu track, 265
Gaiboa, 29
Gaina, 98-9
Gallagher, James, 63
Gallagher, Martin, 168
Gallagher, Mat, 63
Gamundu, 138, 139
Ganai (Map p.184), 186
Ganuganuana, 160
'Gap' (Kokoda Trail), 105, 130, 153, 171
Garaina, 138
Garbutt, H., 230
Gazelle, 52
Geelong, 80
Gelua, 203
Gemaruya, 98
Gensiko, 261
German New Guinea, 53, 101, 109, 138, 139, 141, 162, 167, 219, 221, 223, 256
Germans, 139, 142, 224-8, 254, 256
Gerret, Frank, 9, 31
Gibara (Map p.177), 177-9, 181-2
Giblin, William, 206, 217, 227
Gill, Reverend Wyatt, 76
Gillespie, Andrew, 212, 259
Gio, Constable, 46
Gira Goldfield (Maps pp.3, 91), 176, 185, 192, 193, 194, 211, 259, 265, 269; discovery, 118-19; health, 118, 146-7; population, 121, Table p.124; production, 118, 121, Table p.124; violence, 107, 108-9, 135-7, 172-3; labourers, 144-52; miners, 156-161; dredging, 142

INDEX

Gira River (Map p.91): ascended by MacGregor, 90; Binandere people of, 91, 93, 94, 102-5 *passim*; *see also* Gira Goldfield
Giulianetti, Amedio, 239
Glasson, Dick, 259
Glew, E.J., 69
Goaribari Island, 210
Godaw, 9
Goiye, 116
Goldfields (Map p.3), *see* Gira, Keveri, Kupei, Lakekamu, Laloki, Louisiades (separate entries for Misima and Sudest), Milne Bay, Morobe, Murua, Sepik, Upper Ramu, Waria, Yodda
Goldie, Andrew, 76, 77
Goldie River (Map p.77), 76, 78, 80, 105
Goldmining companies: British New Guinea Goldfields Proprietary, 23; Block 10 Misima Gold Mines (No Liability), 41, 42, 43, 45; Broken Hill Proprietary, 41; New Misima Gold Mines, 41; Cuthbert's Misima Goldmine, 42; Gold Mines of Papua Limited, 42; Woodlark Island Proprietary Goldmining Company, 58-9; Woodlark Ivanhoe Goldmining Company Kulumadau, 58; Kulumadau Woodlark Island Goldmining Company, 58-9; Yodda Goldfields Limited, 171; Tiveri Gold Dredging, 228; Guinea Gold No Liability, 228; Bulolo Gold Dredging, 226; and recruiting, 66, 69
Goldmining laws of New Guinea, Table p.256
Goldmining laws of Papua, Table p.21, 192
Gona, 153
Goodenough Bay (Map p.84), 87, 210
Goodenough Island (Map p.66), 28, 43, 66-7, 149; *see also* Gosiagos
Gors, Walter, 114
Gosiagos, 43, 45, 260
Gosisi, 115
Goulburn, 194
Graham, John, 21
Grant, Alexander (Sandy), 39
Graveyard Hill, 199
Gray, Jack, 176, 179
Gray, John, 169
Greek traders, 6, 25, 40-1
Green, John, 95-104 *passim*, 100-ln., 112, 116, 147, 172, 185
Grey River *Argus*, 150
Griffin, 2, 12
Griffin, Henry, 131, 144, 148, 162, 239; and O'Brien, 164-9
Griffin Point (Map p.8), 2, 22, 37

Guasopa (Map p.51), 49-54 *passim*, 65, 69, 70, 71
Gulewa (Map p.29), 33-4, 37, 38, 45
Gulf Division (Map p.194), 43, 195, 211, 233, 234
Gulf of Papua (Map p.194), 49, 148, 152, 233, 238
Guns, obtained by villagers, 22, 102, 133
Guswei, 140

Hahl, Albert, 141
Hakaia, 157
Hall Sound, 233
Hancock, G.F.B., 23
Hanran, John, 77, 78
Hanuabada (Map p.77), 78
Hara, 123
Hariba (Map p.29), 30, 33, 38, 45
Harimoi, 33
Harrison, Sago Bob, 159
Hasu, Morauta, 237
Hau'ofa, Isikeli, 47
Hawaiu, 239
Haylor, miner, 100, 113, 172
Healesville, 97, 101
Healesville *Guardian*, 97
Health: Sudest, 3, 23, 40-1; Misima, 29, 30, 35, 39, 40-1; Woodlark, 56, 60, 61, 63, 64, 69, 72; Laloki, 80; Keveri, 85; Milne Bay, 87-8; Yodda, 130, 146-7; Gira, 146-7; Trobriands, 164; Lakekamu, 198-207, 209, 217; Morobe, 265
Heath, 8
Heatoare, 232; *see also* Moveave
Heavala, 232; *see also* Moveave
Hely, Bingham, 19, 32, 54
Henderson, James, 57
Henderson, Laurence, 133
Herbert, Justice Charles, 214, 216
Hernsheim, Eduard, 52, 53
Hicks, A.G., 220
Hides, Jack, 230, 249
Higginson, Charles, 164, 195, 234, 239
Higginson, John, 106, 109, 156, 157; and Monckton, 162-3
Higgs, William, 207-8
Highlanders, 261
Hinai Bay (Map p.8), 23
hiri, 232, 238
Hislop, Robert, 127, 164
Hodgkinson Goldfield, 86
Hollingsworth, Walter, 10
Homa Creek, 123
Hopeful, 11
Horan, Dan, 168, 169

Hudson, Gilbert, 116
Hughes Bay (Map p.66), 68
Hula, 29, 80
Humphries, W.R. (Dick), 44, 212, 228, 229, 245-8
Hunt, Atlee, 161
Hunter, Robert, 233
Huon Gulf (Map p.255), 221, 254, 256, 257, 259, 262
Huon Peninsula, 260, 261
Hurley James, 85-7, 259

Iaba, 185
Iadu, 246, 248
Iami, 22
Iariva, 238; see also Kukukuku
Imila River, 223
Inade, 151
Indian, 6
Ingala, 164
Ingham, William, 6, 78
Ingham *Planter*, 114
Ino, 83-4
Ioma, 130, 136, 141, 149, 171
Iova, 219
Ironstone Creek (Map p.200), 197, 198, 202, 209, 211, 219, 235, 237
Isimari, 83
Isurava, 133
Iuoro, 133, 144
Ivanhoe, 55, 176
Ivanhoe prospectors, 95, 96, 99, 112
Ivanhoe Reefs, 58
Ivori, 220
Ivory, Billy, 139, 157, 193, 195, 260
Iworo, 144; see also Iuoro

James, D., 218
Japanese, 25, 47, 71, 105
Jassiack, James (Jimmy the Austrian), 135-6, 168
Jeffrey, Henry, 214-15, 223
Jeneeta, 26
Jiaro, 215-16
Jijingari, 139, 145
Jimmy, of New Caledonia, 76
Johnson, Ross, 99
Jolley, 82
Jones, Edward, 218
Jones, Fleming, 205
Jones, Frank, 80
Jones, George, 59
Jorgensen, Andy, 21
Josephata, 38
Joubert, Les, 167, 227, 255, 259

Juanita, 11, 13
Juries, 170, 216-17

Kabade, 97, 233
Kaili Kaili, 161-2
Kainantu, 261
'Kaindi', 261; see also Mount Kaindi
Kaio, Corporal, 154
Kaioki, 46
Kairuku, 212, 238
Kaiw, Kenneth, 47
Kaisenik, 265
Kakoma, 32, 33
Kalgoorlie, 208
Kambesi, 136
kampani, 261
Kanosia, 221
Kapau language, 238
Kapau River, 249
Kaptin, Constable, 227
Karama, 234, 235
Karavakum (Bonivat) (Map p.51), 55, 59, 65, 192
Karukaru, 135; see also Beda
Kasari, Sergeant, 235
Kasawai, 9
Kataw (Mawatta), 82
Kaurai (Map p.51), 65, 70
Kavatana, 65
Keelan, James, 137, 227
Kelly, T., 157
Kelly, W., 116
Kelly, William, 215-16, 245
Kennedy, Robert, 39
Kess Kess, 101
Kerema (Map p.194), 212, 220, 223, 227, 234, 235, 239
Kerema Bay, 212
Keremahaua, 250
Keveri Goldfield (Maps pp.3, 184), 85, 192, 193, 194, 223, 260, 269; discovery, 185; population, 185-6; production, 185-6; miners and villagers, 186-9
Keveri Valley (Map p.184), 183; see also Keveri Goldfield Keveri village, 218
Kewotai, 93
Kia Ora, 203, 218
Kickbush, Albert, 53
Kikori, 232
Kimberleys, 81
King, John, 122-3, 126, 143
King, Reverend Copland, 109, 150, 159
Kiwai, 98, 210, 212
Klondike, 167
Klotz, George, 185, 195

Koiari, 78-80, 116, 134, 144, 148, 173, 269; see also Mountain Koiari
Koira, 147
Koitapu, 77
Kokoda (Map p.91), 122; station founded, 130-1, 132-4 *passim*, 142-4 *passim*, 153-4 *passim*, 163, 165-71 *passim*, 212
Kokoda Trail, 105, 148
Kokove, 34
Kolinio, 37
Komiatum Hill, 259
Konradt, 226
Koranga Creek, 255, 257, 260
Koromanau, 261
Koropata, 127, 143
Kororo Creek, 220
Kovio, 237-8
Kroos ('Sandfly'), 7
Kropan (Map p.51), 65, 70
Krueger, Walter, 71
Kruger, Fred, 139, 187, 202, 219
Kukipi, 232, 236, 237
Kukipi, 237
Kukukuku: early meetings with prospectors, 195-6, 219-20, 238-42; and government officers from Nepa, 242-50; meet miners from Wau, 265
Kulumadau (Map p.51). 49, 50, 55, 58; town life, 60-2, 65, 67, 70, 192, 196
Kuma, 164
Kumudi, 176
Kumusi River (Map p.91), 90, 91, 93, 104, 106, 107, 119-31 *passim*, 135, 142, 144, 145, 147, 148, 150, 151, 153, 154, 171
Kunimaipa River (Map p.194), 197, 221, 223, 238, 250
Kupei Goldfield (Map p.3), 265
Kurai Hills (Map p.200), 197
Kurnalpi, 265
Kwaiapan Bay (Map p.51), 58, 62, 65
Kwato Island, 185
Kwato Extension Association, 189
Kwarma, 53

Labourers, *see* particular goldfields, D'Entrecasteaux Islands, and Queensland
Lae, 265
Laewomba, 256
Lagalaga, 214
Lagoni, Lance-Corporal, 206
Lahui Vai, 46
Lai, Village Constable, 235
Lai-i-woi, Constable, 245
Lakapu, 63-4

Lakekamu Goldfield (Maps pp.3, 194, 200), 171, 259, 260, 265; prospecting, 192-6, 219-24; development, 196-8; production, 196, 198, Table p.210; dysentery, 198-206; and Commonwealth Parliament, 207-9; labour recruiting, 209-11; population, 196, 197, Table p.210; labourers and miners, 211-19; World War I, 224-8; World War II, 230, 237; decline, 228-30; and Moveave-Toaripi, 232-7; and Kovio, 237-8; and Kukukuku, 238-50
Lakekamu, River (Maps pp.194, 200), 197; *see also* Lakekamu Goldfield
Laloki 'goldfield' (Maps pp.3, 77); discovery, 76-7; population, 77; miners and villagers, 78-80; failure, 80
Laloki River (Map p.77), 260; *see also* Laloki 'goldfield'
La Recherche, 29
Laughlan Islands (Map p.4), 5, 52, 54, 62, 69
Lavai-i, 167
Lawes, Reverend William, 12, 76, 77
Lawrence, James, 144
Lawson, Peter, 158
Legislative Council (Papua), 169, 192
Le Hunte, Sir George: starts crusher, 58; meets Bousimai, 107; and Armit, 122, 153; appoints officers, 125; and Milne Bay, 178, 179, 181
Lett, Lewis, 110-11n.
Lewis, Robert, 63
Lihoiya, 147
Lilydale *Express*, 97
Lindon, James, 176
Lindsay, Bob, 179
Little McKenzie, 59
Little, William, 169, 193, 199
Livesey, G.H., 102
Lizzie, 6, 7, 8, 12
Lobb, Charlie, 54, 88, 113, 114, 259
Logea Island, 39, 186
Lohiki River (Map p.194), 220, 239
London Missionary Society, *see* Missionaries
Lone Hand, 157
Lookout, 144
Louisiade Archipelago (Map p.4), 2, 4, 7; *see also* particular islands
Louisiade Goldfield, 26, 85, 86, 192; *see also* Misima and Sudest
Lovera, 239
Lucy and Adelaide, 2
luluai, 261, 262
Lumley, Charles, 199, 214, 221, 227-8, 229
Lutherans, *see* Missionaries

Lyons, Arthur: at Ioma, 136, 137, 141; at Nepa, 198, 199, 209, 214, 219; and Kukukuku, 240-2

Mabie, 157
MacAlpine, Archie, 221
MacClelland, Moses, 116, 118
McClelland, Samuel, 112, 122-3
McCrann, Thomas, 197, 228
MacDonald, Nurse, 201
MacFarlane, Reverend Samuel, 6
MacGowan, Ernest, 195, 233
MacGregor, Sir William, 131, 173; appointed, 11; on Sudest, 19-23 *passim*; on Misima, 28, 33-4, 39; on Woodlark, 52-7 *passim*, 69; punishes Merani, 83-5; on the Mambare, 90-120 *passim*; on the Musa, 184; on the Lakekamu, 232-3
MacKay, Kenneth, 106, 161-2
MacLaughlin, 112
MacLaughlins Creek, 118, 119, 120, 129, 144, 145; gold strike, 113; miners go overland to, 116; miners and villagers, 127, 134, 257
McLean and Samuelson, reef miners, 22
McLean, Jack, 179, 181
McMurdo, William, 52
McTier, James (Jimmy the Larrikin), 28, 83-4, 183
Madau (Map p.51), 65
Mader, Lt R.G., 46
Maguire, H.R., 109
Mahony, Elizabeth, 25, 26, 42
Mahony, John, 24, 40
Mai, 152
Maiadoma, 68
Mailu Island (Map p.184), 43, 49, 98
Maimera, 218
Maioni, Constable, 129
Maiporo Creek, 220
Makavasi, 52
Malays, 6, 9, 25
Malinowski, Bronislaw, 4
Mamadi, 53-4
Managun Island, 30
Mamba (Mambare) River, 105, 109; *see also* Mambare
Mambare River (Maps pp.91, 98): exploration, 88, 90-1; people living on, 91-4; prospectors on, 94-110; government officers on, 97-110; missionaries on, 109-10; miners on, 113-20, 122-71 *passim*
Mambu, Bishop, 262-3
Manatu (Map p.98), 104, 105

Manawah, Sam, 24
Manboki ('Dixon'), 7
Manewa, 158
Mangrove Island, 142
Manumanu, 212
Manus Island, 6
Mapas Island (Map p.51), 49, 64
Maria, 76
Markham River (Map p.255), 221, 224, 254-5, 257, 260, 263
Martin, Jack, 88
Marx, John, 31
Mary, 5, 50
Mason, Patrick, 131
Massim people, 4, 5, 178
Matunga, 62
Matupit, 52
Mauger, Samuel, 60
Mawai, 93
May, Fred, 135
Mayflower, 95, 225
Mazzucconi, Giovanni, 52
Meek, A.S., 55, 133
Mekeo, 238, 239
Melbourne Cup, 221
Memekowari, 125-6
Merani, 83-5, 184
Mercury, 2
Merrie England, 20, 54, 86, 96, 99, 113, 149, 195, 201, 204
Meteor, 86, 88
Methodists, *see* Missionaries
Miara, 215-16
Mica Creek, 40
Milne Bay (Maps pp.4, 84, 177), 7, 43, 150, 248; prospecting, 87-8, 185; recruiting, 195, 210
Milne Bay Goldfield (Maps pp.3, 84, 177); discovery, 176; early rush, 177; population, 176, Table p.179, 182, 192; violence, 178-82; villagers as miners, 182; companies, 182
mimi, 186
Miners and village women, 18-19, 159-60
Minister, Nicholas, 12, 13, 28, 32
Minnis, Patrick, 78
Mirehea, 232; *see also* Toaripi
Miserie, 23
Misima Goldfield (Maps pp.3, 4, 29), 83, 152, 192; production, Table 2 p.26, 47n., discovery, 28-9; health, 29; population, Table 3 p.26, 29, 39; decline as alluvial field, 39-40; development by companies, 41-3; recruiting by companies, 42-5, Table 5 p.44

INDEX

Misima Islanders, 4, 64; as labour recruits, 9, 30-1, 42-3; population, 29, 41, 45-6; pottery, 30; early contact, 30-1; meet miners, 31-2; and government officers, 32-5, 38-9, 45-7; and missionaries, 35-8, 70; as miners, 40, 43; and World War II, 46-7

Missionaries, 6-7, 156, 177; Methodists (on Misima), 35-7, (on Sudest, 37); relations with state, 37-8; Catholics on Woodlark, 50-2, 71; Methodists on Woodlark, 69-71; United Church, 71, 238; London Missionary Society, 7, 76, 77, 178-82, 232-3, 250; Anglican, 86-7, 109-10, 133, 141, 159-60, 199, 201; Lutherans, 101, 142, 254, 256, 257, 262; *see also* South Sea Islanders

Mollison, Lt P.J., 71

Molnos, 30-1

Monckton, C.A.W., 129, 160; mines on Woodlark, 55-6; and Bousimai, 107; on Green, 111n.; and Orokaiva, 132, 143; arrests deserters at Cape Nelson, 148; and police, 153, 154; on Rochfort, 156; biography, 161-2; and O'Brien, 165-8

Moni River (Map p.184), 183, 184, 185

Montrouzier, Xavier, 70

Morauta, Hasu, 237

Morauta, Mekere, 237

Moree, 221

Moresby, John, 90, 178

Moreton, Matthew, 97, 102, 185; at Milne Bay, 176-81 *passim*

Morley, Harry, 179, 181

Morobe, 141, 142, 246

Morobe Goldfield (Maps pp.3, 255), 250, 268; discovery, 254-8; production, 258-9, Table p.258; development, 259-65; labourers, 260-5; violence, 265; health, 265; companies, 266

Morrison, George, 81

Morrison, J., 19-20

Mosquito Creek, 219, 228

Motorina Island (Map p.8), 46

Motu, Constable, 165

Motu language, 68, 98, 127, 238, 245, 246, 247

Motu people, 77, 85, 232

Motumotu, 218, 232; *see also* Toaripi

Mount Adelaide reef, 23, 155

Mount Albert Edward (Map p.91), 91, 119

Mount Clarence (Map p.184), 183

Mount Hagen, 227

Mount Kabat (Map p.51), 49

Mount Kaindi (Map p.255), 258

Mount Lamington (Map p.91), 128, 129, 130, 154

Mount Lawson (Map p.200), 196, 223

Mount Momoa, 116

Mount Scratchley (Map p.91), 91, 99, 113, 115

Mount Sisa, 41, 42 , 43

Mount Suckling (Map p.184), 183

Mount Victoria (Map p.91), 91, 99

Mountain Koiari, 116, 133, 135, 142

Moveave, 195, 197, 212, 214, 220, 234, 235-8, 240, 248, 249, 250; *see also* Moveave-Toaripi

Moveave-Toaripi, 233-7

Mowata, 106

Mulholland, James, 214, 221

Mullins, Captain T., 9

Mullins Harbour (Map p.177), 125, 177

Murphy, Jack, 171, 223

Murray, Sir Hubert, 163, 260; and miners and crime, 156, 157-8, 169, 214, 216, 217; becomes head of government, 173; and the Lakekamu, 195-204 *passim*, 207, 214, 216, 220-4 *passim*, 227, 229; and Kukukuku, 239, 240, 242-3, 248, 249

Murray, Thomas, 216

Murray, William, 212

Murua Goldfield (Maps pp.3, 4, 51), 114, 118, 158, 176, 192, 193, 194, 260; discovery, 54-5, 88; production, 55, Table p.57, 60; health, 56-7, 60; mining methods, 55-6; miners' wages, 58; development by companies, 58-60; World War I, 62; World War II, 71; recruiting by companies, 65-9, 148, 152, Muruans, 4, 269; attack the *Mary*, 5, 50; and missionaries, 7, 50-2, 69-71; stone axes, 49-50, 51; population, 52, 72; as labour recruits, 52-3, 65-6; and government officers, 53-4, 63-5; health, 50, 63-4, 72; as miners, 64; and World War II, 71

Murua River, 220

Musa River (Map p.184), 96, 114, 147, 184, 194, 260

Muscutt, C.R., 228-30, 247-8

Musgrave, Anthony, 134

Mustar, Pard, 265

Mutiana, 13

Naboko, Constable, 227

Nairoda, 135; *see also* Beda

nambis, 261

Nasai Point, 65

Native Labour Ordinance (New Guinea), 260

Native Labour Ordinance (Papua), 161, 217; labourers not to be taken outside Papua, 141; desertion, 149; recruiting, 151; health, 200, 202, 203-6
Native Plantations Ordinance, 45
Neilson, trader, 53
Nelsson, Gus, 60, 119, 199
Neneba, *see* Beda
Nepa (Maps pp.194, 200); centre for administration of Lakekamu Goldfield and attempts to bring Kukukuku under government control, 198-248 *passim*; established, 198; abandoned, 228-30
Nettle, William, 115
Nettleton, Jack, 257-8
New Britain, 6, 52, 71, 221, 238, 263
New Caledonia, 19
Newcombe, Avard, 220, 227, 240
New Ireland, 238
Newland, W., 210
Newton, Bishop Henry, 199
New South Wales, 194, 221, 223
New Zealand, 35, 161, 163, 194
Nicolas, 68, 151
Nicholls, George, 199, 202
Nigom, 31
Nimoa Island (Map p.8), 9
Nine Mile, Sudest Island, 2
Nivani Island (Map p.29), 23, 34, 38, 39, 40, 47, 56, 63
Normanby Island (Map p.66), 28, 69
North-Eastern Division, 155
Northern Division, 43, 211, 212, 214, 224, 254, 255, 257; government officers, 126; *see also* entries for particular places
Northern Territory, 81, 217
Nowland, Nurse, 201

O'Brien, G., 115
O'Brien, Joe, 157, 164-70, 181, 207
Oia, Corporal, 107, 108
Oivi Ridge, 130
Okavai (Map p.200), 238
Okiduse Range, 49, 55
Oldham, Eric; administration of Lakekamu Goldfield, 215-16, 218, 219, 223; and World War I, 224-7; and Kovio, 238; and Kukukuku, 242-3
Oldorp, Rudolph, 254-5
Olipai River (Maps pp.194, 200), 230, 233; prospecting, 219-24 *passim*; Kukukuku on, 240-50 *passim*
Olsen, miner, 100-2, 113
O'Malley, James, 234
Omdamuda, 53

Ona-Audi, 188
O'Neill, Jack, 268
Ono River (Map p.194), 140
Onombatatu, 104
O'ori', Corporal, 199
Opi River (Map p.91), 43, 90, 104, 106, 107, 119-20, 147
Oreba River (Maps pp.194, 200), 227
Oro Bay (Map p.91), 91, 147
Orokaiva, 43, 104, 210, 269; land and culture, 91-4; as police, 106; relations with Beda, 116; relations with miners, labourers and government officers, 121-73 *passim*; *see also* Binandere
orokaiva, 99, 105
Orokolo, 98, 148, 157, 158, 165, 248
Osborne, D.H. (Harry), 21, 42, 157
Oreya, 147
O'Toole, John, 158
Owen Stanley, Captain, 5-6
Owen Stanley Ranges, 105, 114, 140, 183

Page, William, 57
Paiwa, 87, 210, 216
Pako, 51
Palm Island, 228
Palmer Goldfield, 10, 12, 80, 86, 208
Panaeate Island (Maps pp.2, 29), 4, 9, 30, 34, 36, 38, 40, 46, 50
Panaeate language, 35, 36, 70
Panapompom Island (Map p.29), 32, 33, 38, 63
Pana Tinani Island (Maps pp.4, 8), 2, 5-6, 7, 8-9, 11-13, 28, 32
Panemote Island, 53
Pangari, Constable, 247-8
Pantava, 22, 23, 37
Papaki, 128, 160; violence at, 121-2, 123-4, 129, 133; government station, 127, 130, 143, 145, 153, 154
Papua Bill, 60-1
Papua Hotel, 196
Papuan Courier, 257
Papuan Infantry Battalion, 237
Papuan Times, 188
Parakota, 63
Park, Andy, 171
Park, William (Sharkeye), 171, 221, 255; biography, 257-60, 265
Parke, Halkett, 136, 149, 160
Parkes, William, 160, 169, 171
Partanu, 233
Pearce, Sir George, 265
Pearlers, 21, 53
Pegolo, 214

INDEX

Petari, 101
Peu (Map p.98), 99, 100, 102, 105, 106
Peuma, Village Constable, 238
Phillips, F.B. (Monty), 263-5
Pidgin language, 7, 13, 26, 64, 163, 215-16, 236
Pilhofer, Georg, 257
Pinney, W.S., 202
Pipiri, 124
Piron Island, 2, 7
Playford, Thomas, 169
Police, 23, 34, 58, 85; Laughlan Islanders join, 54; at Merani, 83-4; on Musa, 96; on the Mambare, 99, 102, 104, 105, 108; on Vanapa, 115; in Northern Division, 122, 126, 128, 130, 133, 136, 137, 152-5, 162, 173; and O'Brien, 165, 168, 170; on the Lakekamu, 211, 212, 224, 230, 234, 240, 243, 245, 247-8
Popagania, 13
Population: Sudest, 3; Misima, 29, 46; Woodlark, 52, 62, 71, 72
Port Moresby (Map p.77), 11, 12, 24, 104, 105, 119, 153, 189, 194, 196, 208, 232, 235; centre for Laloki rush, 76-8, 81; start of inland track, 113-15, 120, 130, 171, 212; port for Lakekamu, 197, 200, 203, 204, 214, 217, 218, 227, 228, 229; Kukukuku in, 245-8 *passim*
Port Moresby Hotel, 197
Port Jackson, 5
Poruta, 107, 108
Pottery, 5, 30
Poumbari, 132
Preston, James, 212, 221, 255-6
Preston's, 219
Priddle, Charlie, 219-20
Priddle, Mrs Charlie, 229
Pride of the Logan, 6
Prospect Creek, 145 158
Pryce-Jones, Edwin, 233
Pryke, Dan, 139, 185, 186, 194, 203
Pryke, Frank: on Waria, 139, 140, 145, 146; on Beda track, 144; on O'Brien, 170; prospects Musa, 184-5, 194; Milne Bay, 184-5; at Cloudy Bay and Keveri, 185-8; prospects Lakekamu, 193-8, 238-40; biography, 194; on Lakekamu, 202-3, 204, 209; prospects Vailala, 219-20; prospects Fly, 221-3; at Edie, 259
Pryke, Jim: on Waria, 139, 145; on treatment of labourers, 158; at Keveri, 185; prospects Lakekamu, 193-8; biography, 194; on Lakekamu, 209, 219; on Fly, 221-3; World War I, 227-8

Pumpkin (Nagevagum), 63, 64
Purari River (Map p.194), 210
Pure clan, 93-4

Quartz Mountain (Map p.29), 41, 42
Queensland, 61, 76, 78, 163, 185, 194, 207, 208; north Queensland ports serve goldfields, 2, 28, 55; labour recruiting, 8-9, 13, 22, 30, 53; home of miners, 10-12, 81, 85, 86, 88, 156, 173; Queensland Mining Act, 218
Queen Victoria, 34

Rabaul, 46, 221, 257
Raikupa, 261
Rambuso, 9
Ramsay, Frederick, 159
Ramu River, 256, 266
Rattlesnake, 5, 6
Red Creek people, *see* Seragi
Regan, J., 215, 216
Reid, A.C., 203-4
Reilly, J., 217-18
Rennie, David, 164
Rentoul, Alexander, 43
Retrieve, 6
Reynolds, B., 167
Ribe, 93-4
Rigo, 114, 183
Riley, Pat, 86
Robertson, Gordon, 220, 223, 245, 259
Robertson's Gully, 214, 219
Robinson, Christopher, 60, 129, 130, 154, 160
Robinson, J.B., 9, 13
Rochefort, Frank (Frenchy), 28, 83-4, 183
Rochfort, Frank, 156; on treatment of labourers, 158; and O'Brien, 165, 167, 168; dies, 260
Rockhampton, 115
Rocky Creek, 197, 198, 219
Romilly, Hugh H., 12
Rorke, Frederick, 206
Ross, Ronald, 131, 205
Rossel Island (Maps pp.4, 8), 7, 19, 28, 42; recruiting, 23, 148, 150, 152
Royal Commission: labour recruiting for Queensland, 8; Papua 1906, 61, 158, 165, 169-70, 260
Royal, Mark, 169
Royal, William, 257, 258
Ruatoka, 80
Rulitamu, 7
Runcie River, 11, 13, 21
Rupeni, 43

Russell Goldfield, 86
Ryan, miner, 116
Ryan, Henry, 26, 234
Ryan, Ned, 137, 259

Sagarai Valley (Map p.177), 177, 178, 179, 181
Sago Gully, 26
St Aignan, 29; *see also* Misima
St Andrews, 109
St Joseph (Alabule), 81, 114, 220, 224, 238
St Patricks Creek, 28, 32-3, 43, 83
Salamaua, 93, 259, 261, 265
Samarai (Maps pp.4, 177), 10, 25, 33, 40, 46, 53, 85-6, 88, 95, 100, 101, 113, 114, 118, 141, 148, 150, 151, 152, 156, 171, 185, 186, 193, 196, 197, 203, 205, 234; centre for administration of Milne Bay Goldfield, 176-81 *passim*
Samboga, 131
Samuelson, reef miner, 22
sapi-sapi, 23, 42
Schmitt, miner, 116
Schools: Misima, 37; Woodlark, 62, 65; Gulf of Papua, 233, 235, 238
Scott, Thomas, 63
Sea Breeze, 12
Seagull, 88
Sedu, Corporal, 98, 101, 105
Segaradi, Corporal, 46
Seligman, C.G., 4
semese, 232, 236, 237
Sepik Goldfield (Map p.3), Table p.259, 265
Sepik River, 260, 263, 266
Seragi, 135-8
Shanahan, Michael, 105, 114, 118, 120
Sharp, George, 81
Shaw, Percy, 141
Shearing, William James, *see* Simpson, William
Sheridan, Brinsley, 76
Siagara (Map p.29), 29, 32, 37, 46
Sialum, 265
Simpson, William: leader Ivanhoe prospectors, 95; prospects upper Mambare, 112, 116, 147; in Sydney, 114; on Musa, 184; dies, 118
Simson, C.C., 199, 201
Singapore, 167
Sipiliei, 179, 181
Sisereta, 131, 142
Sisters, 144
Siup, 9

Sloane, Joe: on Aikora, 136; at morality meeting, 159; fights Lumley, 214; on Lakekamu, 216, 218; prospecting Keveri, 223; biography, 259; on Edie, 259
Smith, M. Staniforth C., 169, 193, 208, 224; administers Lakekamu, 204, 205
Smythe, Don, 29
Society of Mary, *see* Missionaries, Catholic
Sogeri, 115, 130
Soich, Luke, 186
Solomon Islands, 6, 14
Sorenson, Neils, 11
South Sea Islanders, 6; as missionaries, 7, 36-7, 41, 70, 76-7, 80, 109, 233
Stanley, 52
Steele, miner, 100-2, 113
Stone-Wigg, Bishop Montague, 153, 154, 156-7, 159, 164
Stores: Sudest, 3, 24; Misima, 29, 32; Woodlark, 60, 61; Yodda, 171; Lakekamu, 197-8, 229
Strachan, John, 14
Streeter, Julius, 200-1, 204, 205
Strickland River, 221
Strikes: Sudest, 23; Woodlark, 58
Stuart-Russell, Henry, 105, 120, 133, 144
Stuart-Russell, John, 154
Suau Island (Map p.177), 7, 43, 46, 151, 178
Sudest Goldfield (Maps pp.3, 4, 8), 150, 152, 155, 192, rush of 1888, 2-3, 12; discovery, 11-12; mining methods, 18; reef mining, 22-3; decline, 23-7; shift to Misima, 28; shipping, 32; planters, 42; population, Table 2, p.26; production, Table 3, p.26, 55
Sudest Islanders, 256; dress, 3; houses, 3; population, 3, 41; trading, 4-5; early contact with traders and labour recruiters, 5-10; meet miners, 11-14; and miners, 18-21; as miners, 21-3, 26-7, 40; and traders, 23-5; attacks on, 32; in World War II, 47; and missionaries, 6-7, 37, 70
Sudest, Jimmy, 22
Suiena, 93
Suloga (Map p.51), 50, 55, 56, 58, 63-4, 66, 70
Suloga Peak, 49, 55
Sunset Creek (Map p.200), 216, 219, 228
Susuina, 35
Swanson, James, 148, 151
Swinger, H.M.S., 28, 31
Sybil, 8
Sydney, 5, 51, 52, 53, 76, 114, 260, 265
Sydney Daily Telegraph, 56

INDEX

Sydney Morning Herald, 12, 150
Symons, Alexander, 178-81 *passim*

Taaffe, John, 62
Tabe, 147
Tacomola, 7
Tagalita, 8
Tai Hill, 93
Taian Dawari, 93
Taian Yabari, 93
Tailend Creek, 219
Tairora language, 261
Talk-Money, 26
Tamana, 20
Tamanabae, 106
Tamata (Maps pp.91, 98), centre for administering Aikora, Gira and Waria goldfields, established, 98, 100-68 *passim*, 185, 196
Tamata Creek (Map p.98), 93, 97, 98, 120
Taniava, 147
Tatoo, David, 109
Taupota (Map p.84), 86, 87, 98, 147; carriers, 88, 112, 113
Tauri River (Map p.194): prospecting, 195, 197, 219-20, 221, 223, 239; Moveave on, 232, 233; Kukukukuon, 242
Tein, 107
Tetebra, 53
Tetzlaff, Wilhelm, 52, 54, 69
Thirty-five Mile Creek, 219
Thursday Island, 97, 196
Tiveri, landing and store (Map p.200), 206, 209, 211, 212, 218, 219, 229, 236, 238, 243
Tiveri River (Maps pp.194, 200): prospecting, 195-8, 220, 224; miners on, 225; government patrols on, 227; named, 233; Kukukuku on, 239, 242, 245, 249-50
Toaripi, 232-7
Toaripi Association, 237
Tomasi, 24
Tomowi-u-ia, 218
Tomu, Sergeant, 153
Torres, Luis Vaez, 5
Torres Strait, 6, 102
Totoadari, 93
Touinsi, 8
Towalla Goldfield, 88
Townsville, 113, 261
Traitors Bay, 90, 93
Tramways, 41, 58, 59
Treachery Bay, 31
Trobriand Islands, 4, 45, 50, 65, 164

Truganini, 9, 12
Tryon, G., 31
Tuari, 83
Tubetube Island, 40
Tudava, 64
Tufi, 43
Tully, 221
Turner, J., 87
Turner, Owen, 150
Twidi, 153
Twisty Creek (Map p.200), 219, 249
Two Mile (Yodda), 165

Uluo, 261
Umauna, 261
Umbogi, 106
Ume (Map p.98), 100, 105, 106, 147
Umuna (Map p.29), 41, 42, 43, 45, 66, 67, 71
Umuta, 108
Unamatana, 64
United Church, *see* Missionaries
United States armed forces, 71
Upaia, 68
Upper Ramu Goldfield (Map p.3), 265, 268
Uritai, 232; *see also* Toaripi
Urulau (Map p.200), 238

Vailala 'madness', 236
Vailala River (Map p.194), 195, 220, 224, 239
Vanapa River (Map p.97), 100, 114, 115
Vaughan, Cecil, 179
Vavasua, 249
Victoria, 193, 194
Victoria, 53
Vilirupu, 223
Village Constables,69; Sudest, 22; Misima, 34, 38; Woodlark, 54, 64, 72; Northern Division, 127, 133, 134, 147, 164, 171; Moveave-Toaripi, 234, 235
Viviga, 53-4
Voura, 6

Wabununu (Map p.51), 72
waga, 4, 30, 50, 71-2
Wagawaga (Map p.177), 178, 182
Wagawaga Dick, 196, 239
Wages: for Australians on Woodlark, 58; for Muruans, 65; on Morobe, 261
Wagima, 32-3
Waibua, 128
Waie, 93
Wai-iupa, 151
Waiko, John, 108, 155
Waima, 234
Wakaia, 140, 141

Wakoia, 53, 54, 65
Wali, Constable, 228
Walker, Archibald, 107, 125-6, 127, 135, 142
Wallace, James, 164, 216
Walsh, Allen, 127, 136, 143
Walsh, James, 60
Walsh, M., 223
Walsh, R.J., 86
Wamana, 54
Wamira, 86-7
Wampit, 266
Wanganui, 32
Ward, J., 157
Ward, William, 230
Waria Goldfield (Maps p.3, 91), 192, 193, 194, 195, 259, 269; discovery, 138; production, Table p.124, 139; miners and villagers, 139-42
Waria River, 254, 255, 257, 260; Binandere on, 93, 94, 102, 108, 135; miners on, 138-42, 144, 145, 146, 159, 172, 268
Wari Island, 4, 7, 12, 30, 40
Warren, Robert, 39
Wasida, 128, 154
Wasilasi, 64
Waterfall Creek, 138
Watson, John, 181
Watut River (Map p.255), 254-5, 256, 263, 266
Wau (Map p.255), 237, 250, 261, 263, 265, 266
Waverley, 35, 36
Wawonga, 144
Weekly, Fred, 59
Western Australia, 61, 113; goldfields of, 81, 115, 156, 173, 176, 193, 194, 257, 265
Western Division, 23, 43, 148, 152, 211, 212
Whalers, 5, 50
White Australia Policy, 181
Whitten Brothers (Robert and William), storekeepers, 11, 185; Samarai, 86; Jijingari, 139, 140; Northern Division, 144, 150, 159, 165, 166; Lakekamu, 197, 199, 203-4, 206, 211, 216, 227, 229, 236
Whitten, William, 147
Whyte, David, 11, 13, 18, 21
Williams, F.E., 183, 236
Wilsoni, 24
Winter, Francis, 20, 119, 180-1
Wio, 148, 150
Woitapi, 115, 116
Wolf, 62
Wonai, 55
Woodlark Island, 30, 40, 206, 212, 220; *see also* Murua Goldfield and Muruans

Woodlark King, 59
Woolf, Steve, 158, 179, 180
World War I, *see* particular goldfields
World War II, *see* particular goldfields
Wriford, George, 115
Wuth, Charles, 188

Yarumeku, 147
Yema, 93
Yeva, 102, 104,105
Yodda Goldfield (Maps pp.3, 91), 185, 193, 259, 269; discovery, 112-13, 118, 119-20, 146-7; early rush, 114-18; health, 114, 130-1; population, 116, 118, 121, Table p.125, 156, 192; violence, 121-9, 133-4, 142-4, 164-70, 172-3; production, 121, Table p.125; tracks, 129-32, 144-5, 149, 161; costs, 160: labourers, 144-52, 160-1; police, 152-5; miners, 156-61; government officers, 161-4; companies, 171-2; World War II, 105, 171
Yukon, 156, 193, 257
Yule Island, 80, 167, 195, 209, 218

Zephyr, 2 ,12
Zia, 93, 108

www.ingramcontent.com/pod-product-compliance
Lightning Source LLC
Chambersburg PA
CBHW050901240426
43671CB00027B/2968